The
Anatomy
Student's
REVIEW WORKBOOK

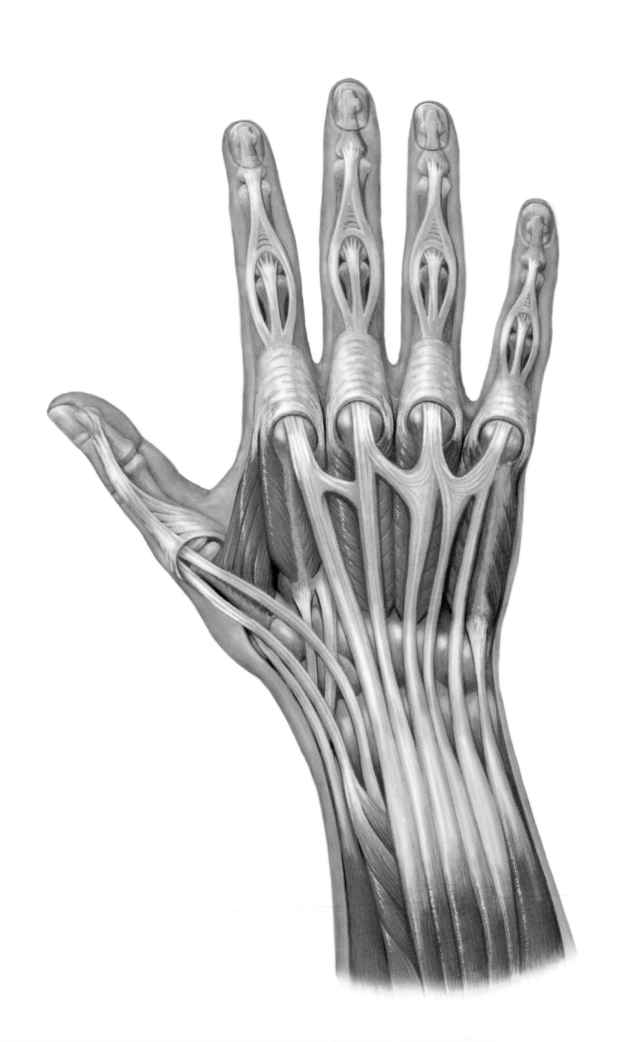

The
Anatomy
Student's
REVIEW WORKBOOK

Test and reinforce your anatomical knowledge

Professor Ken Ashwell, BMedSc, MBBS, PhD

BARRON'S

First edition for North America published in 2018
by Barron's Educational Series, Inc.

© 2018 Quarto Publishing plc

Conceived, designed, and produced by The Quarto Group
The Old Brewery, 6 Blundell Street, London N7 9BH, United Kingdom
T (0)20 7700 6700 **F** (0)20 7700 8066 **www.QuartoKnows.com**

Contributors:
Author: Ken Ashwell, BMedSc, MBBS, PhD
Design: Tony Seddon

All inquiries should be addressed to:
Barron's Educational Series, Inc.
250 Wireless Boulevard
Hauppauge, NY 11788
www.barronseduc.com

ISBN: 978-1-4380-1190-5

Library of Congress Control Number: 2018937022

Printed in China

9 8 7 6 5 4 3 2 1

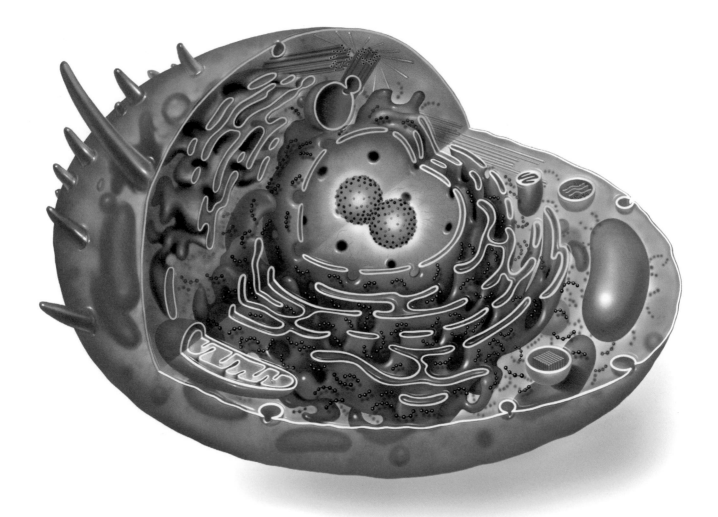

Contents

8   How this book works

10   Chapter 1: Body Systems and Tissues

36   Chapter 2: The Musculoskeletal System

78   Chapter 3: The Nervous System

120   Chapter 4: The Circulatory System

158   Chapter 5: The Respiratory System

184   Chapter 6: The Digestive System

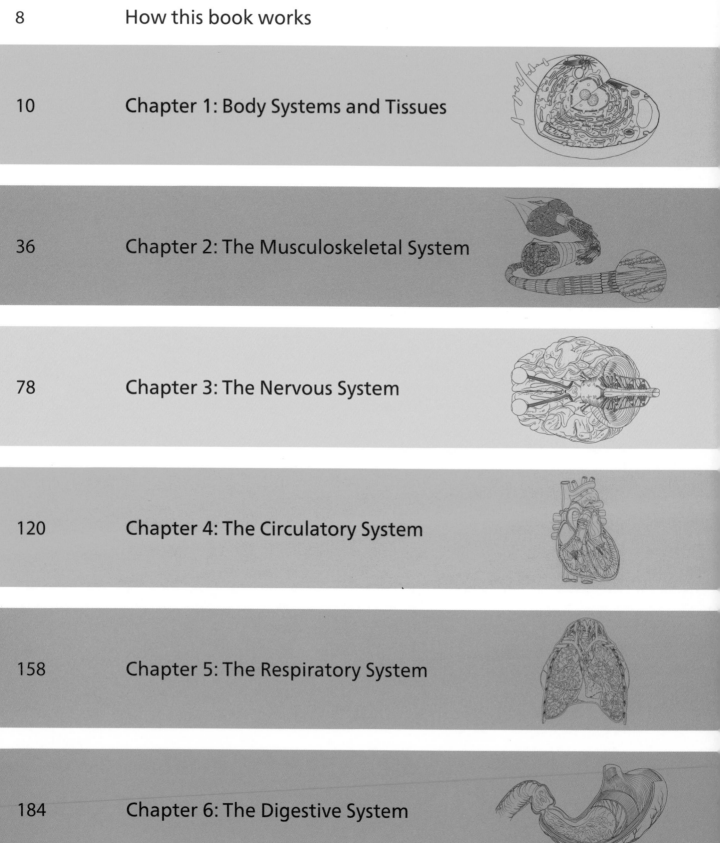

216   Chapter 7: The Urinary System

248   Chapter 8: The Endocrine System

280   Chapter 9: The Reproductive System

306   Chapter 10: The Blood

332   Chapter 11: The Immune System

356   Answers

364   Index

How this book works

Anatomy is a complex subject that incorporates not only visual knowledge—such as the recognition of body features, and their shapes and positions—but also declarative or factual information about the structure and function of body parts. Once a student has reviewed lecture and practical class material, it is time to test that knowledge against well-crafted questions that can reveal weaknesses and focus attention on specific topics for further revision. It is only by repeated sequences of knowledge testing and review of educational material that a student can progress toward a sound understanding of the field.

This compact yet detailed book combines the testing of visual recognition through comprehensive labeling tasks, with the assessment of factual knowledge through multiple choice, true or false, and sentence completion exercises. It also tests deeper understanding of the reasoning behind structure/function relationships in anatomy through the use of matching-statement-to-reason exercises.

Topics in this book are organized by body system, with useful summaries and concise definitions of key terms and concepts. Understanding of the clinical relevance of the content is reinforced by the provision of focused, informative text boxes for each system, so that readers can appreciate the practical importance of the information.

Full answers to all questions are included on pages 356–363.

Summary pages

A brief summary of the broad topic and labeled, full-color illustrations are combined with definitions of the key terms.

Definitions
Arranged in alphabetical order, concise entries define the body parts labeled on the illustrations on the facing page. Each term in bold corresponds to a label.

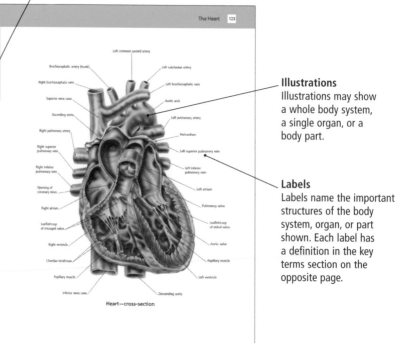

Illustrations
Illustrations may show a whole body system, a single organ, or a body part.

Labels
Labels name the important structures of the body system, organ, or part shown. Each label has a definition in the key terms section on the opposite page.

Exercise pages

Exercises include true-or-false questions, multiple-choice questions, fill in the blanks, and matching statements to the correct reasons. Instructions are given on each page, and full answers are included on pages 356–363.

Chapter name

Subsection

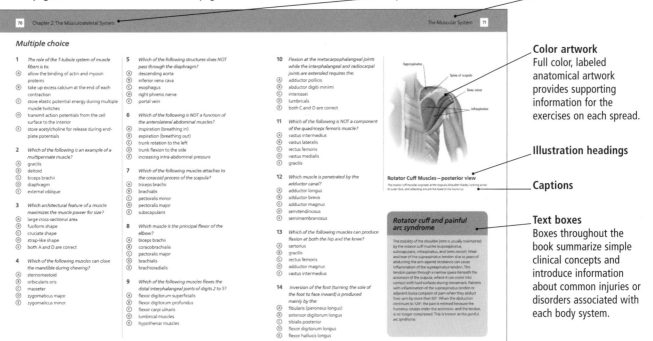

Color artwork
Full color, labeled anatomical artwork provides supporting information for the exercises on each spread.

Illustration headings

Captions

Text boxes
Boxes throughout the book summarize simple clinical concepts and introduce information about common injuries or disorders associated with each body system.

Color and label pages

These pages feature color-and-label activities to help you identify and memorize the location of different body parts.

Blank labels
Numbered blank labels refer to body parts. Fill in the labels to test your anatomical knowledge.

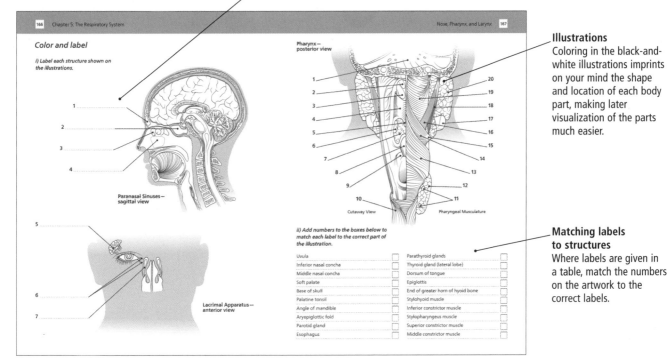

Illustrations
Coloring in the black-and-white illustrations imprints on your mind the shape and location of each body part, making later visualization of the parts much easier.

Matching labels to structures
Where labels are given in a table, match the numbers on the artwork to the correct labels.

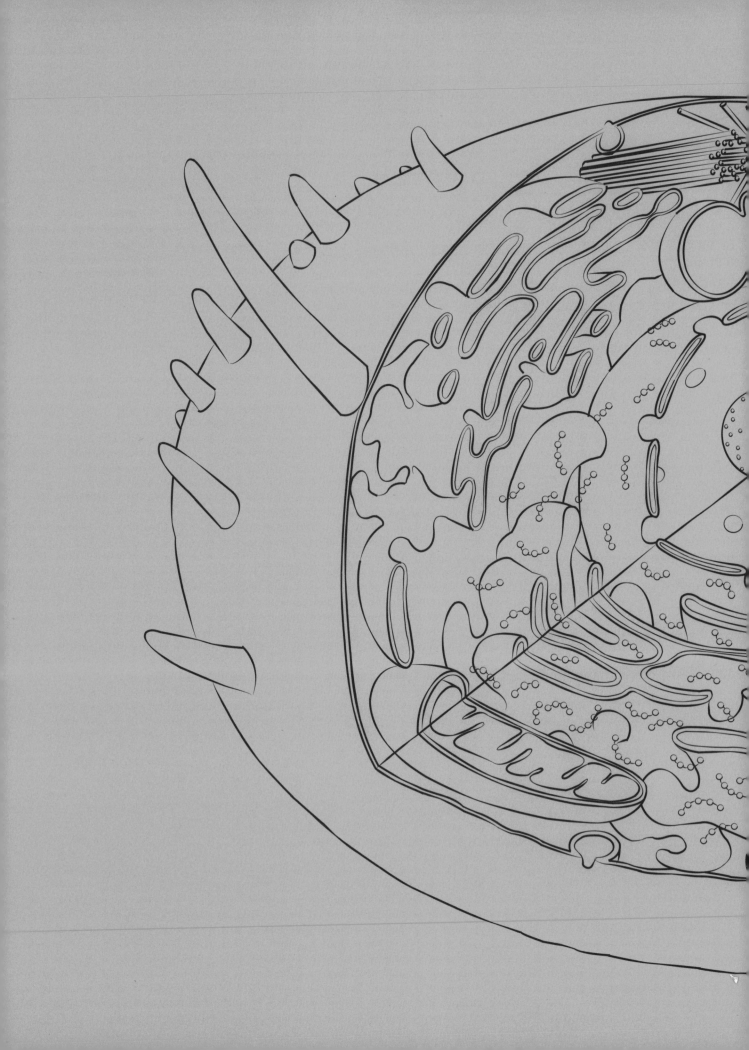

CHAPTER 1:
BODY SYSTEMS
AND TISSUES

Body Systems Overview

◁ Skin

The skin is the outer protective layer of the body. It protects the interior from dehydration, microorganisms, abrasion, and radiation, in addition to serving sensory and metabolic functions.

▷ Skeletal System

The skeletal system comprises the bones and cartilage of the body. It provides rigidity to the body and support for soft internal organs, as well as attachments for skeletal muscles, reservoirs for calcium and fat storage, and a site for blood-cell production.

▷ Muscular System

The muscular system, which brings about body movement, includes the voluntary muscles of the body. Muscle fibers attach either to a bone by a tendon, or to the skin, as is the case with the facial muscles.

◁ Nervous System

The nervous system processes information about the environment and the body's interior, and it initiates movement of the limbs or changes in the body. It comprises the brain, spinal cord, peripheral nerves, and ganglia.

▷ Lymphatic System

The lymphatic system comprises a series of delicate channels that drain tissue fluid through lymph nodes. It contains lymphocytes and macrophages to sweep up foreign proteins, microbes, and cancer cells.

▷ Circulatory System

The circulatory system consists of the heart to pump blood and the arteries, arterioles, capillaries, venules, and veins to carry blood.

◁ **Digestive System**
Concerned with the ingestion, digestion, and absorption of food and water, and the excretion of residual waste, the digestive system extends from the mouth to the anus and has associated glands (salivary, exocrine pancreas, and liver).

▽ **Endocrine System**
The endocrine system is made up of a series of glands that maintain homeostasis—the relatively constant internal state of the body. It includes the pituitary, thyroid, adrenal, and endocrine pancreas glands.

▷ **Respiratory System**
Comprising the nose, larynx, airways, and lungs, the respiratory system is concerned primarily with bringing air into the body for exchange between the air sacs and blood.

▽ **Urinary System**
The kidneys, ureters, urinary bladder, and urethra make up the urinary system, which is primarily concerned with excretion of nitrogenous waste (urea in urine), but is also important for control of the pH and ionic concentrations of the blood.

▽ **Reproductive System**
The male and female reproductive systems are concerned with the production and nurturing of the next generation. They include the gonads, which produce sex cells, as well as associated glands, erectile tissue, the uterus, and breasts.

Body Regions

Key terms:

Abdomen (abdominal) The region of the trunk below the chest and above the pelvis. Abdominal: relating to the abdomen.

Ankle/Tarsus (tarsal) The junction of the leg and foot. Tarsal: relating to the ankle or proximal foot.

Arm/Brachium (brachial) The proximal part of the upper limb consisting of the humerus and surrounding muscles, fascia (connective tissue), nerves, and vessels. Brachial: relating to the arm.

Armpit/Axilla (axillary) The hollow space at the base of the upper limb situated between the arm, chest wall, and scapula. Axillary: relating to the armpit.

Big toe/Hallux Digit 1 of the foot.

Cheek (buccal) Usually refers to the facial region over the maxilla and zygomatic bone but also extends to the lateral wall of the mouth. Buccal: relating to the cheek.

Chest/Thorax (thoracic) The part of the body bounded by the 12 ribs and the sternum. Thoracic: relating to the chest.

Chin (mental) The pointed inferior part of the mandible. Mental: relating to the chin.

Ear (otic) The external ear, pinna, or auricle, or more correctly used for the external, middle, and inner parts of the hearing apparatus. Otic: relating to the ear.

Elbow (antecubital) The junction of the arm and forearm. The antecubital region is the depression (cubital fossa) anterior to the elbow joint.

Eye (ocular or orbital) The organ of sight. The eye and associated glands, muscles, vessels, and nerves lie within the bony orbital cavity. Ocular: relating to the eye itself. Orbital: relating to the bony orbit.

Face (facial) The front of the head. The facial skeleton is covered by facial muscles and highly sensitive skin. Facial: relating to the face.

Fingers/Digits (digital or phalangeal) Digits of the hand. They have phalangeal bones as their core (two in the thumb or digit 1; three in digits 2–5). Digital: relating to fingers or toes. Phalangeal: relating to the phalangeal bones.

Foot/Pes (pedal) The terminal part of the lower limb. The foot consists of tarsal, metatarsal, and phalangeal bones, as well as associated muscles, tendons, ligaments, vessels, and nerves. Pedal: relating to structures in the foot.

Forearm/Antebrachium (antebrachial) The part of the upper limb between the elbow and wrist. Antebrachial: relating to some constituents (e.g., nerves) of the forearm.

Forehead (frontal) The part of the face over the frontal bones of the skull. It has a thin layer of muscle (frontalis) beneath the skin. Frontal: relating to the forehead.

Groin/Inguen (inguinal) The area at the junction of the lower limb and the anterior abdominal wall. Inguinal: relating to the groin region.

Hand The distal extremity of the upper limb. It consists of the metacarpal bones (palm); phalangeal bones (digits); and associated muscles, tendons, vessels, and nerves.

Head The face and braincase. It contains major sensory organs, as well as the entrances to the respiratory and gastrointestinal tracts.

Kneecap/Patella (patellar) The prominence of the knee. It is produced by the patella, a sesamoid bone in the tendon of the quadriceps femoris muscle. Patellar: relating to the kneecap.

Leg/Crus (crural) The part of the lower limb between the knee and ankle. Crural: relating to structures in the region.

Mouth (oral) The opening in the face bounded above by the hard and soft palate, below by the tongue, and on each side by the cheek (buccal wall). Oral: relating to the mouth.

Neck (cervical) The part of the body that connects the head and the trunk. At its core are the cervical vertebrae, which are surrounded by muscles, vessels, and nerves. Cervical: relating to the neck.

Nose (nasal) A protuberance on the face, formed from cartilage and bone, that houses the entrance to the nasal cavity. Nasal: relating to the nose.

Palm (palmar) The proximal part of the hand, formed of the metacarpal bones and associated muscles, tendons, vessels, and nerves. Palmar: relating to the palm.

Pelvis (pelvic) Part of the skeleton formed by two hip bones on each side and the sacrum and coccyx in the midline. Pelvic: relating to the pelvis.

Pubis (pubic) The pubic region at the front of the bony pelvis. Its contents include the pubic part of the hip bone and overlying tissues. Pubic: relating to the pubis.

Skull/Cranium (cranial) Consists of the facial skeleton and the braincase (calvaria). It protects the brain. Cranial: relating to the cranium.

Thigh (femoral) The region of the proximal lower limb that has the femur at its core with the quadriceps femoris muscle anteriorly, and hamstring and adductor muscles posteriorly and medially. Femoral: relating to the thigh.

Thumb/Pollex (pollical) Digit 1 of the hand. Pollical: relating to the thumb.

Toes/Digits (digital or phalangeal) Digits (1–5) of the lower limb. They contain phalangeal bones. Digital: relating to the toes or fingers. Phalangeal: relating to the phalanges.

Trunk/Torso Comprises the thoracic and abdominopelvic cavities and their musculoskeletal walls.

Umbilicus (umbilical) A puckered scar where the umbilical cord was attached during prenatal life. Umbilical: relating to the umbilicus.

Wrist (carpal) Comprises the carpal bones and tendons transmitting muscle power from the forearm to the fingers. Carpal: relating to the wrist.

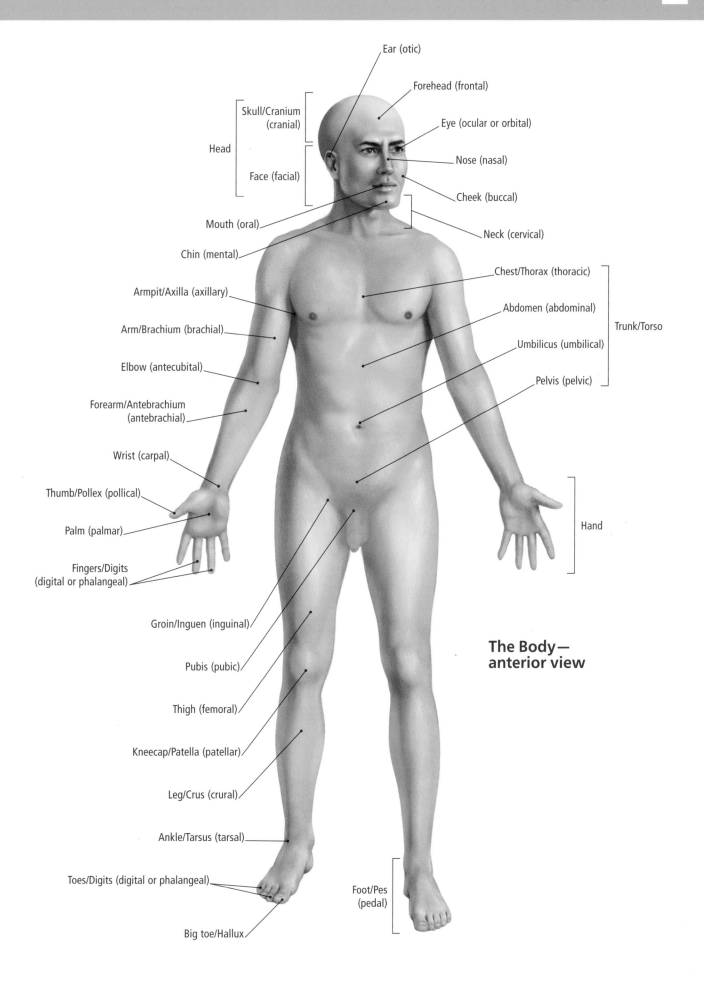

Ear (otic)

Forehead (frontal)

Skull/Cranium (cranial)

Eye (ocular or orbital)

Head

Nose (nasal)

Face (facial)

Cheek (buccal)

Mouth (oral)

Neck (cervical)

Chin (mental)

Chest/Thorax (thoracic)

Armpit/Axilla (axillary)

Abdomen (abdominal)

Arm/Brachium (brachial)

Umbilicus (umbilical)

Trunk/Torso

Elbow (antecubital)

Pelvis (pelvic)

Forearm/Antebrachium (antebrachial)

Wrist (carpal)

Thumb/Pollex (pollical)

Hand

Palm (palmar)

Fingers/Digits (digital or phalangeal)

Groin/Inguen (inguinal)

The Body— anterior view

Pubis (pubic)

Thigh (femoral)

Kneecap/Patella (patellar)

Leg/Crus (crural)

Ankle/Tarsus (tarsal)

Toes/Digits (digital or phalangeal)

Foot/Pes (pedal)

Big toe/Hallux

Body Regions

Key terms:

Abdominal cavity The upper part of the abdominopelvic cavity containing the stomach, intestines, liver, kidneys, and pancreas.

Abdominopelvic cavity The largest ventral cavity of the body, extending from the diaphragm above to the pelvic floor muscles below.

Back (dorsal) The posterior or dorsal part of the body, including the vertebral column and associated muscles.

Buttock/Gluteus (gluteal) The fleshy part of the body posterior to the pelvis and upper femur, composed of gluteal muscles and fat. Gluteal: relating to the buttock.

Calf/Sura (sural) The fleshy part of the posterior leg, largely composed of the triceps surae muscle. Sural: relating to the calf.

Cranial cavity The interior of the skull or cranium. It houses the brain, pituitary gland, and associated blood vessels.

Diaphragm A sheet of muscle and tendon separating the thoracic and abdominal cavities. It is perforated by the aorta, inferior vena cava, and esophagus.

Dorsal cavity The body cavity containing the brain, spinal cord, and their membranes (meninges).

Elbow/Olecranon (olecranal) The elbow is the junction of the arm and forearm. The posterior aspect of the elbow is marked by a bony projection called the olecranon. Olecranal: relating to the olecranon.

Head See pp. 14–15.

Heel/Calcaneus (calcaneal) The core of the heel is the calcaneus bone. Calcaneal: relating to some structures in this region.

Lower back (lumbar) The region of the body that has the lumbar vertebrae and flanking muscles at its core. Lumbar: relating to the lower back.

Lower limb Consists of the thigh (femoral region), leg (crus), foot (pes), and digits.

Mediastinum The space in the chest between the two pleural sacs and their enclosed lungs. It contains the heart, major vessels, and esophagus.

Neck (cervical) See pp. 14–15.

Pelvic cavity The space bounded by the hip bones, sacrum, coccyx, and pelvic diaphragm. It contains the urinary bladder, reproductive organs, rectum, and anus.

Pericardial cavity The sac that lies in the mediastinum of the thorax and encloses the heart. It contains a fluid space to allow free movement of the beating heart.

Shoulder (acromial) The junction of the clavicle, scapula, and humerus. Its rounded shape is produced by the deltoid muscle, arising from the spine and acromial process of the scapula and distal clavicle, and covering the head of the humerus. Acromial: relating to the shoulder.

Sole (plantar) The inferior or plantar surface of the foot. Plantar: relating to the sole.

Spinal canal A hollow cavity that extends the length of the vertebral column. It contains the spinal cord and cauda equina.

Thoracic cavity The upper ventral cavity of the trunk. It is divided into paired pleural sacs enclosing the lungs and the mediastinum in the midline.

Trunk/Torso See pp. 14–15.

Upper limb Comprises the arm (brachium), forearm (antebrachium), hand (manus), and digits.

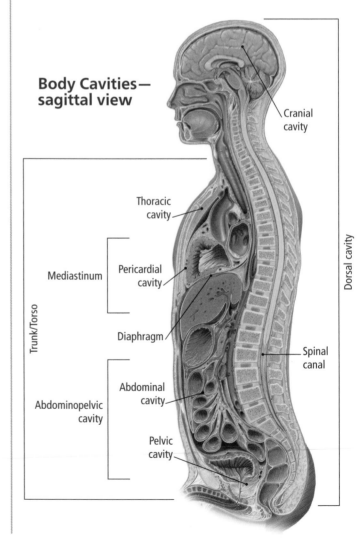

Body Cavities—sagittal view

Cranial cavity

Thoracic cavity

Mediastinum

Pericardial cavity

Diaphragm

Trunk/Torso

Abdominopelvic cavity

Abdominal cavity

Pelvic cavity

Dorsal cavity

Spinal canal

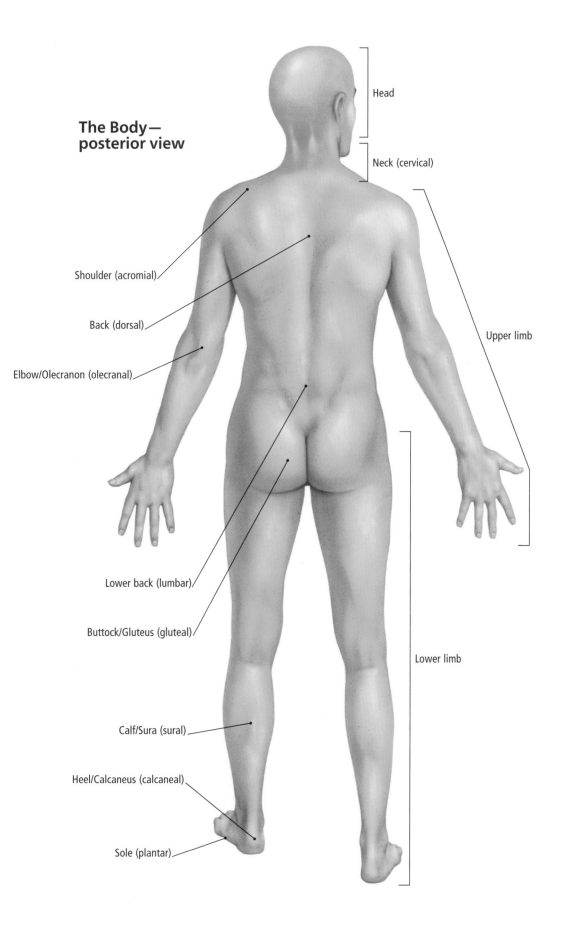

**The Body—
posterior view**

Head

Neck (cervical)

Shoulder (acromial)

Back (dorsal)

Elbow/Olecranon (olecranal)

Upper limb

Lower back (lumbar)

Buttock/Gluteus (gluteal)

Lower limb

Calf/Sura (sural)

Heel/Calcaneus (calcaneal)

Sole (plantar)

View Orientation and Anatomical Planes

The study of anatomy requires clear communication, which is facilitated by the use of a standard body arrangement and agreed-upon descriptive terms. The anatomical position is a standard body position that must be used when describing the location of body parts. In the anatomical position, the subject stands with the two feet together and facing forward. The two hands are by the side, with the palms facing the front and the thumb to the lateral side. The head is erect, and the eyes look forward.

Key terms:

Anterior Toward the front of the body.

Distal Toward the end of an extremity.

Dorsal Toward the back of the body in the trunk but superiorly in the head. The difference is due to the fact that the dorsum is defined relative to the axis of the nervous system, which bends at the midbrain in humans.

Dorsal surface (of foot or hand) The top of the foot or back of the hand.

Frontal (coronal) plane The anatomical plane that divides the body into front and back parts.

Inferior Toward the feet or downward.

Lateral Toward the side of the body.

Medial Toward the midline of the body.

Palmar surface The anterior surface (palm) of the hand.

Plantar surface The inferior surface (sole) of the foot.

Posterior Toward the back of the body.

Proximal Toward the attachment of a limb to the trunk, or toward the beginning of a tubular structure.

Sagittal plane A family of planes that divide the body into left and right parts. One of the planes is in the midline (midsagittal or median) and divides the body into two equal halves. The remaining members of the family (parasagittal or paramedian) divide the body into unequal left and right parts.

Sagittal (midsagittal or median) plane The plane that divides the body into two equal halves.

Sagittal (parasagittal or paramedian) plane One of a family of planes parallel to the midline.

Superior Toward the upper end of the body.

Transverse (axial) plane The family of planes that divide the body into upper and lower parts.

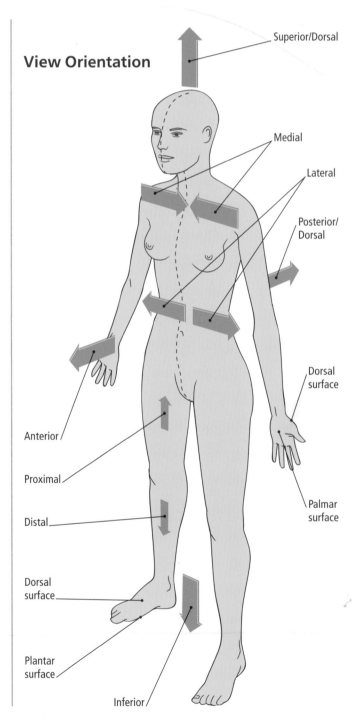

View Orientation

Superior/Dorsal

Medial

Lateral

Posterior/Dorsal

Dorsal surface

Palmar surface

Anterior

Proximal

Distal

Dorsal surface

Plantar surface

Inferior

Anatomical Planes

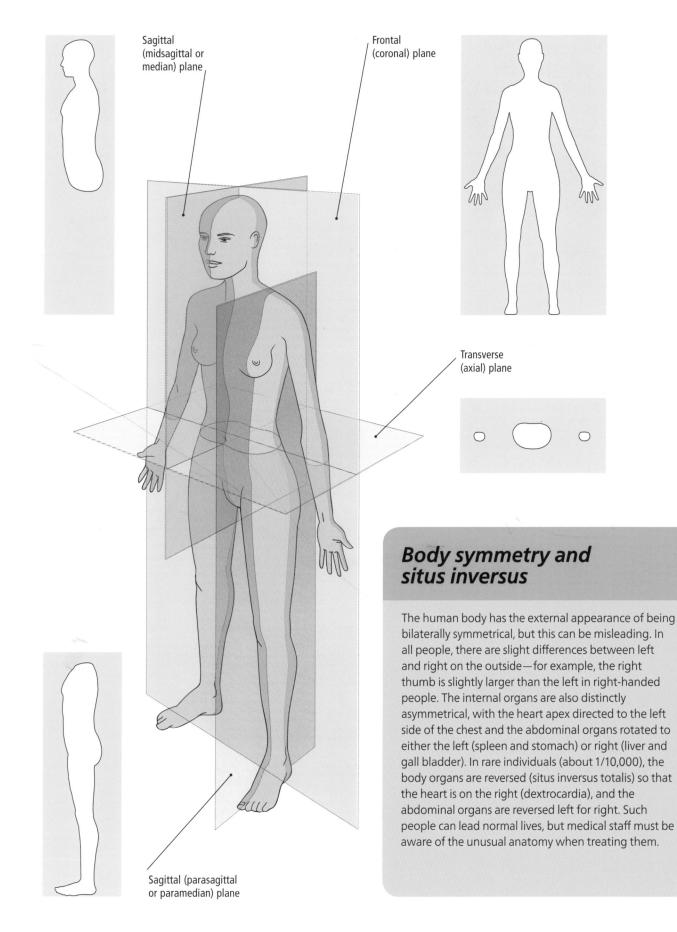

Sagittal
(midsagittal or
median) plane

Frontal
(coronal) plane

Transverse
(axial) plane

Sagittal (parasagittal
or paramedian) plane

Body symmetry and situs inversus

The human body has the external appearance of being bilaterally symmetrical, but this can be misleading. In all people, there are slight differences between left and right on the outside—for example, the right thumb is slightly larger than the left in right-handed people. The internal organs are also distinctly asymmetrical, with the heart apex directed to the left side of the chest and the abdominal organs rotated to either the left (spleen and stomach) or right (liver and gall bladder). In rare individuals (about 1/10,000), the body organs are reversed (situs inversus totalis) so that the heart is on the right (dextrocardia), and the abdominal organs are reversed left for right. Such people can lead normal lives, but medical staff must be aware of the unusual anatomy when treating them.

Tissues

◁ Collagen
This structural protein makes up the bulk of connective tissue. Five different types are recognized. All collagen contains the amino acids hydroxyproline and hydroxylysine, which require vitamin C for their manufacture.

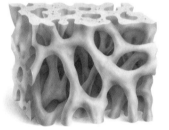

◁ Bone Tissue
A composite material comprising cells (osteocytes) in an extracellular matrix (collagen and mineral salts), bone acts as a structural material and stores calcium and phosphorus.

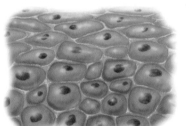

◁ Epithelial Tissue
Epithelial tissue comprises the surface cells of the body exterior (epidermis), as well as the linings and glands of the digestive, respiratory, and genito-urinary tracts. Epithelial cells may be flat (squamous), cuboidal, or columnar and may have specialized features on their apical cell surface to facilitate absorption or movement of mucus.

◁ Loose Connective Tissue
This connective tissue is made up of fibers arranged obliquely to each other. It has minimal structural strength, and its main role is to fill the spaces between organs.

◁ Dense Connective Tissue
Dense connective tissue is designed to withstand high tensile force. Strength is achieved by multiple parallel arrays of collagen fibers. Examples include ligaments and tendons.

◁ Adipose Tissue
This fat tissue includes yellow fat, which is found under the skin in adults and stores energy, and brown fat, which has many mitochondria and is found on the upper back in newborn infants. Brown fat generates heat for thermoregulation.

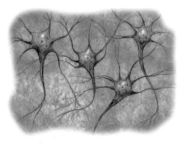

◁ Neural Tissue
Comprising nerve cells (neurons) and their supporting cells, neural tissue is concerned with the processing of sensory information and the rapid control of the body's systems.

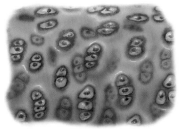

◁ Cartilage Tissue: Hyaline Cartilage
Found on the surfaces of synovial joints, hyaline cartilage has the ability to withstand cycles of compression and relaxation as the joint moves. During these cycles, synovial fluid is exchanged between the hyaline cartilage and the joint space.

◁ Muscle Tissue: Smooth Muscle
This nonstriated involuntary muscle is found in the walls of the respiratory, gastrointestinal, and genitourinary tracts, as well as blood vessels.

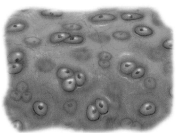

◁ Cartilage Tissue: Elastic Cartilage
Found in parts of the body where flexibility and elasticity are required, such as in the external ear (pinna or auricle), this is a type of cartilage with a high content of elastic fibers.

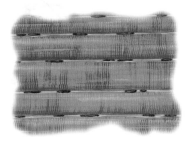

◁ Muscle Tissue: Skeletal Muscle
A type of voluntary striated (striped) muscle, usually with at least one attachment to the skeleton, skeletal muscle produces all voluntary movements either by contraction or by controlled relaxation.

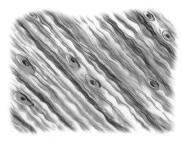

◁ Cartilage Tissue: Fibrocartilage
Found in parts of the body subjected to high compressive forces, such as the edges of intervertebral disks, fibrocartilage has a high content of fibrous tissue (that is, collagen fibers).

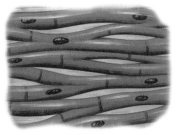

◁ Muscle Tissue: Cardiac Muscle
Cardiac muscle is a type of involuntary striated (striped) muscle tissue and is found solely within the walls of the heart chambers. It is activated by the conducting and pace-making tissues within the heart and is under the influence of the autonomic nervous system.

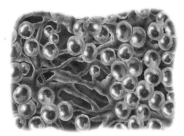

◁ Immune System Tissue
Concerned with the drainage of excess tissue fluid from the extracellular space and the immune surveillance of that fluid for microorganisms and cancer cells, immune system tissue consists of lymphatic vessels. These vessels drain through clusters of lymphoid (immune system) cells known as lymph nodes.

True or false?

1   The median plane divides the body into two roughly equal halves.

2   A plane through the waistline would be a horizontal or transverse plane.

3   The mediastinum is the muscular structure that separates the thoracic and abdominal cavities.

4   The ventral body cavity can be further divided into thoracic, abdominal, and pelvic cavities.

5   The ventral body cavity is continuous from the neck to the pelvic outlet.

6   The dorsal body cavity contains the heart and lungs.

7   The lymphatic system is mainly made up of immune system cells aggregated into large internal organs.

Afferent lymphatic vessels

Follicle of cortex

Trabecula

Capsule

Vein

Artery

Capillary

Efferent lymphatic vessel

8   Homeostasis is the ability of the body to maintain a relatively constant internal state.

9   Both the nervous and endocrine systems are involved with communication within the body.

10 The region of the cell that contains the genetic material (DNA) is the centriole.

Lymph Node

Clustered along the route of the lymphatic vessels, the lymph nodes filter and clean incoming lymph supplied through afferent vessels. Once filtered, efferent vessels carry the lymph to the venous system.

11   *The nucleolus is found in the periphery of the cell body.*   f

12   *Lysosomes contain enzymes that break down waste material, viruses and bacteria, and cellular debris.*   T

13   *Bone is a composite material of the mineral calcium carbonate and the protein collagen.*   F

14   *The ends of long bones are usually covered with a layer of elastic cartilage.*   F

15   *Cilia are motile extensions of the cell surface that are responsible for moving mucus and debris.*   T

16   *The tubular structure of long bones allows minimal strength for a given weight.*   F

17   *Smooth muscle cells are concentrated in the outer membranes of long bones.*   F

18   *The nervous system is concerned primarily with control of body changes over periods of days to months.*   F

19   *Gut movement is mainly due to the actions of smooth muscle.*   T

20   *Glands of the gut are derived from the epithelial tissue.*   T

Defects in cilia function

Cilia are cellular organelles that beat rhythmically, either to move the cell through fluid (such as in paramecia) or to waft fluid past cells (as in the human respiratory epithelium). Kartagener syndrome is a rare genetic condition that leads to defective function of cilia in the respiratory tract and the uterine tube, and defective flagella of sperm. Affected individuals experience recurrent respiratory tract infections, middle ear infections, and sinusitis. In both sexes, there would be reduced fertility because sperm cannot swim and the uterine tube epithelium cannot move the egg.

Multiple choice

1   *The roles of the respiratory system include all of the following EXCEPT:*
(A) gas exchange
(B) thermoregulation
(C) pH regulation
(D) communication
(E) excretion of uric acid

2   *The urinary system includes all of the following organs EXCEPT:*
(A) kidney
(B) seminal vesicle
(C) urinary bladder
(D) ureter
(E) urethra

3   *Which of the following structures would be found in the ventral body cavity?*
(A) brain
(B) spinal cord
(C) pineal gland
(D) spleen
(E) pituitary gland

4   *Which direction in the body is defined with respect to the nervous system axis and is directed toward the back of the body in the trunk, but superiorly in the head?*
(A) anterior
(B) ventral
(C) posterior
(D) caudal
(E) dorsal

5   *A plane that passes through the two eyes and the pubic region would belong to which group?*
(A) parasagittal
(B) median
(C) frontal or coronal
(D) horizontal or transverse
(E) oblique

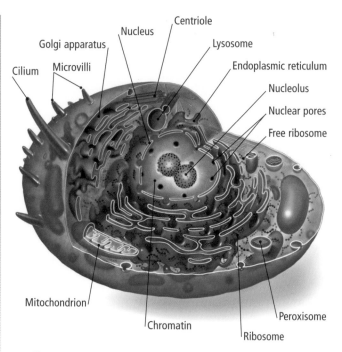

Cell Structure

Cells are the fundamental units of all living things. The human body contains trillions of cells, each containing our unique identity (our DNA).

6   *Which of the following cellular organelles is responsible for the production of protein?*
(A) nucleus
(B) nucleolus
(C) centriole
(D) ribosome
(E) mitochondrion

7   *Which structure would be found within the nucleus?*
(A) ribosome
(B) nucleolus
(C) mitochondrion
(D) microvillus
(E) Golgi apparatus

8   *The role of microvilli is to:*

(A)   increase cellular surface area for absorption

(B)   allow the cell to move freely in the body

(C)   transmit nuclear material between adjacent cells

(D)   allow cells to bind together to form tissue layers

(E)   provide a surface for energy production by metabolizing glucose

9   *The role of the Golgi apparatus is to:*

(A)   store and retrieve genetic information to control cellular activities

(B)   produce lipids and proteins

(C)   package protein and lipid for transport and secretion

(D)   transmit information around the cell body

(E)   produce energy in the form of adenosine triphosphate (ATP)

10   *The role of the centriole is to:*

(A)   produce protein and lipid for the cell membrane

(B)   generate energy for the cell's activities

(C)   regulate the cellular activities using information stored in DNA

(D)   facilitate chromosome movement during cell division

(E)   break down lipid and foreign protein from ingested debris

11   *A defect in mitochondrial DNA is most likely to cause:*

(A)   problems with energy production in the cell

(B)   problems with cellular division

(C)   defective protein synthesis

(D)   abnormal degradation of cellular debris

(E)   no problem at all

12   *Cells with cilia are usually found in the:*

(A)   epidermis of skin

(B)   gut lining

(C)   bladder lining

(D)   paranasal sinuses

(E)   ventricles of the brain

13   *Cells with microvilli are usually found in the:*

(A)   cerebral cortex

(B)   lining of the small intestine

(C)   lining of the trachea

(D)   interior of the spleen

(E)   wall of the urinary bladder

14   *Which of the following cell types would have a flagellum?*

(A)   respiratory epithelium

(B)   choroid plexus of lateral ventricle

(C)   kidney epithelial cell

(D)   lining of duodenum

(E)   spermatozoon

Mitochondrial disease

Mitochondria are the powerhouses of the cell and may be derived from microorganisms that took up symbiotic residence inside our ancestral cells during early evolution. Mitochondria have their own DNA, and defects in this can cause significant disease. Mitochondria are present in all cells except red blood cells, so the effects of mitochondrial disease are widespread and serious: muscle weakness, neurological problems, learning disorders, early-onset dementia, heart dysfunction, and liver and kidney disease. Treatment is difficult, but future avenues could involve transfer of normal genes to affected individuals (gene therapy).

Color and label

i) Label each structure shown on the illustrations.

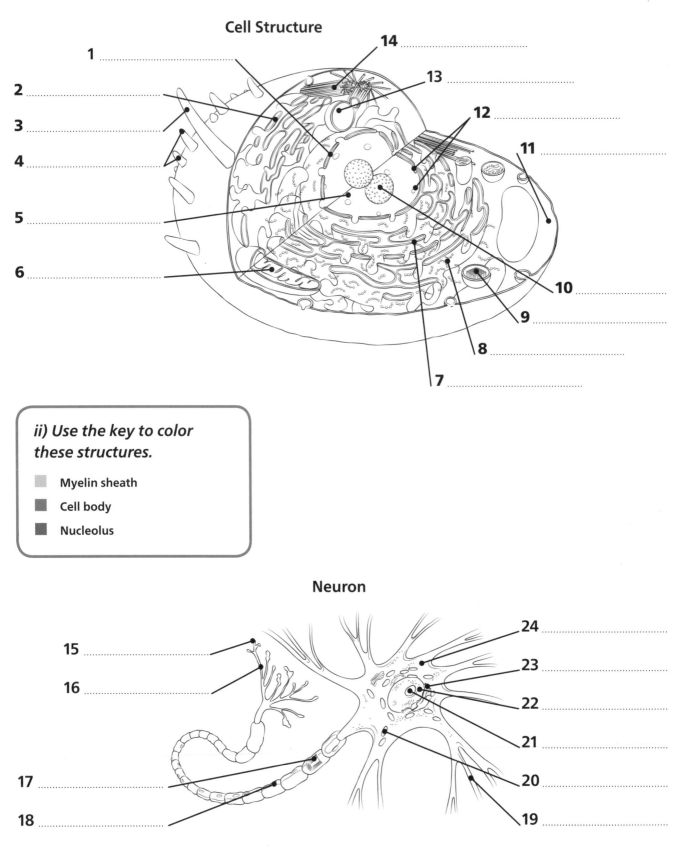

Cell Structure

1 ....

2 ....

3 ....

4 ....

5 ....

6 ....

14 ....

13 ....

12 ....

11 ....

10 ....

9 ....

8 ....

7 ....

ii) Use the key to color these structures.

☐ Myelin sheath

■ Cell body

■ Nucleolus

Neuron

15 ....

16 ....

17 ....

18 ....

24 ....

23 ....

22 ....

21 ....

20 ....

19 ....

Blood Cells

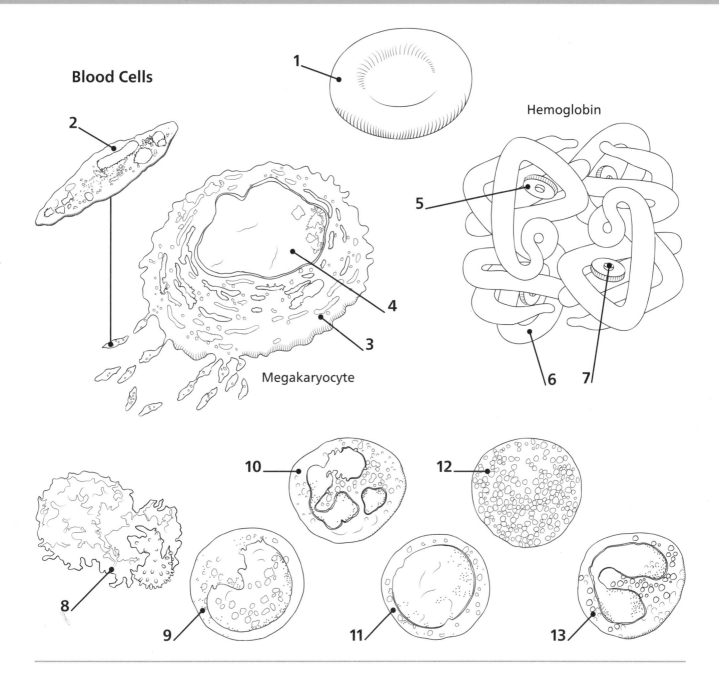

Hemoglobin

Megakaryocyte

---

iii) *Add numbers to the boxes below to match each label to the correct part of the illustrations.*

Globin protein strand	☐	Macrophage	☐
Cytoplasm	☐	Nucleus	☐
Eosinophil	☐	Iron ion	☐
Monocyte	☐	Neutrophil	☐
Red blood cell	☐	Basophil	☐
Heme	☐	Platelet	☐
Lymphocyte	☐		

Multiple choice

1   *Which of the following is NOT a type of connective tissue?*

(A)   bone
(B)   tendon
(C)   blood
(D)   sweat gland
(E)   adipose tissue (fat)

2   *Which of the following is NOT a type of epithelial tissue?*

(A)   sweat gland
(B)   sebaceous gland
(C)   adipose tissue
(D)   cornea
(E)   olfactory mucosa

3   *Which of the following is a defining feature of epithelial tissue?*

(A)   it is exquisitely sensitive to touch
(B)   it is found only on the exterior of the body
(C)   it always consists of a single layer of cells
(D)   it has no secretory function
(E)   it covers the internal and external body surfaces

4   *Which of the following would be an example of loose connective tissue?*

(A)   tendon
(B)   ligament
(C)   subcutaneous tissue
(D)   an aponeurosis
(E)   blood serum

5   *The role of dense connective tissue is to:*

(A)   resist tensile forces and store potential energy
(B)   provide a storage site for lipids and calcium
(C)   produce voluntary movement under command from the brain
(D)   resist compressive forces during jumping
(E)   both A and D are correct

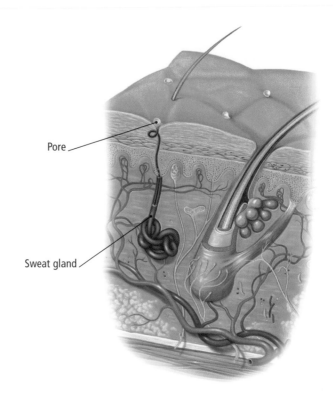

Pore

Sweat gland

Sweat Gland

Involved in temperature regulation, the sweat glands release sweat to the surface of the skin, which cools the skin as it evaporates.

6   *Which of the following is an example of peripheral nervous system tissue?*

(A)   dorsal root (spinal) ganglion
(B)   cerebral cortex
(C)   autonomic ganglion
(D)   spinal cord gray matter
(E)   both A and C are correct

7   *Which of the following statements is true concerning fat?*

(A)   it consists of brown and yellow types
(B)   it can develop directly from muscle cells
(C)   it is often deposited inside the fluid of the peritoneal cavity
(D)   it is richly supplied with blood
(E)   none of the above is correct

8 *Which of the following is true concerning muscle tissue?*

Ⓐ it includes voluntary smooth muscle

Ⓑ smooth muscle can beat rhythmically at up to 150 times per minute

Ⓒ the heart contains voluntary muscle

Ⓓ both skeletal and cardiac muscle are striated

Ⓔ smooth muscle has a richer blood supply than cardiac muscle

9 *Elastic cartilage is found in which parts of the body?*

Ⓐ external ear

Ⓑ synovial joint surface

Ⓒ external nose

Ⓓ pubic symphysis

Ⓔ both A and C are correct

10 *Fibrocartilage is found in which parts of the body?*

Ⓐ head of femur

Ⓑ intervertebral disk

Ⓒ heart

Ⓓ external ear

Ⓔ skull

Where does cancer come from?

Cancer is more likely to come from some tissue types than others. Most cancers arise from epithelial tissue because it lies at the body surface (e.g., skin, airways, and gut lining) and is constantly exposed to cancer-causing agents (carcinogens). Normal epithelial tissues are also actively dividing, so any cancer-causing gene (oncogene) is likely to lead to a rapidly growing tumor. Cancers may also arise from cells in bone, muscle, or fat (osteosarcoma, myosarcoma, liposarcoma), but this is less common.

11 *Where is brown fat usually found?*

Ⓐ between the shoulder blades of newborns

Ⓑ in the medullary cavities of bone

Ⓒ in the mesenteries of the abdomen

Ⓓ in the capsule of the spleen

Ⓔ between the lymph nodes of the mediastinum

12 *Approximately what percentage of body weight is made up of skeletal muscle in a person of healthy body mass index?*

Ⓐ 10 percent

Ⓑ 20 percent

Ⓒ 30 percent

Ⓓ 40 percent

Ⓔ 50 percent

13 *Which of the following tissue types is a good thermal insulator?*

Ⓐ skeletal muscle

Ⓑ nervous tissue

Ⓒ adipose tissue

Ⓓ epidermal tissue

Ⓔ endocrine tissue

14 *Which of the following tissue has the highest metabolic demand when the body is at rest?*

Ⓐ adipose tissue

Ⓑ gut wall

Ⓒ lung tissue

Ⓓ tongue muscle

Ⓔ brain tissue

Color and label

i) Label each structure shown on the illustrations.

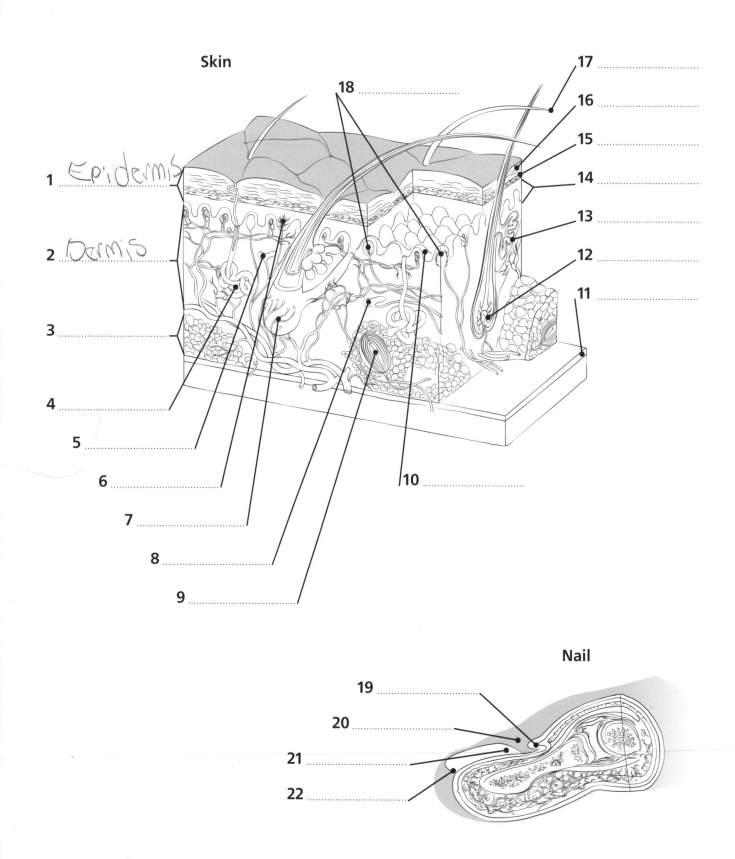

Skin

1 Epidermis

2 Dermis

3

4

5

6

7

8

9

10

11

12

13

14

15

16

17

18

Nail

19

20

21

22

ii) Add numbers to the boxes below to match each label to the correct part of the illustrations.

Nerve	
Medulla	
Internal root sheath	
Cuticle	
Dermal hair papilla	
Epidermis	1
Follicle sheath	
Hair bulb	
Melanocyte	
Internal root sheath	
Erector pili muscle (arrector pili)	13
Follicle sheath	
Skull bone	
Sebaceous gland	
Precuticular epithelium	
Cortex	
External root sheath	
Hair shaft	
External root sheaths	
Aponeurosis	
Loose areolar tissue	
Pericranium	
Hair	
Hair follicle	
Skin	

Hair

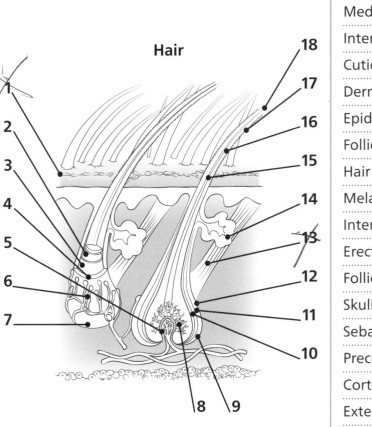

1
2
3
4
5
6
7

18
17
16
15
14
13
12
11
10

8   9

Scalp

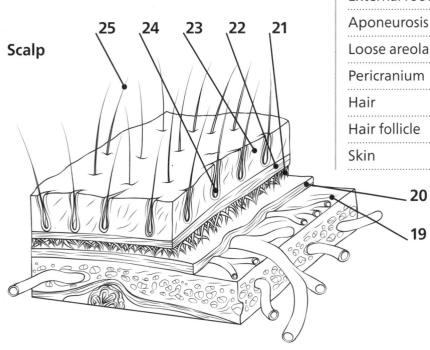

25 24 23 22 21

20

19

Fill in the blanks

1   The correct anatomical term for the region between the shoulder and elbow is the

     _____ .

2   The _____ is the region where the umbilical cord was attached during prenatal life.

3   The _____ region is at the junction of the anterior abdominal wall and the
     lower limb.

4   The correct anatomical term for the region between the elbow and the wrist is the

     _____ .

5   The _____ is a standard position of the body and its parts for use when
     describing anatomical structures.

6   In anatomical terminology, the _____ is the region between the knee
     and the ankle.

7   The _____ is the region between the proximal arm and the chest wall.

8   The term _____ is used to refer to the neck.

9   The depression posterior to the knee is called the _____ fossa.

10   The anatomical term for the thumb (digit 1) is the _____ .

11   The body cavity that encloses the heart is called the _____ .

12   The _____ system is made up of a series of glands that secrete mainly into the
     bloodstream.

13   The inferior surface of the foot is called the _____ surface.

14 The _Skin_ protects the rest of the body against radiation, heat loss, abrasion, and microorganisms.

15 The _Cubital_ fossa is the space anterior to the elbow when the body is in the anatomical position.

16 Free _ribosomes_ have the ability to use messenger RNA to construct protein molecules.

17 Nerve cells are characterized by having an _axon_ to convey nerve impulses away from the cell body.

18 Between cell divisions, the cell's DNA and associated proteins are dispersed within the nucleus as _Chromatin_ .

19 The body fiber type that has optimal elastic properties is _____ .

20 The protein _collagen_ makes up the bulk of connective tissue and is a key component of bone.

Osteogenesis imperfecta

For its strength and resilience, bone tissue depends on the blending of biological fiber (mainly collagen) and mineral crystals (hydroxyapatite). Osteogenesis imperfecta is a rare but serious genetic disease (affecting around 7 people per 100,000) in which bone strength is impaired due to defective production of collagen. In milder cases, the condition leads to increased risk of fractures, hearing loss, and a blue or gray tint to the sclera of the eyeball. In severe cases, the fractures may be so common as to cause short stature, serious disability, dental abnormalities, and respiratory problems.

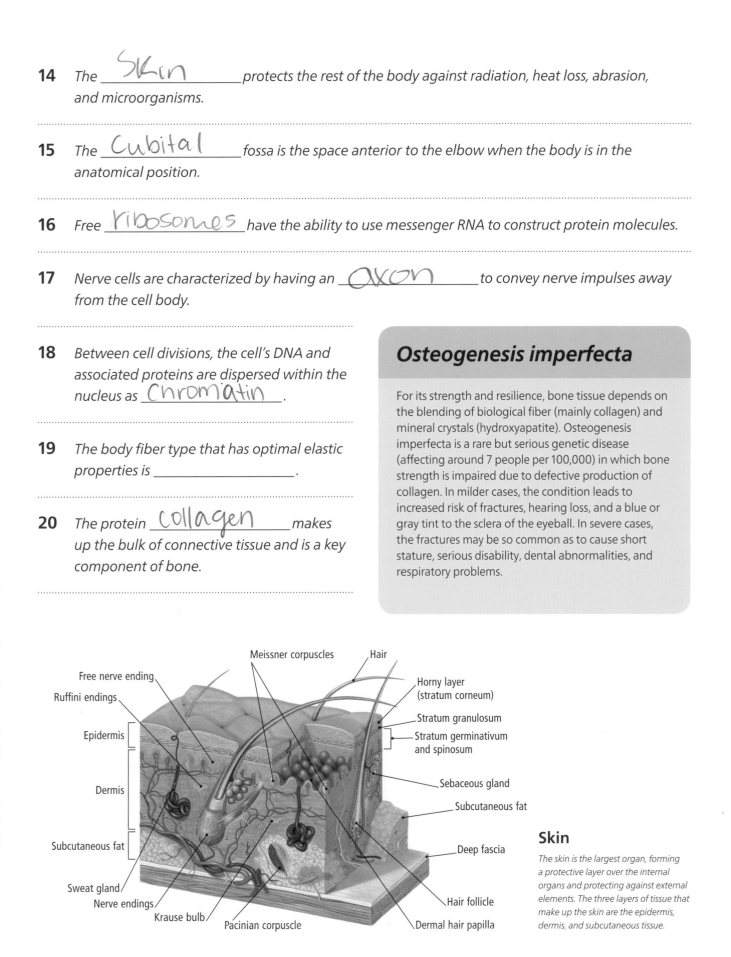

Meissner corpuscles
Hair
Free nerve ending
Ruffini endings
Horny layer (stratum corneum)
Stratum granulosum
Epidermis
Stratum germinativum and spinosum
Dermis
Sebaceous gland
Subcutaneous fat
Subcutaneous fat
Deep fascia
Sweat gland
Nerve endings
Krause bulb
Pacinian corpuscle
Hair follicle
Dermal hair papilla

Skin

The skin is the largest organ, forming a protective layer over the internal organs and protecting against external elements. The three layers of tissue that make up the skin are the epidermis, dermis, and subcutaneous tissue.

Match the statement to the reason

1   *Scholars of anatomy can always reliably describe the position of body parts to other anatomists because…*

C

a   *component glands secrete hormones into the bloodstream for distribution throughout the body.*

2   *The endocrine system can act on diverse parts of the body because…*

A

b   *lubricating fluid from the joint space can enter or leave the tissue depending on joint loading.*

3   *Cardiac muscle has a striated appearance under the microscope because…*

e

c   *they use a standard anatomical position and agreed-upon terminology when communicating their findings.*

4   *Scars often become white with time because…*

d

d   *collagenous connective tissue is formed as an important stage in tissue repair.*

5   *Hyaline cartilage is an ideal tissue to cover the surfaces of joints because…*

b

e   *component cells have regularly arranged contractile proteins that can slide past each other.*

Components of Blood

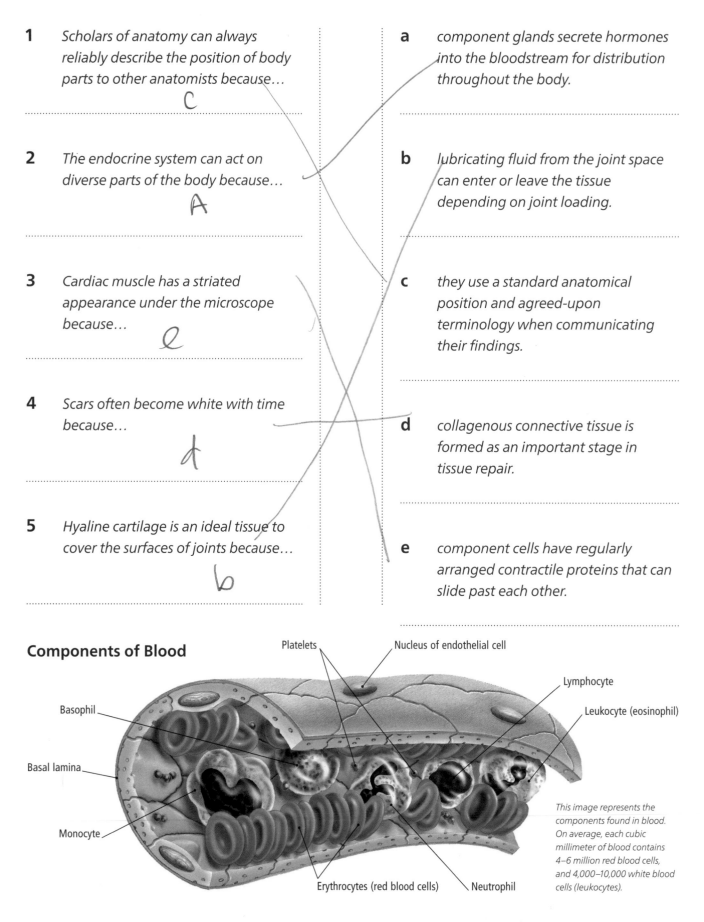

Platelets

Nucleus of endothelial cell

Lymphocyte

Leukocyte (eosinophil)

Basophil

Basal lamina

Monocyte

Erythrocytes (red blood cells)

Neutrophil

This image represents the components found in blood. On average, each cubic millimeter of blood contains 4–6 million red blood cells, and 4,000–10,000 white blood cells (leukocytes).

1   Bone combines the features of mechanical strength and resistance to cracking because…

a   it is a highly flexible type of tissue that allows these body parts to deform but return to their original shape and size.

2   Blood is regarded as a connective tissue because…

b   it consists of cellular components (red and white blood cells, platelets) embedded in a fluid matrix (plasma) containing dissolved proteins, lipoproteins, nutrients, hormones, and gases.

3   Elastic cartilage is found in the external ear (auricle or pinna) and the nose because…

c   it consists of nerve cells (neurons) connected to each other by axons that are coated with a fatty layer (myelin) that increases the speed of conduction.

4   The cells of the immune system are in an excellent position to provide surveillance of the body's tissues for invading microorganisms and cancer because…

d   it is a composite material consisting of organic fibers (mostly the protein collagen) embedded in a mineral matrix (hydroxyapatite).

5   Neural (nervous) tissue provides a means of rapid internal communication within the body because…

e   they are distributed in nodular structures (lymph nodes) along the lymphatic channels that drain excess tissue fluid back toward the venous side of the circulation.

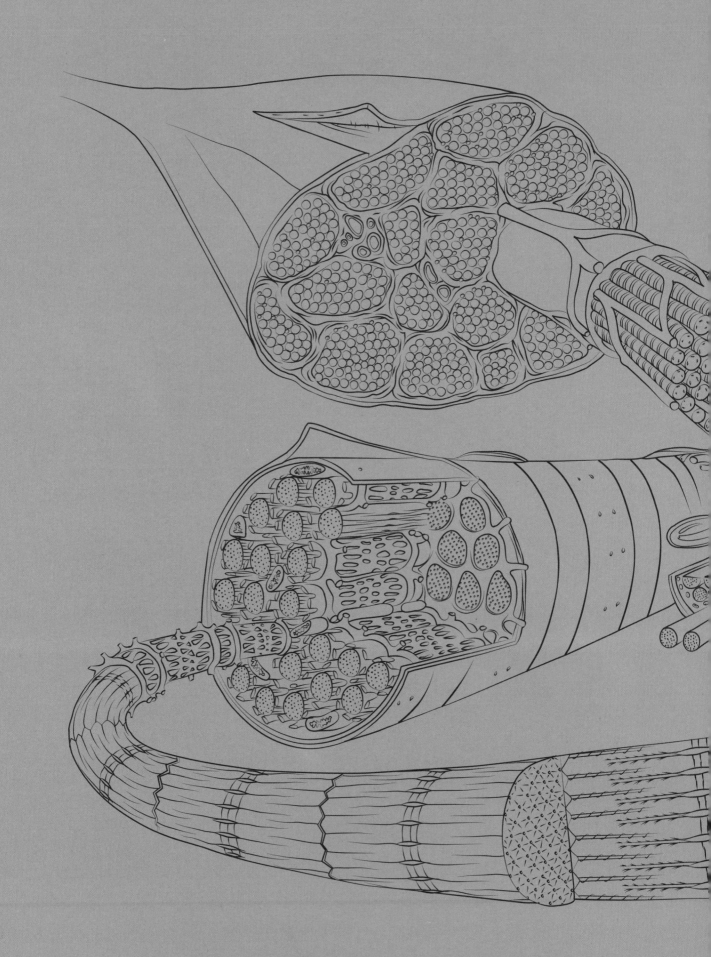

CHAPTER 2: THE MUSCULOSKELETAL SYSTEM

Axial Skeleton and Joints

The axial skeleton consists of the skull, hyoid bone, vertebral column, ribs, and sternum. The role of the axial skeleton is primarily to protect the nervous system and other visceral organs. The limb bones (appendicular skeleton) attach to the axial skeleton by only the sternoclavicular joint for the upper limb and the strong sacroiliac joint for the lower limb. Joints between components of the axial skeleton may be synovial in type but are of relatively low mobility.

Key terms:

Acromion The distal end of the spine of the scapula. Forms a synovial joint with the distal clavicle and provides attachment for the coracoacromial ligament and deltoid muscle.

Anterior nasal (piriform) aperture The front opening of the bony nasal cavity of the skull.

Atlas (C1) The first cervical vertebra.

Axis (C2) The second cervical vertebra.

Cervical vertebrae Region of the vertebral column formed by seven vertebrae (C1–C7). The upper two in order are named the atlas and axis, respectively.

Clavicles The paired bones forming the anterior parts of each shoulder girdle. The clavicles form struts, which allow the scapulae to move around the thoracic cage.

Coccyx The terminal part of the vertebral column. It is a vestigial tail with only four or five segments.

Costal cartilage Bars of hyaline cartilage that attach to the anterior ends of the ribs. Cartilages 1–7 connect directly to the sternum; 8–10 articulate with the cartilage above; 11 and 12 have no connection with the sternum at all.

False ribs (pairs 8–10) Ribs that have only an indirect connection with the sternum.

Floating ribs (pairs 11 & 12) Ribs that have no connection with the sternum, either directly or indirectly.

Frontal bone The bone underlying the forehead.

Humerus The bone of the upper arm.

Ilium One of the three component bones of the hip. It has a flared crest or wing superiorly and a body inferiorly.

Lower teeth In an adult, each side has two incisors, one canine, two premolars, and three molars.

Lumbar vertebra Vertebrae of the lower back. Each lumbar vertebra has a large body; short, thick laminae; and pedicles and mammillary processes extending backward from the superior articular processes.

Mandible The bone of the lower jaw. It has an alveolar margin to support the teeth, paired condylar processes for articulation with the temporal bone, and paired coronoid processes.

Maxilla The bone of the medial cheek. It contributes to the inferior margin of the orbit and support of the upper teeth.

Occipital bone The large bone on the underside of the skull. It provides attachment for the postvertebral muscles of the neck and bears occipital condyles for articulation with the atlas (cervical vertebra 1).

Orbit The bony cavity containing the eye, lacrimal gland, extraocular muscles, nerves, vessels, and orbital fat. Bounded by the maxilla, frontal, sphenoid, ethmoid, and lacrimal bones.

Parietal bone A flat bone of the braincase (calvaria). It articulates with the frontal, temporal, sphenoid, and occipital bones.

Sacrum The large fused part of the vertebral column that attaches to the ilium of the hip bone.

Scapula The posterior bone of the shoulder girdle. It has anterior and posterior surfaces; medial, lateral, and superior borders; and coracoid and acromial processes. The glenoid fossa makes a joint with the humerus.

Spinous processes The posteriorly projecting processes of vertebrae.

Sternum The bone in the front wall of the chest forming part of the skeleton of the thorax. It consists of the manubrium, body, and xiphoid process.

Temporal bone The temporal bone has squamous, tympanic, styloid, mastoid, and petrous parts.

Thoracic vertebra A vertebra of the thoracic region. Typical thoracic vertebrae have superior and inferior demifacets on the body for articulation with the rib heads and long transverse processes with facets for articulation with the tubercles of ribs.

Transverse processes The transverse processes of vertebrae extend laterally from the junction of the laminae and pedicles of the neural arch.

True ribs (pairs 1–7) Ribs that have a direct connection (through their costal cartilages) to the sternum.

Upper teeth In an adult, each side has two incisors, one canine, two premolars, and three molars.

Zygomatic bone The bone of the upper lateral cheek. It articulates with the maxilla, temporal, and frontal bones.

Axial Skeleton—anterior view

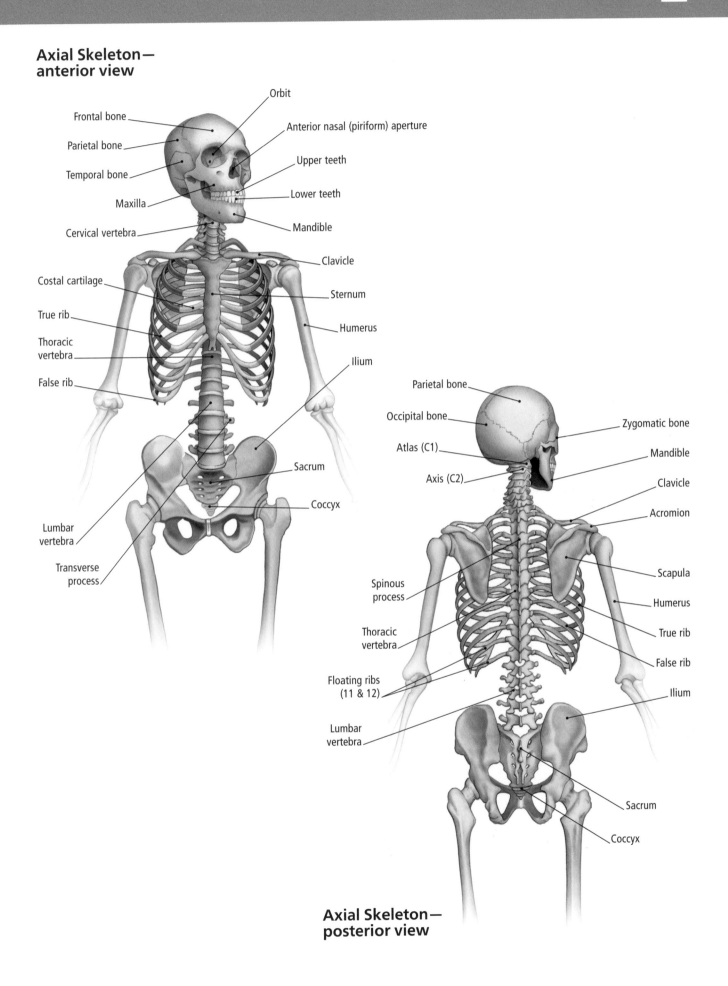

Frontal bone

Parietal bone

Temporal bone

Maxilla

Cervical vertebra

Orbit

Anterior nasal (piriform) aperture

Upper teeth

Lower teeth

Mandible

Clavicle

Costal cartilage

True rib

Thoracic vertebra

False rib

Sternum

Humerus

Ilium

Lumbar vertebra

Transverse process

Sacrum

Coccyx

Parietal bone

Occipital bone

Atlas (C1)

Axis (C2)

Zygomatic bone

Mandible

Clavicle

Acromion

Spinous process

Thoracic vertebra

Floating ribs (11 & 12)

Lumbar vertebra

Scapula

Humerus

True rib

False rib

Ilium

Sacrum

Coccyx

Axial Skeleton—posterior view

Bones and Joints of the Upper Limb

The upper limb bones consist of a pectoral girdle (scapula and clavicle), which has a relatively loose attachment to the axial skeleton by way of the shoulder muscles and the synovial sternoclavicular joint; the single bone of the arm (humerus); bones of the forearm (radius and ulna); wrist bones (the eight carpal bones); palm bones (metacarpals); and finger bones or phalanges (two only in digit one, but three in each of the other digits).

Key terms:

Acromioclavicular joint The plane joint between the acromion of the scapula and the distal end of the clavicle. The articular surfaces are predominantly fibrocartilaginous.

Acromion The distal end of the spine of the scapula. It forms a synovial joint with the distal clavicle and provides attachment for the coracoacromial ligament and the deltoid muscle.

Carpal bones The eight bones of the wrist. The proximal row (from lateral to medial) comprises the scaphoid, lunate, triquetrum, and pisiform. The distal row comprises the trapezium, trapezoid, capitate, and hamate.

Clavicle See pp. 38–39.

Coracoid process An anterior process of the scapula. It provides attachment for the short head of the biceps brachii and pectoralis minor muscles.

Glenoid cavity The cavity of the glenohumeral synovial joint (shoulder joint) between the glenoid fossa of the scapula and the head of the humerus.

Glenoid fossa The shallow articular surface of the lateral angle of the scapula. It engages in a synovial joint with the head of the humerus (glenohumeral or shoulder joint).

Greater tubercle of humerus The elevation on the lateral humerus that provides attachment for the supraspinatus, infraspinatus, and teres minor muscles.

Head of humerus The articular surface of the proximal humerus. It is slightly less than a hemisphere and is surrounded by the anatomical neck.

Humerus The bone of the arm. It consists of a shaft and proximal and distal articular surfaces.

Lateral border of scapula The lateral edge of the scapula. It provides attachment for the long head of the triceps brachii (at the infraglenoid tubercle), teres minor, and teres major muscles.

Lesser tubercle A roughened elevation of the anterior proximal humerus.

Medial border of scapula The medial edge of the scapula. It provides attachment for the levator scapulae, rhomboid minor, and rhomboid major muscles.

Metacarpal bones The bones of the palm. They are numbered from 1 to 5, from the thumb to the little finger. The space between the metacarpals is filled by muscles.

Phalanges The bones of the digits. There are two (proximal and distal) in digit 1 (the thumb) and three (proximal, middle, and distal) in digits 2–5. Singular: phalanx.

Radius The lateral bone of the forearm. It engages in a pivot synovial joint with the ulna, allowing rotational movements of the radius around the ulna (pronation and supination).

Scapula The posterior bone of the shoulder girdle. It has anterior and posterior surfaces; medial, lateral, and superior borders; and coracoid and acromial processes. The glenoid fossa makes a joint with the humerus.

Shoulder joint The glenohumeral (shoulder) synovial joint. It is a freely mobile ball-and-socket joint that is stabilized by the joint capsule and muscles of the rotator cuff group (including supraspinatus, infraspinatus, and teres minor).

Spine of scapula The laterally running sharp elevation of the posterior scapula. It separates the supraspinous and infraspinous fossae.

Subscapular fossa The concave, anterior surface of the scapula. It provides attachment for the subscapularis, a muscle that medially rotates the arm.

Superior border of scapula The thin and sharp upper border of the scapula. It is interrupted at its junction with the coracoid process by the scapular notch, which transmits the suprascapular nerve.

Ulna The medial bone of the forearm. It articulates with the humerus proximally (humeroulnar synovial joint of the elbow) and engages in a pivot joint with the radius (proximal and distal radioulnar synovial joints).

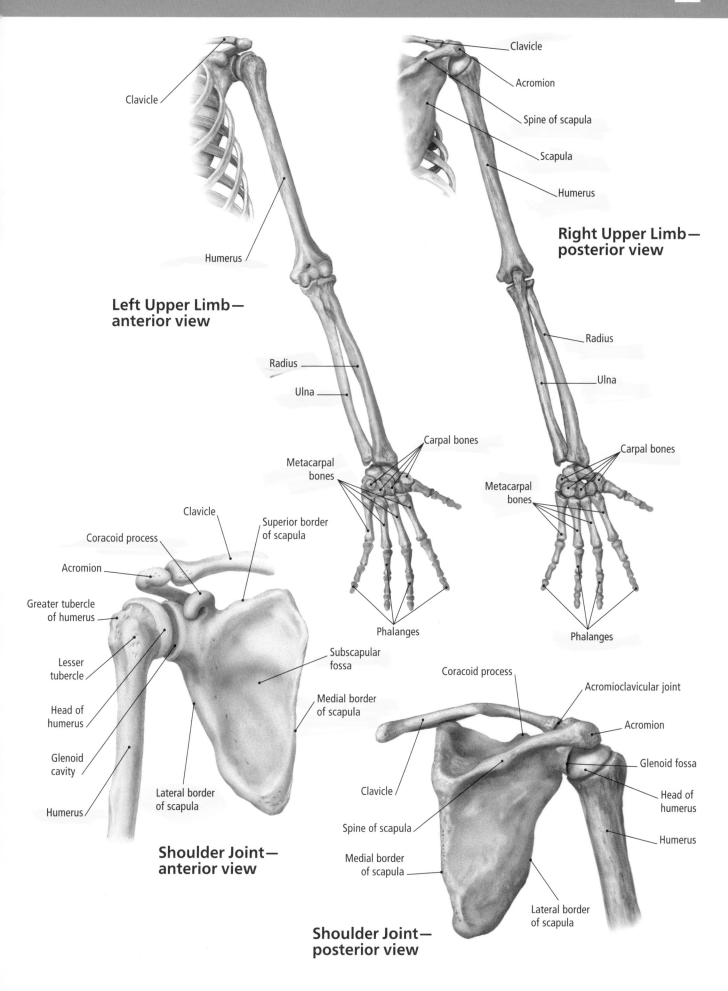

Clavicle

**Left Upper Limb—
anterior view**

Humerus

Clavicle
Acromion
Spine of scapula
Scapula
Humerus

**Right Upper Limb—
posterior view**

Radius
Ulna

Radius
Ulna

Carpal bones
Metacarpal bones
Phalanges

Carpal bones
Metacarpal bones
Phalanges

Clavicle
Coracoid process
Acromion
Superior border of scapula
Greater tubercle of humerus
Lesser tubercle
Head of humerus
Glenoid cavity
Humerus
Subscapular fossa
Medial border of scapula
Lateral border of scapula

**Shoulder Joint—
anterior view**

Coracoid process
Acromioclavicular joint
Acromion
Glenoid fossa
Head of humerus
Humerus
Clavicle
Spine of scapula
Medial border of scapula
Lateral border of scapula

**Shoulder Joint—
posterior view**

Bones and Joints of the Lower Limb

The lower limb bones include the hip bone or os coxa (ilium, ischium, and pubis together), joining at the acetabulum, which attaches to the sacrum at the sacroiliac joint; the single bone of the thigh (femur); the paired bones of the leg (robust tibia and pin-like fibula); the seven tarsal bones (talus, calcaneus, cuboid, navicular, and three cuneiforms); five metatarsals; and 14 toe bones (two in the big toe, but three in each of the other digits).

Key terms:

Adductor tubercle A prominent elevation on the uppermost part of the medial femoral condyle.

Anterior border The ridge of the tibia at the front of the shin.

Anterior intercondylar area The region between the paired articular facets of the anterior tibial plateau.

Apex of fibula The upper end of the fibula. It has one of the ligaments that runs alongside the knee joint attached to it.

Articular facet for talus The fibula contributes to the lateral part of the ankle joint. Its lower end articulates with the lateral surface of the trochlea of the talus.

Articular facet of medial malleolus The distal tibia forms the medial part of the ankle joint and has articular cartilage.

Articular surface with head of fibula A flat, circular facet on the lateral condyle of the tibia for articulation with the head of the fibula.

Calcaneus The bone of the heel, also known as os calcis. Has articular surfaces on the upper surface for the talus and cuboid.

Femur The bone of the thigh.

Fibula A slender bone on the lateral leg that bears no weight.

Fibular notch The notch on the lateral surface of the distal tibia for articulation with the fibula.

Fovea capitis A depression in the head of the femur for attachment of the ligament of the head of the femur.

Greater trochanter A large protrusion on the proximal femur. It provides attachment for the gluteus medius and gluteus minimus muscles.

Head of femur The ball-shaped proximal end of the femur that fits into the hip joint socket (acetabulum).

Head of fibula The expanded proximal part of the fibula. Has a circular articular facet for the lateral condyle of the tibia.

Inferior articular surface Inferior surface of the tibia for articulation with superior surface of trochlea of the talus.

Intercondylar eminence An elevation in the center of the tibia. The cruciate ligaments within the knee and the menisci, which help the joint surfaces conform, attach to the tibia immediately in front and behind it.

Intercondylar fossa The depression between the medial and lateral condyles of the femur.

Interosseous border The sharp lateral border of the tibia.

Lateral condyle The lateral part of the expanded upper end of the tibia.

Lateral epicondyle A prominence on the lateral surface of the lateral femoral condyle.

Lateral malleolus The inferior end of the fibula forms the lateral malleolus of the ankle.

Lateral surface The lateral surface of the tibia has the tibialis anterior muscle attached to it.

Lesser trochanter An elevation from the back of the junction of the neck and shaft of the femur.

Medial condyle The medial part of the expanded upper end of the tibia. Articulates with the medial condyle of femur.

Medial epicondyle A prominence on the medial surface of the medial femoral condyle.

Medial malleolus The medial prominence of the ankle formed by the distal tibia.

Metatarsal bones The five bones of the forefoot.

Neck of femur The part of the femur between the head of the femur and the greater trochanter.

Neck of fibula The narrow upper part of the fibula immediately distal to the head.

Patella The kneecap. It is a triangular sesamoid bone embedded in the tendon of the quadriceps femoris muscle.

Patellar surface The articular surface on the distal femur for the patella.

Phalanges The bones of the digits.

Shaft (diaphysis) of femur Main tubular part of the femur.

Superior articular surfaces (medial and lateral facets) The paired articular facets of the upper tibia.

Talus The most superior tarsal bone. It comprises a head, neck, and body.

Tarsal bones The tarsal bones include the talus, calcaneus, navicular, cuboid, and three cuneiforms.

Tibia The larger weight-bearing bone of the leg.

Tibial tuberosity An elevation on the anterior margin of the proximal tibia.

Right Lower Limb— anterior view

Left Lower Limb— posterior view

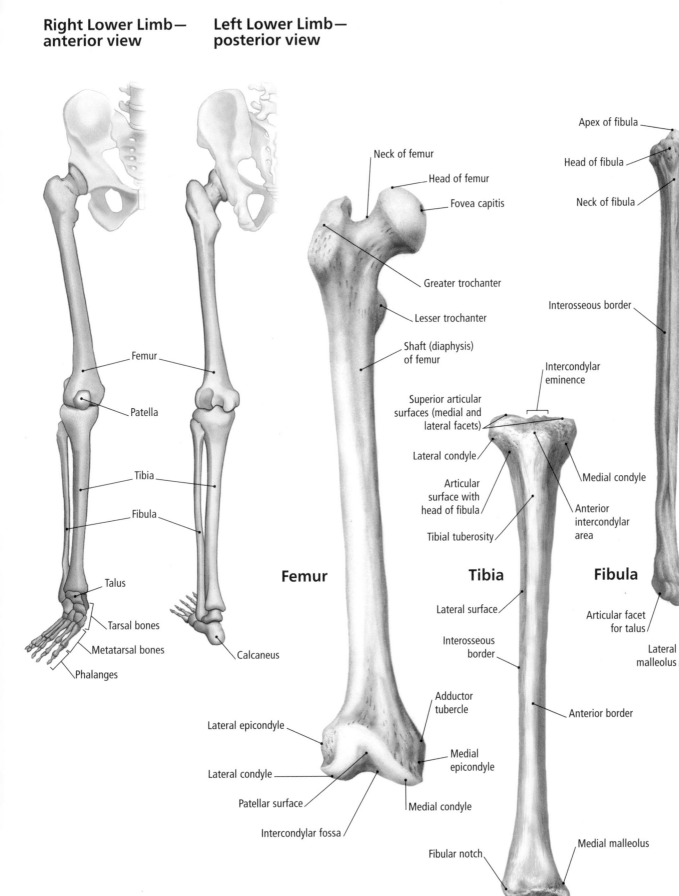

Neck of femur

Head of femur

Fovea capitis

Greater trochanter

Lesser trochanter

Shaft (diaphysis) of femur

Femur

Lateral epicondyle

Lateral condyle

Patellar surface

Intercondylar fossa

Adductor tubercle

Medial epicondyle

Medial condyle

Femur

Patella

Tibia

Fibula

Talus

Tarsal bones

Metatarsal bones

Phalanges

Calcaneus

Apex of fibula

Head of fibula

Neck of fibula

Interosseous border

Intercondylar eminence

Superior articular surfaces (medial and lateral facets)

Lateral condyle

Articular surface with head of fibula

Tibial tuberosity

Medial condyle

Anterior intercondylar area

Tibia

Lateral surface

Interosseous border

Anterior border

Fibular notch

Inferior articular surface

Medial malleolus

Articular facet of medial malleolus

Fibula

Articular facet for talus

Lateral malleolus

True or false?

1   The axial skeleton includes the skull, vertebral column, ribs, sternum, and hyoid bone.

2   The hyoid bone is in direct contact with the vertebral column.

3   There are three pairs of floating ribs that have no contact with the sternum.

4   The five lumbar vertebrae allow free rotation of the trunk.

5   Fusion of the five sacral vertebrae occurs in early adult life.

6   Bones of the pectoral girdle include the scapula, clavicle, and humerus.

7   The humerus is expanded distally as the capitulum and trochlea.

8   The radial tuberosity lies toward the proximal end of the radius.

9   The bones in the proximal row of the carpus are the scaphoid, lunate, triquetrum, and pisiform.

10   The bones of the palm are collectively called the metatarsals.

11   The three bones that make up the hip bone (os coxa) are the ilium, ischium, and pubis.

12   The components of the hip bone meet at the pubic symphysis.

13   The structure on which we sit is the ischial spine.

14   The femur has a proximal rounded head that fits into the acetabulum.

15   The shaft of the femur angles inward toward the knee joint.

16   *The distal end of the femur has two condyles, each with an elliptical cross-section.*

17   *The thickest bone of the leg is the tibia.*

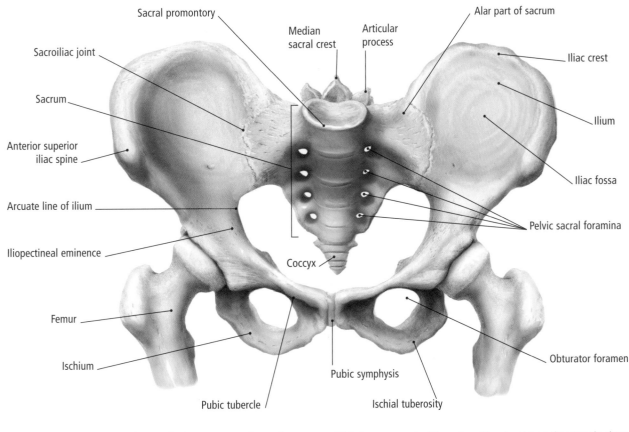

Female Pelvis—anterior view

While the components of the male and female pelvis are the same, the shape and dimensions differ. The male pelvis is generally larger to meet the demands of heavier weight. The female pelvic inlet is wider to allow the passage of the baby's head during childbirth.

Fractured neck of femur

The proximal femur has an almost spherical head supported on a narrow neck. Fractures of the neck of the femur are common, particularly in the osteoporotic elderly and in automobile accidents. There is often compression of the fracture site, so limb shortening and rotation are common. The femoral head is also vulnerable to necrosis in this type of fracture because much of its blood supply passes through the femoral neck. Femoral neck fractures may require complete surgical replacement of the femoral head, neck, and acetabulum (prosthetic hip) or pinning of the fracture site.

Multiple choice

1 *Which of the following is NOT a component bone of the skull?*

(A) temporal
(B) sphenoid
(C) hyoid
(D) vomer
(E) lacrimal

2 *Which of the following groups of bones contribute to the walls of the cranial cavity?*

(A) frontal, sphenoid, and vomer
(B) parietal, temporal, and zygomatic
(C) occipital, parietal, and nasal
(D) axis, frontal, and parietal
(E) mandible, sphenoid, and occipital

3 *Which of the following bones carries the upper teeth?*

(A) palatine
(B) sphenoid
(C) ethmoid
(D) maxilla
(E) none of the above is correct

4 *Which of the following statements is correct concerning the adult (permanent) teeth?*

(A) there are 20 total
(B) there are 8 incisors total
(C) the last premolars are called the wisdom teeth
(D) canines are located between premolars and molars
(E) molars have only one root

5 *The number of ribs that directly connect with the sternum, i.e., the true ribs, is:*

(A) 3
(B) 5
(C) 7
(D) 9
(E) 12

6 *Which of the following is correct concerning the false ribs?*

(A) they connect to the costal cartilages of ribs above
(B) they protect the kidneys
(C) they directly attach to the xiphoid process
(D) they are purely cartilaginous in adults
(E) both A and D are correct

7 *The humerus has a smooth, rounded head that is surrounded by a rim called the:*

(A) surgical neck
(B) capitulum
(C) glenoid labrum
(D) spiral line
(E) anatomical neck

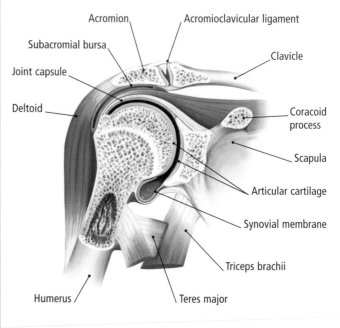

Acromion
Subacromial bursa
Joint capsule
Deltoid
Humerus
Acromioclavicular ligament
Clavicle
Coracoid process
Scapula
Articular cartilage
Synovial membrane
Triceps brachii
Teres major

Shoulder—cross-section

The smooth movements achieved by the shoulder are the result of several factors. A small contact area between the articulation surfaces and a generous, well-lubricated joint capsule provide friction-free movement.

8 *The surgical neck of the humerus is the place where:*

(A) the radial nerve passes posterior to the humeral shaft

(B) the humerus is highly prone to fracture

(C) the brachial artery is a close relation of the humeral shaft

(D) surgeons can remove bone for grafting

(E) most of the triceps major muscle attaches

9 *The ulna articulates with which part of the humerus?*

(A) capitulum

(B) trochlea

(C) head

(D) medial epicondyle

(E) none of the above is correct

10 *The bone that directly articulates with the distal radius is the:*

(A) scaphoid

(B) triquetrum

(C) trapezium

(D) capitate

(E) hamate

Shoulder dislocation

The stability of the shoulder (glenohumeral) joint depends mainly on surrounding muscles because the apposing joint surfaces are small and the ligaments are relatively weak. Dislocation of the shoulder is a common injury, particularly when the subject abducts and extends the arm, as in throwing the arm back vigorously to catch a ball. The weak ligaments at the front of the glenohumeral joint get torn, and the humeral head pops downward into the space inferior to the coracoid process and anterior to the glenoid lip. Recurrent dislocation may require surgical repair of the ligaments.

11 *The part of the femur that is palpable at the hip is the:*

(A) lesser trochanter

(B) medial condyle

(C) linea aspera

(D) greater trochanter

(E) femoral neck

12 *A prominent feature of the femur immediately superior to the medial condyle is the:*

(A) greater trochanter

(B) adductor tubercle

(C) lesser trochanter

(D) femoral head

(E) intercondylar eminence

13 *The proximal end of the tibia is characterized by the presence of:*

(A) a single articular surface with an eminence laterally

(B) a dish-shaped articular surface that conforms to the femoral condyles

(C) paired, flat articular surfaces on either side of an eminence

(D) paired supracondylar ridges for articulation with the femur

(E) none of the above is correct

14 *Long bones have internal medullary cavities where:*

(A) hematopoiesis (blood cell production) takes place

(B) calcium and phosphate are stored

(C) immune system surveillance cells predominate

(D) energy can be stored in the form of yellow fat

(E) both A and D are correct

Color and label

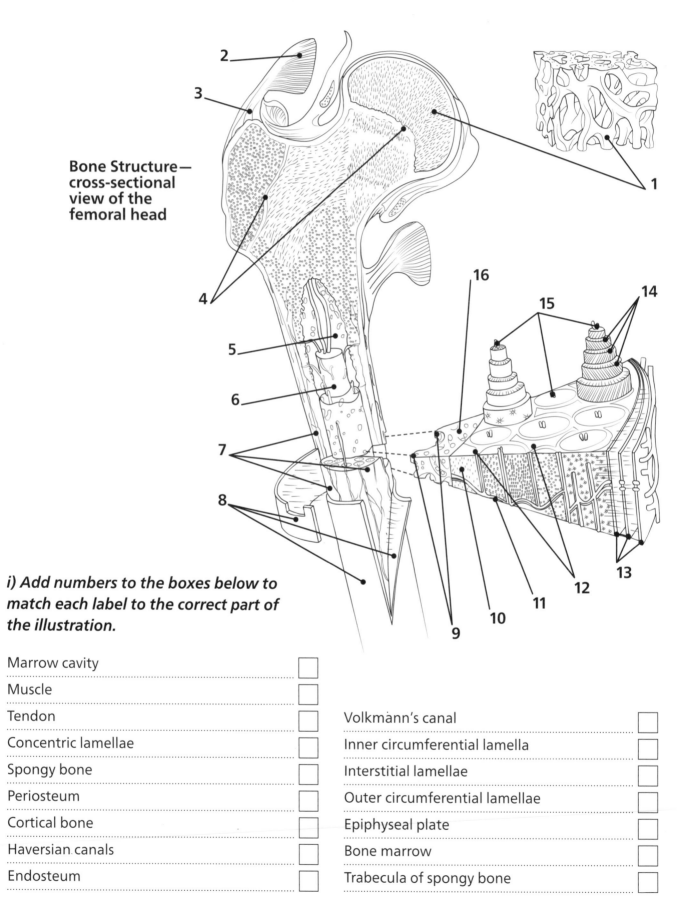

Bone Structure—cross-sectional view of the femoral head

i) Add numbers to the boxes below to match each label to the correct part of the illustration.

Marrow cavity	☐
Muscle	☐
Tendon	☐
Concentric lamellae	☐
Spongy bone	☐
Periosteum	☐
Cortical bone	☐
Haversian canals	☐
Endosteum	☐

Volkmann's canal	☐
Inner circumferential lamella	☐
Interstitial lamellae	☐
Outer circumferential lamellae	☐
Epiphyseal plate	☐
Bone marrow	☐
Trabecula of spongy bone	☐

ii) Label each structure shown on the illustrations.

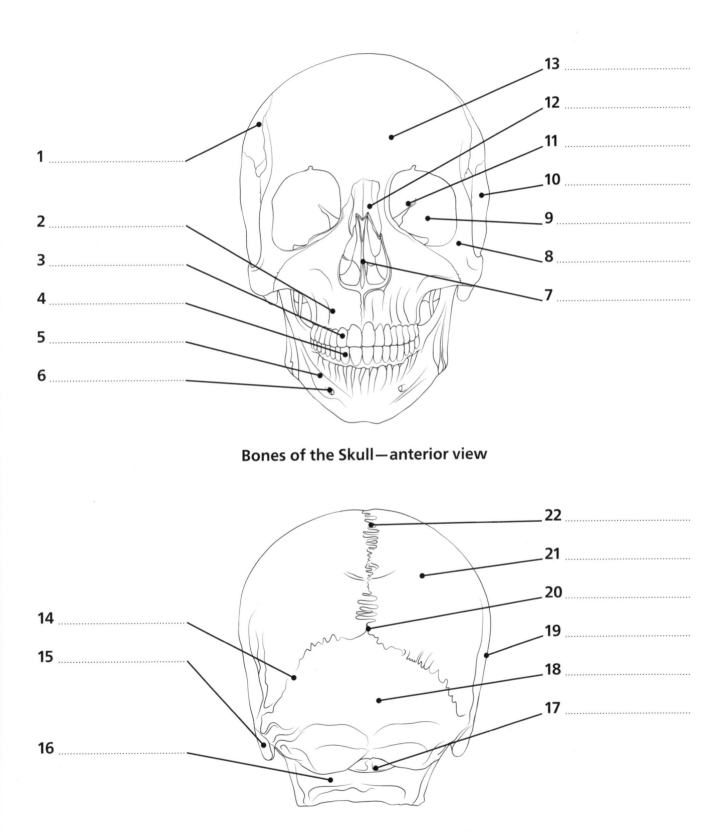

1 ....

2 ....

3 ....

4 ....

5 ....

6 ....

13 ....

12 ....

11 ....

10 ....

9 ....

8 ....

7 ....

Bones of the Skull—anterior view

14 ....

15 ....

16 ....

22 ....

21 ....

20 ....

19 ....

18 ....

17 ....

Bones of the Skull—posterior view

Fill in the blanks

1   Vascular supply to the bone tissue is provided by vessels running in the _____, which interconnect the canals of the Haversian system.

2   Longitudinal canals running through the bone tissue and surrounded by concentric lamellae of woven bone are called the _____.

3   The highly sensitive layer of connective tissue on the outside of bone is called the _____.

4   Growth in the circumference of bones occurs by deposition of bone tissue under the _____ on the outside of the bone and by (usually) removal of bone tissue at the _____ of the medullary cavity.

5   The skull is divided into two main regions: _____ and _____.

6   Blood vessels to the bone interior penetrate the surface of the bone through the _____.

7   The vertebral column includes _____ cervical vertebrae, _____ thoracic vertebrae, _____ lumbar vertebrae, _____ fused sacral vertebrae, and _____ separate or fused coccygeal vertebrae.

Traumatic neck injury

The cervical vertebral column is the most fragile part of the vertebral column because it allows the highest degree of mobility. Unfortunately, this makes it extremely vulnerable to high acceleration injuries, such as blows to the head or automobile accidents. The upper parts of the cervical vertebral column are the most vulnerable, and injuries such as fractures or dislocation of the C1 and C2 vertebrae can have fatal consequences if the cervical spinal cord is severed and control of respiratory muscles such as the diaphragm is lost. Fractures and dislocation of the lower cervical vertebrae can cause quadriplegia.

8   The _____ provides attachment for muscles of the tongue and protects the airway from collapse.

9   The sacrum has _____ and _____ sacral foramina for the passage of nerve fibers.

10   The proximal end of the humerus has two anterolateral elevations, the _____ and _____, separated by a sulcus.

11 The proximal ulna has a _____ notch to articulate with the humerus.

12 The bones of the digits include _____ and _____ phalanges in digit 1 (the thumb) and _____, _____, and _____ phalanges in digits 2 to 5.

13 The _____ is a space in the hip bone formed by the junction of the superior and inferior rami of the pubic bone with the _____ and _____, respectively.

14 The ischial spine separates the greater and lesser _____ notches.

15 The part of the hip bone palpable at the high side of the hip region is the _____ crest.

16 The _____ is a prominent ridge on the posterior surface of the femur that provides attachment for muscles and intermuscular septa.

17 The space between the two femoral condyles is known as the _____.

18 The _____ is the distal part of the tibia that forms the bony bump on the medial side of the ankle.

19 The navicular bone lies between the _____ and the three _____.

20 The _____ is the bone supported on the calcaneus.

Metacarpophalangeal joint
Carpometacarpal joint
Proximal phalanges
Carpal bones
Proximal interphalangeal joint
Middle phalanx
Capsule
Articular cartilage
Palmar ligament
Distal phalanx
Nail
Proximal phalanx
Distal phalanx of thumb
Distal interphalangeal joint

Bones of the Hand

Each of the four fingers contains three bones, the phalanges, while the thumb contains two phalangeal bones.

Match the statement to the reason

1 Bone microarchitecture is able to resist routine tensile and compressive forces because…

a the cartilaginous epiphyseal growth plates at the ends of long bones are progressively converted into bone during childhood and adolescence.

2 Growth of long bones occurs exclusively before adulthood because…

b progressively lower vertebral bodies must be able to withstand greater compressive forces due to the weight of the trunk, head, and upper limbs above them.

3 The sizes of vertebral bodies increase as one descends the vertebral column because…

c the bones of children are more flexible than those of adults.

4 The shaft of the femur is angled inward toward the knee because…

d the trabeculae or spicules of bone are arranged along the lines of force transmission through the bone for normal weight-bearing function.

5 Children are more prone to greenstick fractures than adults because…

e this keeps the point of contact between the body and the ground as close as possible to the midline during walking, thereby reducing the swaying of the trunk and the energy necessary to balance the body.

1 The ability to breathe deeply declines after the fourth decade of life because…

a this fine balancing reduces the muscular effort of holding the head up when walking upright.

2 The atlas (C1) vertebra lies directly beneath the center of the skull because…

b much of the blood supply of the femoral head passes through the femoral neck.

3 The cervical lordosis (posterior concavity of the cervical vertebral column) appears only after 3 months of age because…

c the fracture is often due to a fall onto the outstretched hand when one's balance is lost.

4 The femoral head is vulnerable to avascular necrosis following femoral neck fracture because…

d ossification ("turning to bone") of the costal cartilages reduces the flexibility of the chest wall.

5 Fracture of the distal radius is a common injury in the elderly because…

e the infant begins to lift his/her head from that age.

Bone Formation

Bone is continually being formed, enclosing each blood vessel, (a) and (b). As the vessel becomes completely surrounded, and more bone is laid down around it, an osteon is formed (c). The process is continually repeated, resulting in the bone growing in width (d).

Periosteum
Artery
Ridge

a

b

c

New osteon

d

Color and label

i) Label each structure shown on the illustrations.

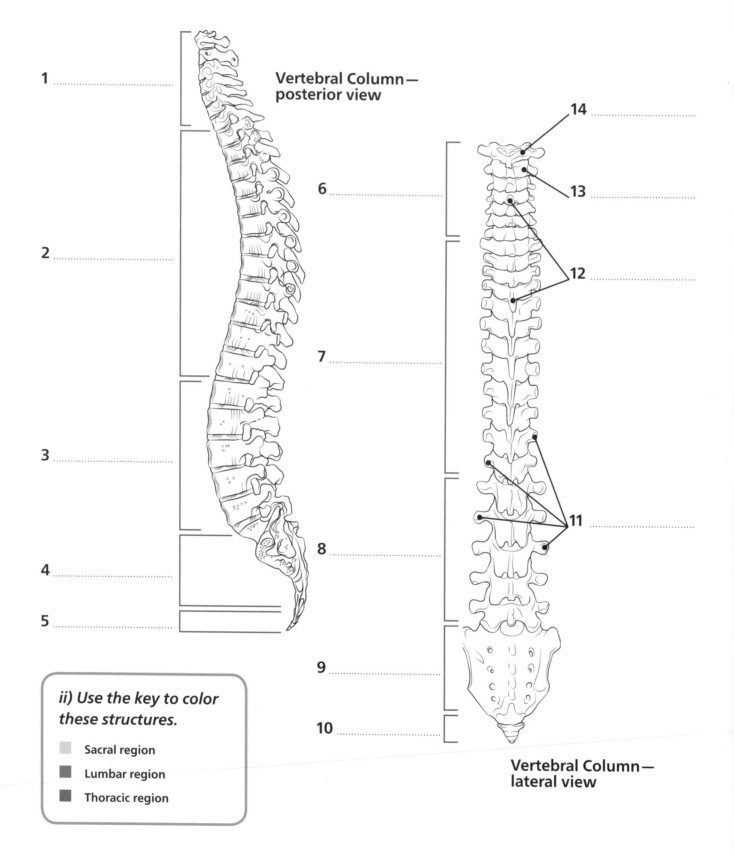

1 ....

Vertebral Column— posterior view

2 ....

3 ....

4 ....

5 ....

6 ....

7 ....

8 ....

9 ....

10 ....

14 ....

13 ....

12 ....

11 ....

ii) Use the key to color these structures.

Sacral region

Lumbar region

Thoracic region

Vertebral Column— lateral view

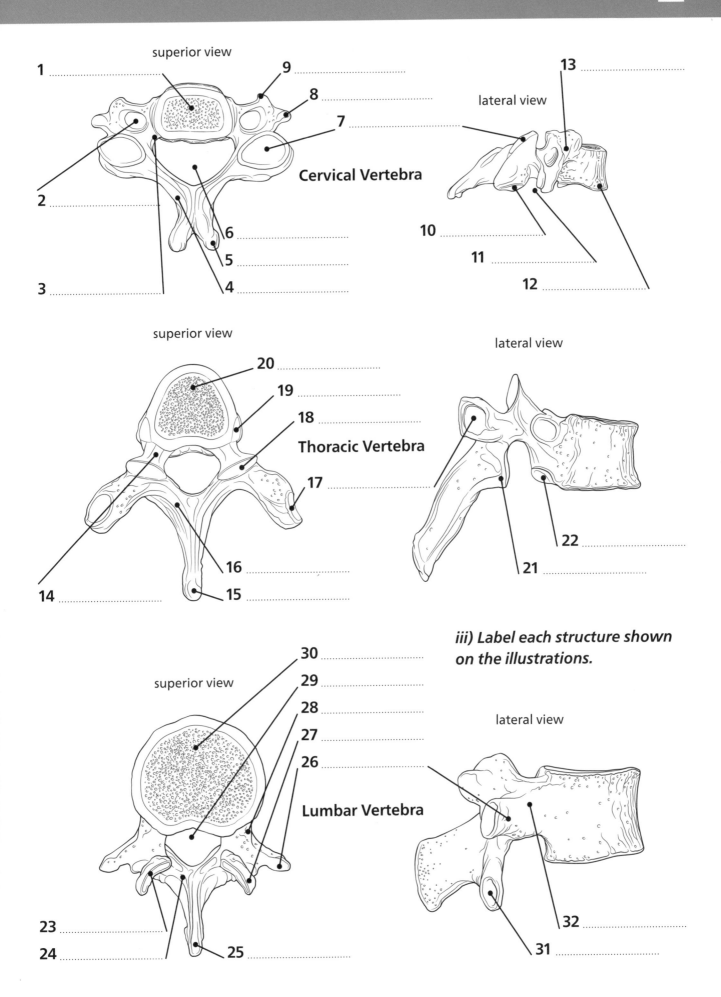

superior view

1 ....

9 ....

8 ....

7 ....

Cervical Vertebra

2 ....

6 ....

5 ....

3 ....

4 ....

13 ....

lateral view

10 ....

11 ....

12 ....

superior view

20 ....

19 ....

18 ....

Thoracic Vertebra

17 ....

16 ....

14 ....

15 ....

lateral view

22 ....

21 ....

iii) Label each structure shown on the illustrations.

superior view

30 ....

29 ....

28 ....

27 ....

26 ....

Lumbar Vertebra

lateral view

23 ....

24 ....

25 ....

32 ....

31 ....

True or false?

1   *The most mechanically stable joints are synovial joints.*

2   *Examples of fibrous joints would include the sutures of the skull and the gomphoses of the teeth.*

3   *Ball-and-socket joints are the most mobile in the body, allowing movement around three axes.*

4   *Ellipsoidal joints, like the radioscaphoid joint in the wrist, allow movement around only one axis.*

5   *Fibrocartilaginous joints consist of a fibrocartilage disk sandwiched between two joint surfaces.*

6   *Intervertebral disks consist of a central annulus fibrosus surrounded by a nucleus pulposus.*

7   *The joint between the trapezium and the base of the first metacarpal is a planar joint.*

8   *The knee joint is a good example of a hinge joint because it allows only flexion and extension.*

9   *The joints between the tarsal bones are all essentially planar joints.*

10   *The ankle joint is formed by the tibia, fibula, and talus.*

Gliding (Plane) Joint

Saddle Joint

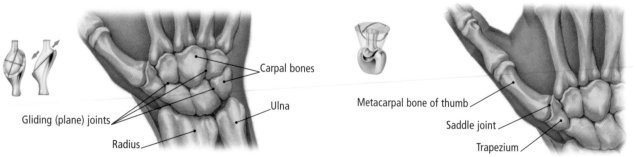

Carpal bones

Ulna

Gliding (plane) joints

Radius

Metacarpal bone of thumb

Saddle joint

Trapezium

Multiple choice

1 *The joint between the scapula and the humerus is highly mobile, but mechanically unstable, because:*

(A) the glenoid cavity of the scapula is a relatively shallow depression

(B) there are no bony projections over the superior aspect of the joint

(C) there are very few muscles providing stability for the joint

(D) there are no ligaments protecting the joint

(E) the joint capsule can be easily stripped from the scapula posteriorly

2 *The first carpometacarpal joint (at the base of the thumb) allows which movements?*

(A) rotation

(B) inversion

(C) flexion

(D) extension

(E) both C and D are correct

3 *Good example(s) of pivot joints would include the:*

(A) atlantoaxial joint

(B) proximal and distal radioulnar joints

(C) thoracic intervertebral joints

(D) midtarsal joints

(E) both A and B are correct

4 *Condylar joints are found in the:*

(A) elbow joint

(B) ankle joint

(C) metacarpophalangeal joints

(D) intercarpal joints

(E) wrist joint

5 *Stability of the hip joint is maintained by:*

(A) a ligamentum teres joining the femoral head to the acetabular fossa

(B) close matching of the acetabular and femoral head surfaces

(C) strong muscles surrounding the hip joint

(D) strong iliofemoral, pubofemoral, and ischiofemoral ligaments

(E) all of the above are correct

6 *Stability of the knee joint during standing is achieved by:*

(A) tension of the surrounding muscles

(B) strong anterior and posterior longitudinal ligaments

(C) rotation of the distal femur on the tibia to ensure close-packing of the joint

(D) tension in the patellar ligament

(E) none of the above is correct

7 *Interphalangeal joints are examples of which joint type?*

(A) condylar

(B) ellipsoidal

(C) ball-and-socket

(D) hinge

(E) saddle

Ellipsoidal Joint

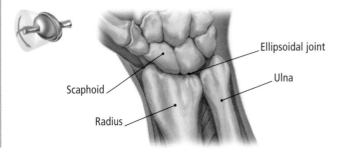

Fill in the blanks

1　Sutures visible on the top (dorsum) of the skull include the _____ and _____ sutures.

2　The joint of the jaw is a bicondylar joint between the _____ of the _____ and the _____ bone.

3　The joint that allows us to shake our heads from side to side is the _____ joint.

4　The joint that allows us to nod our heads is the _____ joint.

5　Joints between vertebrae in the lumbar part of the vertebral column are arranged in a series of _____ planes, so that trunk _____ are possible, but not _____.

6　The paired proximal and distal radioulnar joints allow the forearm movements of _____ and _____.

7　The condyloid metacarpophalangeal joints allow the movements of finger _____, _____, and some limited _____.

8　The knee joint allows the movements of _____, _____, _____, and _____.

9　Movements at the midfoot include _____ to turn the sole to face medially and _____ to turn the sole to face laterally.

10　Spraining of the ankle usually involves tearing of the fibers of the _____ and _____ ligaments.

Pivot Joint

Atlas (C1)

Pivot joint

Axis (C2)

Match the statement to the reason

1 The fibrous joints of the skull sutures are very mechanically stable because…

a the socket of the acetabulum is deep and encircles the femoral head, and strong encircling ligaments stabilize the joint.

2 Rotational movements of the trunk mainly occur in the thoracic part of the vertebral column because…

b the intervertebral joint planes in that region are aligned on the circumference of a circle centered on the nucleus or core of the intervertebral disk.

3 The hip joint is very mechanically stable because…

c there is a thin film of lubricating synovial fluid between the articular cartilage surfaces.

4 Synovial joints allow free movement of the joint surfaces against each other because…

d the glenohumeral ligaments are most easily stripped from the anterior surface of the glenoid lip of the scapula.

5 Dislocation of the shoulder usually involves abnormal movement of the humeral head inferiorly and anteriorly because…

e the bone edges are locked together by the complex interdigitation of the bone margin.

Ball-and-Socket Joint

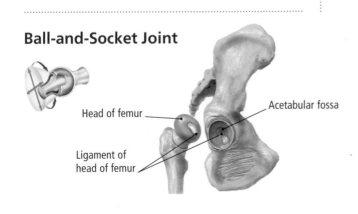

Head of femur

Acetabular fossa

Ligament of head of femur

Hinge Joint

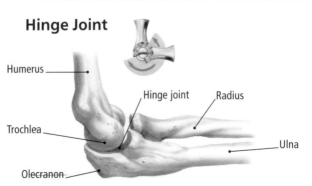

Humerus

Hinge joint

Radius

Trochlea

Ulna

Olecranon

Muscles of the Head, Neck, and Trunk

Muscles of the head and neck can be divided into groups concerned with facial movements (muscles of facial expression), moving the eyes (extraocular muscles), chewing (muscles of mastication), and moving the head and neck (cervical muscles). Muscles of the trunk include groups concerned with flexion, extension, and rotation of the vertebral column; muscles of lung ventilation (mainly the diaphragm, with assistance from intercostal muscles); muscles of the anterolateral abdominal wall; and muscles of the pelvic and urogenital diaphragms.

Key terms:

Deltoid A triangular muscle arising from the lateral clavicle, acromion, and spine of the scapula and inserting into the deltoid tuberosity of the humerus. It abducts, flexes, or extends the arm depending on which components are activated.

Depressor anguli oris Inserts into the corner of the mouth. It pulls the angle of the mouth downward.

External abdominal oblique The outermost muscle layer of the lateral abdominal wall. When both sides are contracted, they raise intra-abdominal pressure.

Frontalis A sheet-like muscle on the forehead. It attaches to the anterior end of the galea aponeurotica and, by acting alternately with the occipitalis, can move the scalp backward and forward.

Galea aponeurotica A dense, but mobile, connective tissue sheet that supports the scalp.

Iliac crest The large superior ridge of the ilium. It has attachments for the lateral abdominal wall muscles and the quadratus lumborum.

Latissimus dorsi The broadest muscle of the back. It arises from the lower thoracic, lumbar, and sacral vertebrae and inserts into the humerus. It is a powerful adductor and extensor of the humerus during swimming and climbing.

Levator labii superioris One of the muscles of the mouth. It inserts into the upper lip, close to the midline, and elevates the upper lip.

Masseter One of the muscles of mastication. It is a quadrilateral muscle that arises from the zygomatic arch and inserts into the ramus of the mandible. It elevates the mandible.

Occipitalis The facial muscle at the back of the scalp that inserts into the galea aponeurotica. When it contracts, it pulls the scalp posteriorly.

Orbicularis oculi A facial muscle that encircles the eye. It closes the eye tightly when it contracts fully.

Orbicularis oris A facial muscle that encircles the lips. It purses or puckers the mouth when it contracts.

Pectoralis major A muscle arising from the medial clavicle and upper six costal cartilages to insert into the crest of the greater tubercle of the humerus. It adducts, medially rotates, and flexes the arm.

Rectus abdominis A long, thin, sheet-like muscle running from the xiphisternum and fifth to seventh costal cartilages to insert into the pubic crest. It flexes the trunk anteriorly.

Serratus anterior Arises from the upper eight ribs and curves around the chest to insert into the medial border of the scapula. It pulls the scapula forward (protraction).

Sternocleidomastoid Extends from the sternoclavicular joint to the mastoid process of the skull. When the two sternocleidomastoids act together, they flex the head on the neck. Acting singly causes rotation of the head to the other side.

Sternohyoid A muscle band from the manubrium of the sternum to the body of the hyoid bone. It lowers the larynx, hyoid, and floor of the mouth.

Temporalis A muscle of mastication (chewing) that arises from the temporal fossa and inserts into the coronoid process and anterior border of the ramus of the mandible. It elevates the mandible.

Teres major Arises from the dorsal scapula near the inferior angle and inserts into the crest of the lesser tubercle of the humerus. It adducts the arm.

Teres minor One of the rotator cuff group. It arises from the dorsal surface of the scapula and inserts into the greater tubercle of the humerus. It laterally rotates the arm.

Thoracolumbar fascia A broad sheet of fascia extending from the vertebral spines of the lower thoracic and lumbar regions to the iliac crest and sacrum. It encloses the latissimus dorsi muscle.

Trapezius Arises from the occipital bone and the cervical and upper thoracic vertebrae. It inserts into the lateral clavicle and the spine and acromion of the scapula. It elevates and rotates the scapula.

Triceps brachii A three-headed muscle on the posterior aspect of the arm. It arises from the infraglenoid tubercle of the scapula and the back of the humerus and inserts into the olecranon. It extends the forearm.

Zygomaticus major A muscle band extending from the zygomatic bone of the cheek to the angle of the mouth. It pulls the angle of the mouth laterally.

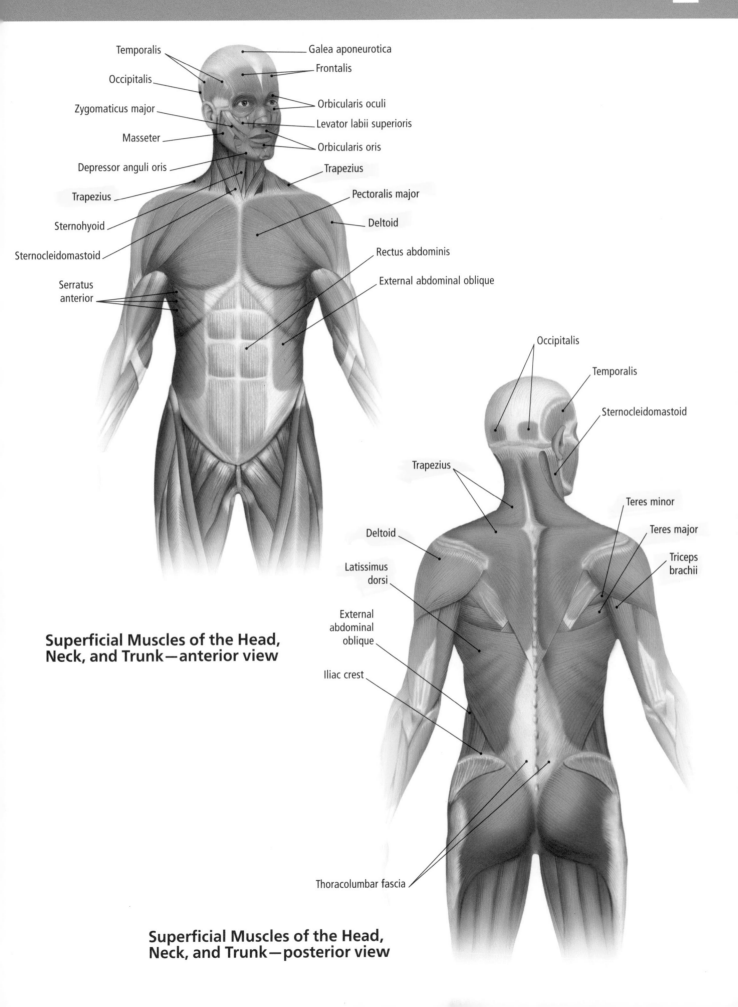

Temporalis

Galea aponeurotica

Occipitalis

Frontalis

Zygomaticus major

Orbicularis oculi

Masseter

Levator labii superioris

Depressor anguli oris

Orbicularis oris

Trapezius

Trapezius

Pectoralis major

Sternohyoid

Deltoid

Sternocleidomastoid

Rectus abdominis

Serratus
anterior

External abdominal oblique

**Superficial Muscles of the Head,
Neck, and Trunk—anterior view**

Occipitalis

Temporalis

Sternocleidomastoid

Trapezius

Teres minor

Deltoid

Teres major

Latissimus
dorsi

Triceps
brachii

External
abdominal
oblique

Iliac crest

Thoracolumbar fascia

**Superficial Muscles of the Head,
Neck, and Trunk—posterior view**

Muscles of the Upper Limb

Muscles of the upper limb can be divided into muscles of the shoulder girdle (those that move the scapula and humerus), muscles of the arm (mainly biceps brachii, brachialis, and triceps brachii), muscles of the forearm (flexor and extensor groups), muscles for forearm rotation, intrinsic muscles of the hand (located in the thenar and hypothenar eminences at the base of the thumb and little finger, respectively), and muscles in the palm.

Key terms:

Abductor pollicis longus Forearm muscle that has a long tendon extending to the first metacarpal. It abducts the thumb.

Anconeus Arises from the lateral epicondyle of the humerus and inserts into the lateral surface of the olecranon. It stabilizes the elbow joint during forearm rotation.

Biceps brachii Arises by two heads from the coracoid process and a tubercle immediately above the glenoid cavity (supraglenoid tubercle) and inserts into the forearm at the radial tuberosity and bicipital aponeurosis. It flexes and supinates the forearm.

Brachialis Arises from the humerus and inserts into the coronoid process and tuberosity of the ulna. It flexes the forearm.

Brachioradialis Arises from the lateral supracondylar ridge of the humerus and inserts into the lateral surface of the distal radius. It flexes the forearm.

Deltoid See pp. 60–61.

Extensor digiti minimi Arises from the lateral supracondylar ridge of the humerus, with a tendon running to the little finger. It allows independent extension of the little finger.

Extensor digitorum Arises from the lateral epicondyle of the humerus and gives rise to multiple tendons, which extend digits 2–5.

Extensor pollicis brevis Arises from the distal radius and inserts into the proximal phalanx of the thumb. It extends the thumb.

Extensor retinaculum A thickening of deep fascia running across the back of the wrist. It holds the extensor tendons to the wrist during the activation of extensor muscles.

Fibrous flexor sheath Fibrous sheaths enclosing the deep and superficial digital flexor tendons. The tendons are surrounded by synovial membranes to reduce friction.

Flexor carpi ulnaris Arises from the medial epicondyle of the humerus and inserts into the pisiform. It flexes and adducts the hand.

Flexor digitorum superficialis Arises from the medial epicondyle of the humerus. It flexes the middle phalanges of digits 2–5 on the proximal phalanges.

Flexor retinaculum A tough fibrous band running across the wrist, turning the carpal arch into a tunnel. It prevents the bowing of the flexor tendons during activation of the forearm flexor muscles.

Hypothenar muscles Muscles at the base of the little finger. These small muscles abduct or flex the little finger, or oppose it with the thumb.

Lateral head of triceps brachii Arises from the posterior surface of the upper humerus.

Long head of triceps brachii Arises from a tubercle immediately below the glenoid cavity of the scapula.

Palmaris brevis A small muscle that protects the ulnar artery and nerve and deepens the concavity of the palm when it contracts.

Pectoralis major See pp. 60–61.

Pronator teres Runs from the medial supracondylar ridge of the humerus to the radius. It pronates the forearm.

Tendons of extensors of the digits Inserts into the extensor hoods over the dorsum of each finger and on to the distal phalanges.

Tendon of flexor carpi radialis Inserts into the bases of the second and third metacarpals. It produces flexion and abduction of the hand.

Tendon of flexor carpi ulnaris Inserts into the pisiform bone and exerts its force to other carpal bones by the ligaments between the pisiform and the hamate and fifth metacarpal. It produces flexion and adduction of the hand.

Tendon of palmaris longus Attaches to the flexor retinaculum and the palmar aponeurosis to produce tension in the latter.

Tendon of triceps brachii Inserts into the olecranon to produce extension of the forearm.

Thenar muscles Muscles at the base of the thumb. These muscles flex or abduct the thumb, or oppose it with the other digits.

Triceps brachii A three-headed muscle on the posterior aspect of the arm. It arises from the infraglenoid tubercle of the scapula and the back of the humerus and inserts into the olecranon. It extends the forearm.

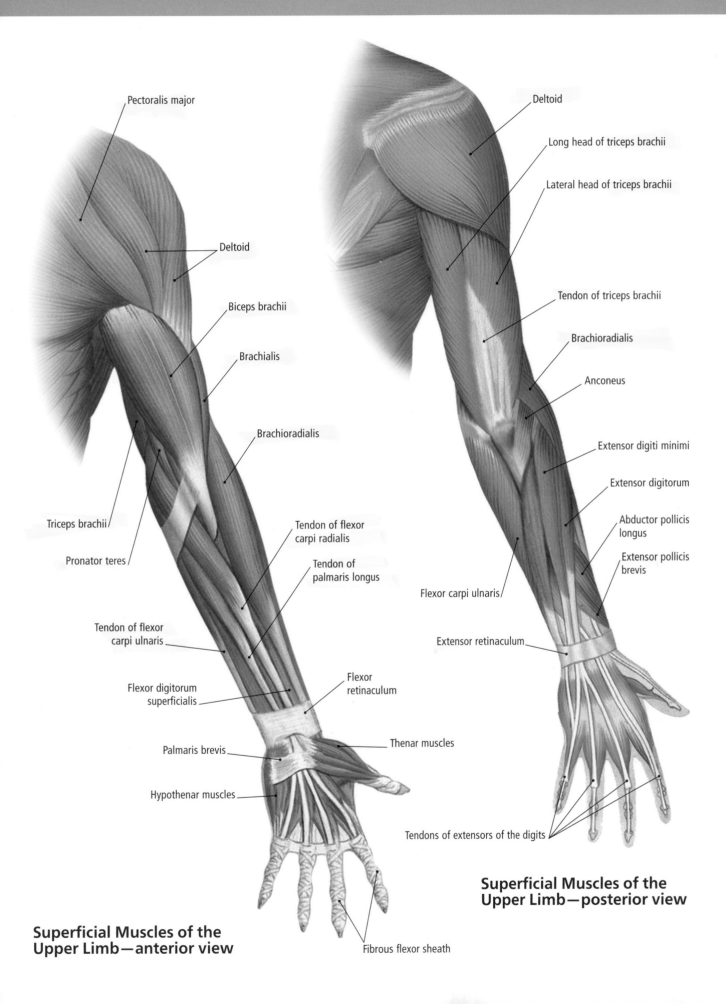

Pectoralis major

Deltoid

Biceps brachii

Brachialis

Brachioradialis

Triceps brachii

Pronator teres

Tendon of flexor
carpi radialis

Tendon of
palmaris longus

Tendon of flexor
carpi ulnaris

Flexor digitorum
superficialis

Flexor
retinaculum

Palmaris brevis

Thenar muscles

Hypothenar muscles

Fibrous flexor sheath

**Superficial Muscles of the
Upper Limb—anterior view**

Deltoid

Long head of triceps brachii

Lateral head of triceps brachii

Tendon of triceps brachii

Brachioradialis

Anconeus

Extensor digiti minimi

Extensor digitorum

Abductor pollicis
longus

Extensor pollicis
brevis

Flexor carpi ulnaris

Extensor retinaculum

Tendons of extensors of the digits

**Superficial Muscles of the
Upper Limb—posterior view**

Muscles of the Lower Limb

Muscles of the lower limb are divided into the gluteal or buttock group (mainly for moving the femur laterally and posteriorly); muscle groups of the anterior, medial, and posterior thigh (quadriceps femoris, thigh adductors, and hamstrings, respectively); muscles of the leg split into anterolateral and posterior groups; and intrinsic muscles of the foot in four layers. Nerves and vessels often run in the connective tissue planes that separate the muscle compartments.

Key terms:

Adductor longus This muscle arises from the body of the pubis and inserts into the medial lip of a ridge on the back of the femur (linea aspera). It adducts the thigh and stabilizes it during flexion and extension.

Adductor magnus Has two parts: a large adductor part running from the ischiopubic ramus to the femur and an extensor part running from the ischial tuberosity to the adductor tubercle of the femur.

Biceps femoris This hamstring muscle has a long head from the ischial tuberosity and a short head from the femur. The common tendon from both heads forms the lateral boundary of the depression behind the knee.

Extensor digitorum longus Runs from the lateral condyle of the tibia and splits into four tendons, which pass under the extensor retinacula of the ankle before inserting into digits 2–5. It extends the toes.

Fibularis (peroneus) longus Arises from the lateral condyle of the tibia and the fibula. Its tendon crosses under the foot before inserting into the medial cuneiform and base of the first metatarsal. It everts and plantarflexes the foot.

Gastrocnemius A two-headed muscle arising from the femur. The two bellies converge to form a membranous sheet, which fuses with the tendon of the underlying soleus muscle. The gastrocnemius flexes the leg and plantarflexes the foot.

Gluteus maximus This large muscle arises from the ilium, dorsal sacrum, and coccyx. It inserts into the back of the femur and the iliotibial tract of the fascia lata. It is a powerful extensor of the thigh.

Gluteus medius A muscle from the gluteal (posterolateral) surface of the ilium that inserts into the lateral surface of the greater trochanter. It abducts the thigh and supports the pelvis during the stance phase of walking.

Gracilis A slender muscle running from the body of the pubis to the upper medial tibia. It is a flexor, adductor, and medial rotator of the thigh.

Iliopsoas The tendons of the psoas major and iliacus combine to insert into the lesser trochanter of the femur. This combined muscle produces flexion of the thigh.

Iliotibial tract A band of tough connective tissue running from the iliac crest to a sheet of dense connective tissue attached to the side of the patella (lateral retinaculum).

Inferior extensor retinaculum One of two extensor retinacula that hold the extensor tendons in place as they pass anterior and lateral to the ankle joint.

Inguinal ligament A ligament joining the anterior superior iliac spine with the pubic tubercle. It forms the floor of the inguinal canal, which transmits the ductus deferens from the testis.

Lateral head of gastrocnemius Arises from the lateral surface of the lateral condyle of the femur. It may contain a sesamoid bone called the fabella.

Medial head of gastrocnemius Arises from the popliteal surface of the femur and the upper medial condyle of the femur.

Pectineus Arises from the pectineal line of the pubis and inserts into the femur. It adducts the thigh.

Semimembranosus Arises from the ischial tuberosity by a flattened tendon (hence the name) and inserts into the medial tibia. It flexes the leg and extends the thigh.

Semitendinosus Arises in common with the long head of the biceps femoris. Forms a long, thin tendon that inserts into the upper medial tibia. Flexes the leg and extends the thigh.

Soleus Arises from the posterior fibula and the posterior intermuscular septum. It receives the tendinous insertion of the gastrocnemius to form a thick, strong tendon inserting into the calcaneus. It plantarflexes the foot.

Superior extensor retinaculum One of two extensor retinacula that hold the extensor tendons in place as they pass anterior and lateral to the ankle joint.

Tibialis anterior Arises from the lateral condyle of the tibia and the anterior interosseous membrane and inserts into the medial cuneiform and base of the first metatarsal. It dorsiflexes and inverts the foot.

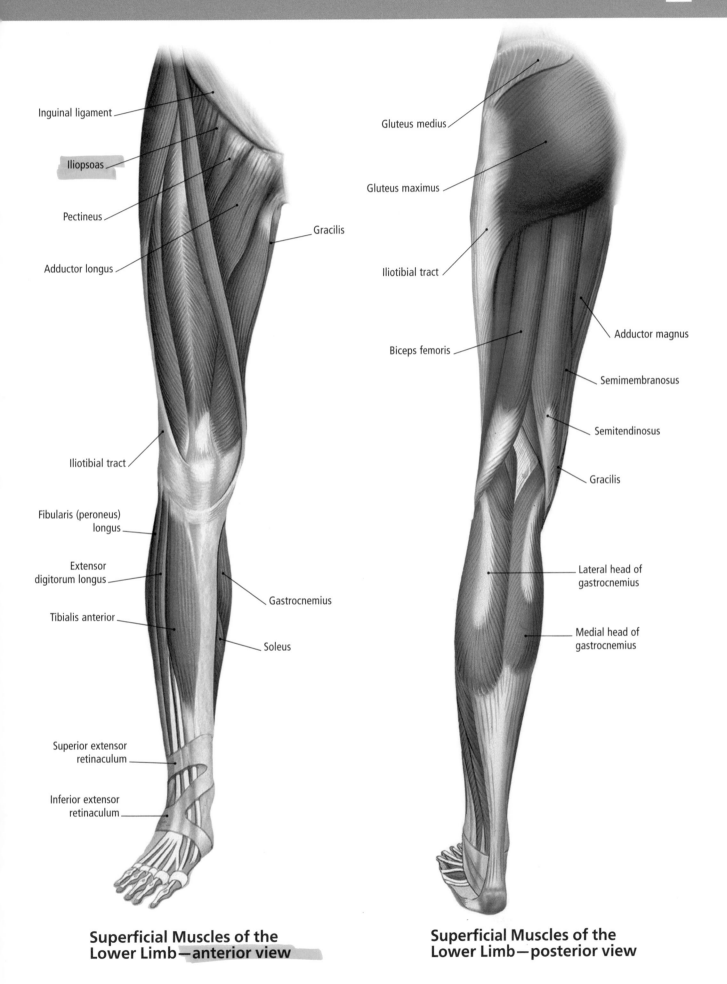

Inguinal ligament

Iliopsoas

Pectineus

Adductor longus

Gracilis

Iliotibial tract

Fibularis (peroneus) longus

Extensor digitorum longus

Tibialis anterior

Gastrocnemius

Soleus

Superior extensor retinaculum

Inferior extensor retinaculum

Gluteus medius

Gluteus maximus

Iliotibial tract

Biceps femoris

Adductor magnus

Semimembranosus

Semitendinosus

Gracilis

Lateral head of gastrocnemius

Medial head of gastrocnemius

Superficial Muscles of the Lower Limb—anterior view

Superficial Muscles of the Lower Limb—posterior view

True or false?

1   Facial muscles usually have one attachment to bone and the other to subcutaneous tissue.

2   Muscle contraction is fundamentally achieved by interactions between the troponin and tropomyosin proteins.

3   Digastric muscles are so called because they have two muscle bellies.

4   Concentric, circular, or sphincteric muscles are usually found surrounding the entrances to body orifices.

5   Muscles of the anterolateral abdominal wall are (from outside to inside): transversus abdominis, external oblique, and internal oblique.

6   In females, the pelvic diaphragm is traversed by the urethra, vagina, and anorectal canal.

7   The medial fibers of the pelvic diaphragm play a key role in supporting the bladder neck and controlling the onset of urination (micturition).

8   The posterior muscles of the rotator cuff group (infraspinatus/teres minor) play a key role in lateral rotation of the humerus.

9   The two heads of the biceps brachii attach to the supraglenoid tubercle and coracoid process of the scapula.

Transverse fibers of extensor expansions (hoods)

Dorsal interosseus muscles

Extensor indicis

Extensor carpi radialis longus

Extensor carpi radialis brevis

Extensor digiti minimi

Extensor pollicis longus

Extensor pollicis brevis

Extensor digitorum

Abductor pollicis longus

Muscles and Tendons of the Hand—dorsal view

Muscles within the hand combine with those of the forearm to enable the fine motor capabilities and dexterity of the hand and fingers.

10   *The triceps brachii attaches to the supraglenoid tubercle of the scapula.*

11   *The role of the biceps brachii muscle is confined to flexion at the elbow.*

12   *The triceps brachii muscle attaches to the coronoid process of the ulna.*

13   *Muscles of the anterior compartment of the forearm are concerned with flexion of the digits and wrist.*

14   *There is a separate extensor indicis muscle that allows the index finger to be independently extended, as in pointing.*

15   *All the intrinsic muscles of the hand are confined to the thenar and hypothenar eminences.*

16   *The principal role of the gluteus maximus is to support the hip when standing on one lower limb.*

17   *Extension of the knee is produced by the quadriceps femoris muscle.*

Thigh muscle injury

Muscles of the thigh, particularly the quadriceps and adductors, are prone to injury from rapid stretching during vigorous sports. Tears of the quadriceps femoris present as sudden pain in the front of the thigh, with swelling and bruising in some cases. Tears of the adductor muscles present as sudden onset of groin pain. Swelling may be present but is so deep that it is often not visible. Strains are usually treated by the principles of protection, rest, ice, compression, and elevation (PRICE).

18   *Muscles for everting the foot are located on the lateral side of the leg.*

19   *The gastrocnemius muscles can produce movement only at the ankle joint.*

20   *The muscles of the foot are arranged in five layers.*

Color and label

*i) Label each structure shown on
the illustration.*

Muscle Fiber—microstructure

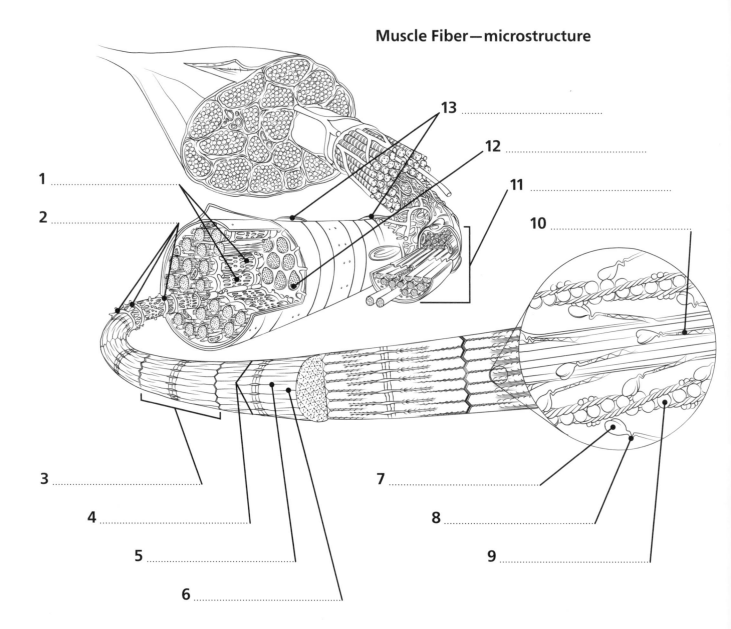

1 ....

2 ....

13 ....

12 ....

11 ....

10 ....

3 ....

4 ....

5 ....

6 ....

7 ....

8 ....

9 ....

Muscle Shapes

1

2

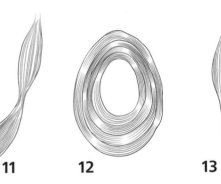
3

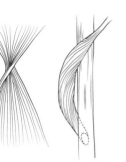
4

5

6

7

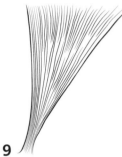
8

9

10

11   **12**   **13**   **14**   **15**   **16**

ii) Add numbers to the boxes below to match each label to the correct illustration.

Multipennate	☐	Multicaudal	☐
Radial	☐	Triangular	☐
Unipennate	☐	Digastric	☐
Quadricipital	☐	Strap	☐
Bipennate	☐	Fusiform	☐
Quadrate	☐	Bicipital	☐
Circular	☐	Tricipital	☐
Cruciate	☐	Spiral	☐

Multiple choice

1 *The role of the T-tubule system of muscle fibers is to:*

Ⓐ allow the binding of actin and myosin proteins

Ⓑ take up excess calcium at the end of each contraction

Ⓒ store elastic potential energy during multiple muscle twitches

Ⓓ transmit action potentials from the cell surface to the interior

Ⓔ store acetylcholine for release during end-plate potentials

2 *Which of the following is an example of a multipennate muscle?*

Ⓐ gracilis

Ⓑ deltoid

Ⓒ biceps brachii

Ⓓ diaphragm

Ⓔ external oblique

3 *Which architectural feature of a muscle maximizes the muscle power for size?*

Ⓐ large cross-sectional area

Ⓑ fusiform shape

Ⓒ cruciate shape

Ⓓ strap-like shape

Ⓔ both A and D are correct

4 *Which of the following muscles can close the mandible during chewing?*

Ⓐ sternomastoid

Ⓑ orbicularis oris

Ⓒ masseter

Ⓓ zygomaticus major

Ⓔ zygomaticus minor

5 *Which of the following structures does NOT pass through the diaphragm?*

Ⓐ descending aorta

Ⓑ inferior vena cava

Ⓒ esophagus

Ⓓ right phrenic nerve

Ⓔ portal vein

6 *Which of the following is NOT a function of the anterolateral abdominal muscles?*

Ⓐ inspiration (breathing in)

Ⓑ expiration (breathing out)

Ⓒ trunk rotation to the left

Ⓓ trunk flexion to the side

Ⓔ increasing intra-abdominal pressure

7 *Which of the following muscles attaches to the coracoid process of the scapula?*

Ⓐ triceps brachii

Ⓑ brachialis

Ⓒ pectoralis minor

Ⓓ pectoralis major

Ⓔ subscapularis

8 *Which muscle is the principal flexor of the elbow?*

Ⓐ biceps brachii

Ⓑ coracobrachialis

Ⓒ pectoralis major

Ⓓ brachialis

Ⓔ brachioradialis

9 *Which of the following muscles flexes the distal interphalangeal joints of digits 2 to 5?*

Ⓐ flexor digitorum superficialis

Ⓑ flexor digitorum profundus

Ⓒ flexor carpi ulnaris

Ⓓ lumbrical muscles

Ⓔ hypothenar muscles

10 *Flexion at the metacarpophalangeal joints while the interphalangeal and radiocarpal joints are extended requires the:*

- Ⓐ adductor pollicis
- Ⓑ abductor digiti minimi
- Ⓒ interossei
- Ⓓ lumbricals
- Ⓔ both C and D are correct

11 *Which of the following is NOT a component of the quadriceps femoris muscle?*

- Ⓐ vastus intermedius
- Ⓑ vastus lateralis
- Ⓒ rectus femoris
- Ⓓ vastus medialis
- Ⓔ gracilis

12 *Which muscle is penetrated by the adductor canal?*

- Ⓐ adductor longus
- Ⓑ adductor brevis
- Ⓒ adductor magnus
- Ⓓ semitendinosus
- Ⓔ semimembranosus

13 *Which of the following muscles can produce flexion at both the hip and the knee?*

- Ⓐ sartorius
- Ⓑ gracilis
- Ⓒ rectus femoris
- Ⓓ adductor magnus
- Ⓔ vastus intermedius

14 *Inversion of the foot (turning the sole of the foot to face inward) is produced mainly by the:*

- Ⓐ fibularis (peroneus longus)
- Ⓑ extensor digitorum longus
- Ⓒ tibialis posterior
- Ⓓ flexor digitorum longus
- Ⓔ flexor hallucis longus

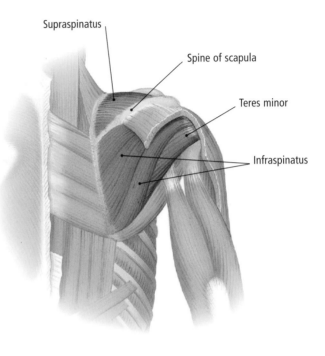

Supraspinatus

Spine of scapula

Teres minor

Infraspinatus

Rotator Cuff Muscles—posterior view

The rotator cuff muscles originate at the scapula (shoulder blade), running across its outer face, and attaching around the head of the humerus.

Rotator cuff and painful arc syndrome

The stability of the shoulder joint is usually maintained by the rotator cuff muscles (supraspinatus, subscapularis, infraspinatus, and teres minor). Wear and tear of the supraspinatus tendon due to years of abducting the arm against resistance can cause inflammation of the supraspinatus tendon. This tendon passes through a narrow space beneath the acromion of the scapula, where it can come into contact with hard surfaces during movement. Patients with inflammation of the supraspinatus tendon or adjacent bursa complain of pain when they abduct their arm by more than 60°. When the abduction continues to 120°, the pain is relieved because the humerus rotates under the acromion, and the tendon is no longer compressed. This is known as the painful arc syndrome.

Fill in the blanks

1   Muscles of mastication include the _____, _____, _____, and _____.

2   The sternocleidomastoid muscle produces _____ against resistance when acting bilaterally and turns the head _____ when acting unilaterally.

3   The diaphragm has attachments to three groups of bones: _____, _____, and _____.

4   The pelvic diaphragm has attachments to the _____, _____, _____, and _____ bones.

5   Component muscles of the pelvic diaphragm include the _____, _____, _____, and _____.

6   The two principal muscles used in push-ups are the _____ and _____.

7   Muscles producing pronation of the forearm include the _____ and _____.

8   The biceps brachii muscle can produce the movements of elbow _____, shoulder _____, and forearm _____ thanks to its attachments above the shoulder and below the elbow.

9   The two muscles responsible for supination of the forearm are the _____ and _____.

10   Precision grip mainly depends on the actions of the _____ muscles of the hand.

11   The power grip is achieved mainly by application of the _____ muscles of the forearm.

12 Muscles of the _____ group are critically important for the ability to support the body weight on one leg during the stance phase of walking.

13 The thigh muscles are arranged in three groups: quadriceps femoris, _____, and _____.

14 The _____ provides attachment for the hamstrings and adductor magnus.

15 The only muscle of the quadriceps femoris group that crosses two joints is the _____.

16 Muscles of the hamstring group include the _____, _____, and _____.

17 External or lateral rotators of the hip include the _____, _____, _____, and _____ femoris.

18 Muscles of the adductor group of the thigh include the _____, _____, _____, _____, _____, and _____.

Pelvic Floor Muscles (Female)— anterior view

Anterior sacrococcygeal ligament

Sacral promontory

Psoas minor

Psoas major

Sacral ventral nerve roots

Iliacus

Piriformis

Obturator internus

Ischiococcygeus (coccygeus)

Inguinal ligament

Pelvic diaphragm

Iliococcygeus

Tendinous arch of levator ani

Levator ani

Pubococcygeus

Rectum

Vagina

Puborectalis

Lacunar ligament

Obturator foramen

Bladder

The pelvic floor muscles straddle the pelvis, forming a muscular cradle for the pelvic organs.

Pubic symphysis

Pubic tubercle

Color and label

**Superficial and
Deep Muscles of
the Head and Neck—
anterior view**

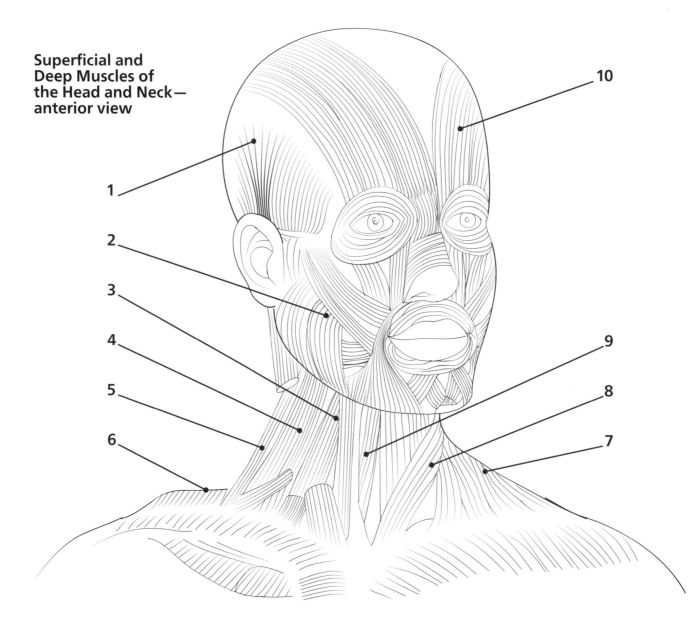

1

2

3

4

5

6

10

9

8

7

**i) Add numbers to the boxes below to
match each label to the correct part of
the illustration.**

Scalenus medius ☐

Frontalis ☐

Scalenus anterior ☐

Sternocleidomastoid ☐

Levator scapulae ☐

Trapezius (cut) ☐

Temporalis ☐

Trapezius ☐

Sternohyoid ☐

Masseter ☐

ii) Label each structure shown on the illustration.

iii) Color the trapezius red and the latissimus dorsi orange.

Superficial Muscles of the Back— posterior view

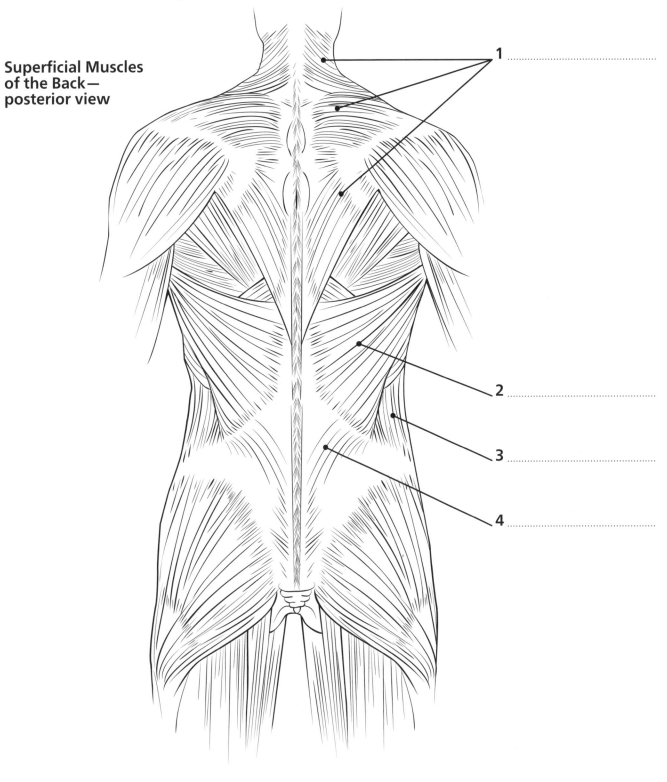

1 ....

2 ....

3 ....

4 ....

Match the statement to the reason

1   *The biceps brachii muscle is able to produce supination of the forearm because…*

a   *the medial and lateral gastrocnemius attach to the femur above the knee joint, and all three muscles (the two gastrocnemius and the soleus) insert onto the calcaneus below the ankle joint.*

2   *The fibers of the vastus medialis (quadriceps femoris group) approach the patella obliquely because…*

b   *the almost horizontal orientation of these muscle fibers allows them to stabilize the patella against the intercondylar surface of the distal femur.*

3   *The triceps surae group is able to produce both plantar flexion and flexion of the knee because…*

c   *when the forearm is in the anatomical position, the bicipital tendon curves around the radius to insert into the radial tuberosity.*

Muscles of the Neck— transverse section

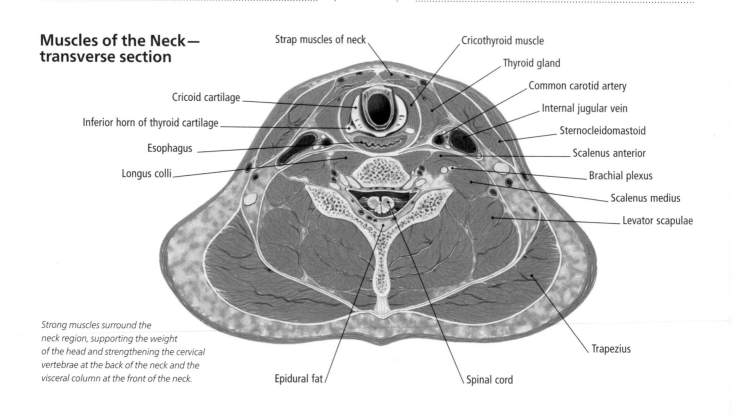

Strap muscles of neck
Cricothyroid muscle
Thyroid gland
Common carotid artery
Internal jugular vein
Sternocleidomastoid
Scalenus anterior
Brachial plexus
Scalenus medius
Levator scapulae
Trapezius
Cricoid cartilage
Inferior horn of thyroid cartilage
Esophagus
Longus colli
Epidural fat
Spinal cord

Strong muscles surround the neck region, supporting the weight of the head and strengthening the cervical vertebrae at the back of the neck and the visceral column at the front of the neck.

1 The neck muscles of modern humans are relatively weak compared with our nearest relatives because...

a the medial fibers of the pelvic diaphragm (levator prostatae) in males and pubovaginalis in females support the urinary bladder neck and regulate initiation of bladder emptying.

2 The diaphragm assists with the return of venous blood from the lower body to the heart because...

b it is a powerful hip extensor.

3 Pelvic diaphragm exercises are important to maintain urinary continence because...

c modern humans rely on balancing of the skull more than muscular effort to keep the head stable.

4 The gluteus maximus is an important muscle for climbing stairs because...

d it is a powerful adductor and extensor of the humerus when the arm is above the head.

5 The latissimus dorsi is well developed in freestyle swimmers because...

e it compresses the abdomen, raising pressure in the abdomen above that in the thorax.

Muscles of ventilation and lung disease

Most of the muscular effort required for normal lung ventilation is during inspiration (breathing in) because the elastic potential energy built up during inspiration can be harnessed to force air out of the lungs. The main muscle of inspiration is the diaphragm, which increases the vertical dimension of the chest when it contracts. Additional inspiratory force is applied by the external intercostal muscles and by those shoulder girdle muscles that have attachment to the ribs (pectoralis major and minor, and serratus anterior). Use of the latter is seen in chronic obstructive pulmonary disease, where the patient struggles to draw air into the lungs. Patients struggling to breathe out—for example, in asthma—will make use of the anterolateral abdominal wall muscles.

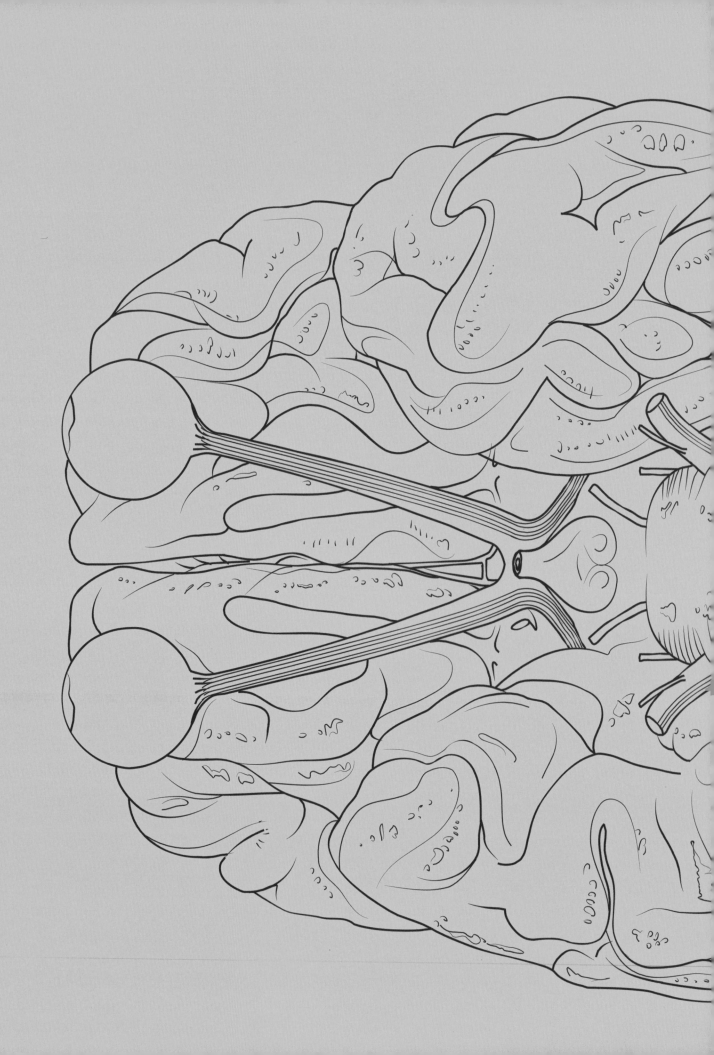

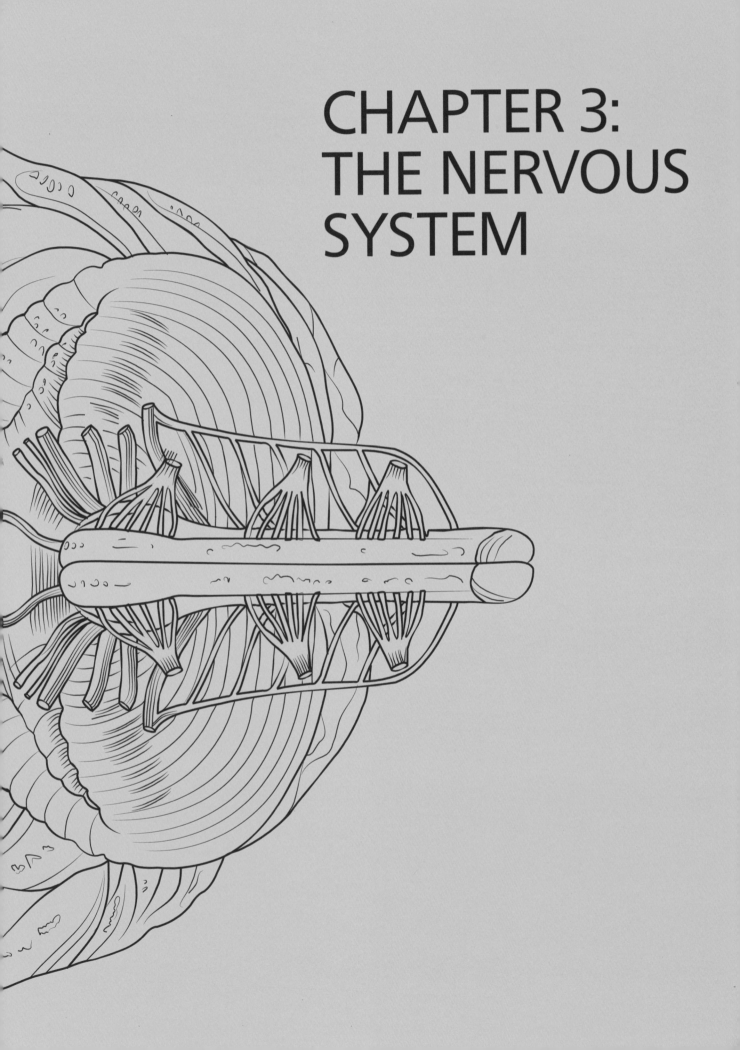

CHAPTER 3:
THE NERVOUS SYSTEM

The Peripheral Nervous System

The peripheral nervous system is that part of the nervous system outside the brain and spinal cord. It includes groups of nerve cell bodies (ganglia) and the peripheral nerves. Ganglia may be sensory in function—e.g., the cranial nerve sensory ganglia and the spinal or dorsal root ganglia—or they may be autonomic (part of the sympathetic or parasympathetic nervous systems). There is also a nerve network within the gut tube (enteric nervous system).

Key terms:

Axillary (circumflex) nerve One of the branches of the brachial plexus that supplies the deltoid and teres minor muscles, shoulder joint, and skin on the back of the arm.

Brachial plexus A network of nerves formed from spinal nerves cervical 5 to thoracic 1 to supply the upper limb. Major nerves arising from it are the radial, ulnar, median, axillary (circumflex), and musculocutaneous.

Cauda equina The roots of the lumbar, sacral, and coccygeal nerves in the vertebral canal beyond the end of the spinal cord. The name derives from the similarity to a horse's tail.

Cervical enlargement of spinal cord Enlargement of the spinal cord at spinal segments cervical 5 to thoracic 1. It supplies the upper limb and houses the neurons concerned with sensory input from, and motor control of, that limb.

Cervical nerve Branches of the cervical plexus supply muscles of the neck and skin of the neck and posterior head.

Common fibular nerve Branch of the sciatic nerve that descends through the popliteal fossa to the neck of the fibula and branches into deep and superficial fibular nerves.

Deep fibular nerve Supplies the tibialis anterior, extensor hallucis longus, extensor digitorum longus, and fibularis (peroneus) tertius muscles, as well as the tarsal joints and dorsal skin of the first and second toes.

Digital nerve Runs along either side of a digit (hand or foot), supplying sensation to that side.

Femoral nerve The largest branch of the lumbar plexus that emerges behind the psoas major and passes deep to the inguinal ligament to enter the femoral triangle at the front of the thigh. Supplies the quadriceps femoris, pectineus, sartorius, and iliopsoas, as well as the skin of the front of the thigh and anteromedial leg.

Intercostal nerve Runs in the space between adjacent ribs. Upper intercostal nerves supply the intercostal muscles and skin of the chest. Lower intercostal nerves cross into the abdomen (thoraco-abdominal nerves) to supply abdominal muscles and the skin of the abdomen.

Lateral femoral cutaneous nerve Branch of the lumbar plexus that supplies the skin of the lateral surface of the thigh.

Lumbosacral enlargement of spinal cord Enlargement of the spinal cord at spinal segments lumbar 2 to sacral 3.

Lumbosacral plexus Network of nerves formed from spinal nerves lumbar 2 to sacral 3 to supply lower limb and perineum. Can be divided into a lumbar plexus on the posterior abdominal wall and a sacral plexus in the pelvic cavity.

Median nerve Branch of the brachial plexus that supplies all muscles of the front of the forearm except the flexor carpi ulnaris and the medial half of the deep flexors of the fingers. Also supplies the thumb muscles and the skin of the lateral palm and digits.

Medulla oblongata The lowest part of the brainstem. It connects the spinal cord with the pons.

Musculocutaneous nerve Branch of the brachial plexus that supplies the biceps brachii muscle and skin of the forearm.

Obturator nerve Branch of the lower lumbar plexus that supplies the adductor group of thigh muscles and the skin of the upper medial thigh.

Radial nerve Branch of the brachial plexus that supplies the triceps brachii, anconeus, brachioradialis, and muscles of the extensor compartment of the forearm. Also supplies the skin of the lateral dorsum of the hand.

Sciatic nerve This branch of the sacral plexus is the largest nerve in the body. It divides into tibial and common fibular nerves, which supply the muscles of the posterior thigh and all of the leg and foot.

Superficial fibular nerve Branch of the common fibular nerve that supplies the fibularis (peroneus) longus and brevis muscles and the skin of the lateral leg and dorsum of the foot and toes.

Sural nerve A cutaneous branch of the tibial nerve that supplies the back of the calf, foot, and heel.

Tibial nerve Branch of the sciatic nerve that passes through the popliteal fossa to give muscular branches to soleus, tibialis posterior, flexor hallucis longus, and flexor digitorum longus.

Ulnar nerve A branch of the brachial plexus that supplies the flexor carpi ulnaris, medial flexor digitorum profundus, muscles of the little finger, interosseus muscles of the palm, the adductor pollicis muscles, and the skin of the medial one-and-a-half digits.

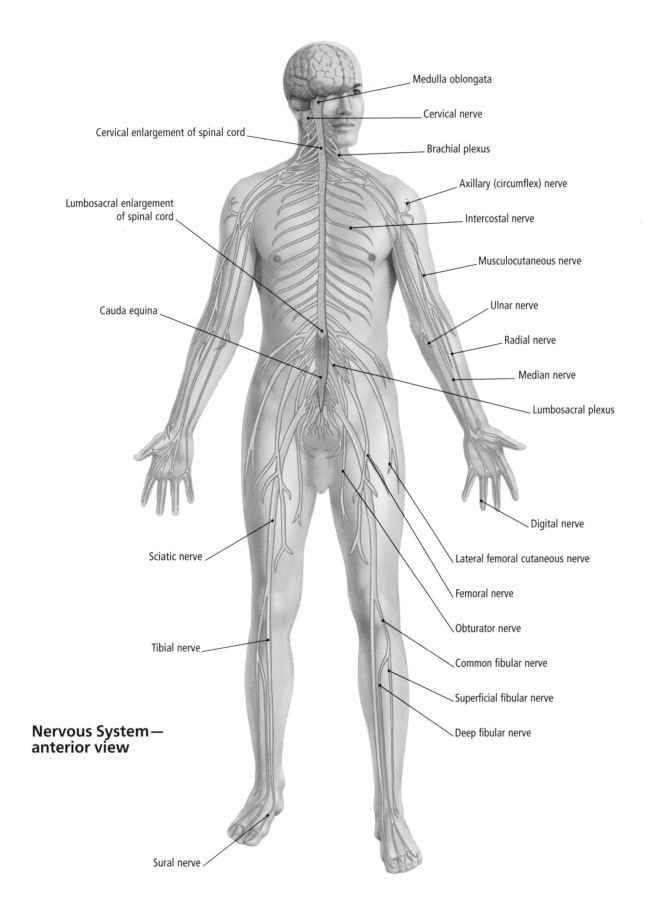

Medulla oblongata

Cervical nerve

Cervical enlargement of spinal cord

Brachial plexus

Axillary (circumflex) nerve

Lumbosacral enlargement of spinal cord

Intercostal nerve

Musculocutaneous nerve

Ulnar nerve

Cauda equina

Radial nerve

Median nerve

Lumbosacral plexus

Digital nerve

Sciatic nerve

Lateral femoral cutaneous nerve

Femoral nerve

Obturator nerve

Tibial nerve

Common fibular nerve

Superficial fibular nerve

Deep fibular nerve

Nervous System— anterior view

Sural nerve

The Spinal Cord

The spinal cord is that part of the central nervous system that lies within the vertebral column. It connects with the brain above and has more than 32 pairs of spinal nerves attached. The spinal cord has dorsal and ventral roots attached to its posterolateral and anterolateral margins, respectively. The dorsal roots carry sensory information from the skin, limbs, body wall, and internal organs, whereas the ventral roots carry motor and autonomic commands to muscles and glands.

Key terms:

Anterior corticospinal tract Descending spinal cord pathway from cerebral cortex to almost all levels of the spinal cord.

Anterior median fissure A deep groove in the front of the spinal cord.

Anterior radicular artery A fine artery that runs along and supplies the ventral rootlets of the spinal cord.

Anterior ramus of spinal nerve Branch of the spinal nerve. Supplies muscles, bones, joints, and skin of the limbs and anterolateral body wall, and organs of ventral body cavities.

Anterior spinal artery An artery running longitudinally in the anterior median fissure of the spinal cord.

Anterior spinal vein A vein running longitudinally in the anterior median fissure of the spinal cord.

Arachnoid The delicate spider's weblike membrane underlying the dura mater.

Axon The long process of a nerve cell that transmits the impulse to another part of the brain or body.

Central canal A delicate fluid-filled tube in the center of the spinal cord.

Cuneate fasciculus Bundle of ascending axons carrying info about touch, vibration, and conscious proprioception from the upper trunk and limbs to the medulla.

Dorsal funiculus The region of white matter on the dorsal or posterior aspect of the spinal cord.

Dorsal horn The region of the spinal cord gray matter concerned with sensory input.

Dorsal rootlets Fine nerve fibers carrying sensory information from the skin surface, muscles, and internal organs.

Dorsal spinocerebellar tract Spinal cord pathway carrying nonconscious proprioceptive information from the upper spinal cord to the cerebellum.

Dorsolateral sulcus A groove running longitudinally down the posterolateral surface of the spinal cord.

Dura mater The tough outer layer of the meninges.

Endoneurium Delicate connective tissue holding together the individual nerve fibers in a fascicle (fine nerve bundle).

Epineurium The connective tissue sheath around an entire spinal nerve or trunk.

Gracile fasciculus Bundle of ascending axons carrying info about touch, vibration, and conscious proprioception from the lower trunk and limbs to the medulla.

Gray ramus communicans A nerve strand joining the sympathetic trunk to the spinal nerve.

Lateral corticospinal tract A descending spinal cord pathway from the cerebral cortex to almost all levels of the spinal cord.

Lateral funiculus The region of white matter on the lateral aspect of the spinal cord.

Lateral reticulospinal tract A descending spinal cord pathway from the reticular formation of the brainstem to the spinal cord.

Lateral vestibulospinal tract A descending spinal cord pathway from the lateral vestibular nucleus of the brain stem to all levels of the spinal cord.

Medial reticulospinal tract A descending spinal cord pathway from the reticular formation of the brainstem to the spinal cord.

Medial vestibulospinal tract A descending spinal cord pathway from the vestibular nuclei of the brainstem to the cervical levels of the spinal cord.

Myelin sheath of Schwann cell Each Schwann cell wraps its cytoplasm around a peripheral axon to form one internodal segment of the myelin sheath.

Node of Ranvier The region of a myelinated nerve fiber where the axon is naked.

Perineurium The connective tissue sheath around a fascicle or primary bundle of nerve fibers.

Pia mater The most delicate layer of the meninges, it is in direct contact with the brain surface.

Posterior radicular artery An arterial branch that runs along and supplies the dorsal or posterior roots of the spinal cord.

Posterior ramus of spinal nerve A branch of the spinal nerve that supplies the muscles, bones, joints, and skin of the back.

Posterior spinal artery One of a pair of arteries running longitudinally along the grooves of the spinal cord where the dorsal roots attach.

Posterior spinal vein A vein running longitudinally along the posterior median fissure of the spinal cord.

Spinal Cord—cross-sectional view

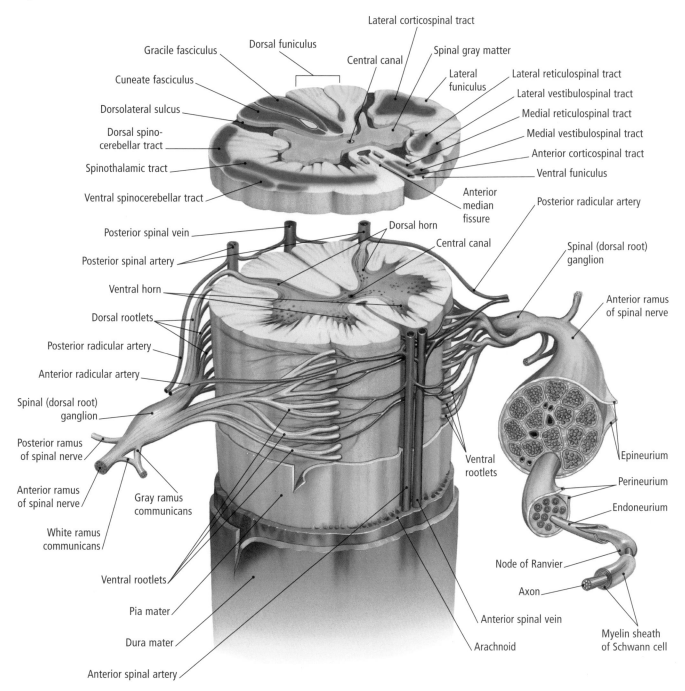

Lateral corticospinal tract

Gracile fasciculus

Dorsal funiculus

Central canal

Spinal gray matter

Cuneate fasciculus

Lateral funiculus

Lateral reticulospinal tract

Lateral vestibulospinal tract

Dorsolateral sulcus

Medial reticulospinal tract

Dorsal spino-cerebellar tract

Medial vestibulospinal tract

Anterior corticospinal tract

Spinothalamic tract

Ventral funiculus

Ventral spinocerebellar tract

Anterior median fissure

Posterior radicular artery

Posterior spinal vein

Dorsal horn

Central canal

Spinal (dorsal root) ganglion

Posterior spinal artery

Anterior ramus of spinal nerve

Ventral horn

Dorsal rootlets

Posterior radicular artery

Anterior radicular artery

Epineurium

Spinal (dorsal root) ganglion

Perineurium

Posterior ramus of spinal nerve

Endoneurium

Anterior ramus of spinal nerve

Gray ramus communicans

Ventral rootlets

White ramus communicans

Node of Ranvier

Ventral rootlets

Axon

Pia mater

Anterior spinal vein

Myelin sheath of Schwann cell

Dura mater

Arachnoid

Anterior spinal artery

Spinal (dorsal root) ganglion A collection of sensory neurons lying along the dorsal root.

Spinal gray matter An H-shaped region in the center of the spinal cord.

Spinothalamic tract A spinal cord pathway that carries information about pain, temperature, and simple touch from all levels of the spinal cord to the thalamus.

Ventral funiculus The region of white matter on the ventral or anterior aspect of the spinal cord.

Ventral horn The region of the spinal cord gray matter where

the cell bodies of motor neurons and their associated interneurons are located.

Ventral rootlets Fine nerve fibers carrying motor neuron axons from the ventral surface of the spinal cord to the spinal nerve.

Ventral spinocerebellar tract A spinal cord pathway carrying nonconscious proprioceptive information from the lower spinal cord to the cerebellum.

White ramus communicans A nerve strand joining the sympathetic ganglia to the spinal nerve.

True or false?

1   The ascending and descending pathways within the spinal cord are located within the gray matter.

2   The dorsal horn of the spinal cord contains neurons concerned with sensory function from the skin, muscles, and joints.

3   The ventral roots of the spinal cord contain outgoing (efferent) axons of skeletal muscle motor neurons and preganglionic visceral autonomic neurons.

4   The dorsal root ganglion cells are autonomic visceromotor in function.

5   Blood supply to the spinal cord is provided by a single midline anterior spinal artery and paired posterior spinal arteries.

6   The radial nerve supplies the triceps brachii and the forearm extensor compartment muscles.

7   The largest nerve in the body is the femoral nerve that supplies the quadriceps femoris muscle.

8   The median nerve passes through the carpal tunnel in the wrist, where it can be compressed in carpal tunnel syndrome.

9   The ulnar nerve is vulnerable to damage where it runs posterior to the lateral epicondyle.

10 The sciatic nerve divides into the posterior tibial and common peroneal (fibular) nerves.

Multiple choice

1 *Which of the following tissues or cell types would NOT be found inside the spinal cord?*

Ⓐ choroid plexus
Ⓑ microglia
Ⓒ astrocyte
Ⓓ oligodendrocyte
Ⓔ vascular endothelial cell

2 *Where are the autonomic preganglionic neurons located?*

Ⓐ the dorsal horn
Ⓑ the ventral horn
Ⓒ the lateral horn
Ⓓ the white matter
Ⓔ the ganglia of the abdominal cavity

3 *Concerning the human spinal nerves, there are:*

Ⓐ 7 cervical, 12 thoracic, 5 lumbar, and 5 sacral
Ⓑ 8 cervical, 11 thoracic, 5 lumbar, and 5 sacral
Ⓒ 8 cervical, 12 thoracic, 5 lumbar, and 5 sacral
Ⓓ 8 cervical, 12 thoracic, 4 lumbar, and 5 sacral
Ⓔ 7 cervical, 11 thoracic, 5 lumbar, and 5 sacral

4 *Which of the following muscles are supplied by the median nerve?*

Ⓐ biceps brachii
Ⓑ short muscles of the thumb (thenar eminence)
Ⓒ brachioradialis
Ⓓ extensors of the wrist
Ⓔ muscles of the little finger (hypothenar eminence)

5 *Where is the radial nerve most likely to be damaged?*

Ⓐ on the dorsum of the wrist at the base of the thumb
Ⓑ in the cubital fossa near the brachial artery
Ⓒ in the medial forearm near the flexor digiti minimi
Ⓓ as it passes posterior to the middle part of the humerus
Ⓔ in the axilla against the teres major tendon

6 *Where is the femoral nerve closest to the skin surface?*

Ⓐ in the femoral triangle
Ⓑ immediately above the adductor tubercle of the femur
Ⓒ as it passes anterior to the iliacus
Ⓓ near the anterior superior iliac spine
Ⓔ deep to the gluteus medius

7 *Which of the following muscles is/are supplied by the sciatic nerve?*

Ⓐ gluteus maximus
Ⓑ adductor magnus
Ⓒ semitendinosus
Ⓓ biceps femoris
Ⓔ both C and D are correct

Neuron

The unique structure of neurons comprises several dendrites and a single axon projecting from the cell body. Nerve impulses are received by the dendrites and carried to the cell body, while the axon carries impulses away from the cell body.

Fill in the blanks

1   The _____ of the spinal cord is a fluid-filled remnant of the embryonic neural tube.

2   The _____ of the spinal cord contains the nerve cell bodies and most of their dendritic processes.

3   The brachial plexus for nerve supply to the upper limb is derived usually from the _____ to _____ spinal nerves, inclusive.

4   The axillary (circumflex) nerve supplies the _____ muscle and a patch of skin over the _____ .

5   The biceps brachii and brachialis muscles are supplied by the _____ nerve.

6   The sympathetic outflow from the central nervous system is through the _____ to _____ spinal nerves.

7   The parasympathetic supply to the pelvic organs is through the _____ nerves.

8   The _____ nerve supplies the adductor muscles of the thigh.

9   Sensory supply to the medial side of the leg is by the _____ nerve, a branch of the _____ nerve.

10 All the intrinsic muscles of the foot are supplied by branches of the _____ nerve, which is in turn a branch of the _____ nerve.

Paraplegia

Paraplegia is the result of spinal cord injury between the T1 and L2 vertebrae. It results in loss of movement in parts of the body below the injured area.

T1

Spinal cord

L1

L2

Spine

Match the statement to the reason

1 *Damage to the wall of the descending aorta can cause an infarct (death of tissue) of the caudal segments of the spinal cord because…*

a *the sciatic nerve runs in the medial and inferior quadrants of the buttock and can be damaged by injection there.*

2 *Fractures of the midshaft of the radius can cause wrist drop because…*

b *the radial nerve that passes posterior to the humeral shaft supplies the extensors of the wrist.*

3 *Injections into the gluteus maximus of the buttock should always be made into the center of the superior and lateral quadrant because…*

c *the median nerve that supplies the thenar muscles and skin of the lateral palm passes anterior to the distal humerus.*

4 *Fractures of the distal humerus can cause paralysis of thumb movement and loss of sensation over the lateral palm because…*

d *arterial supply to the lumbosacral spinal cord is derived from the abdominal aorta.*

Paraplegia

Damage to the spinal cord below the T1 segmental level will lead to paralysis of only the trunk and two lower limbs (paraplegia), as well as loss of direct cortical control of bowel and urinary bladder function. If the site of injury is large, all ascending and descending transfer of information will be blocked. Spinal cord damage may be congenital—for example, spina bifida—or due to injury such as that sustained from an autombile or sporting accident, a gunshot, or a knife wound. Muscles of the affected limbs will develop spastic paralysis, where muscle tone and reflexes are increased.

Color and label

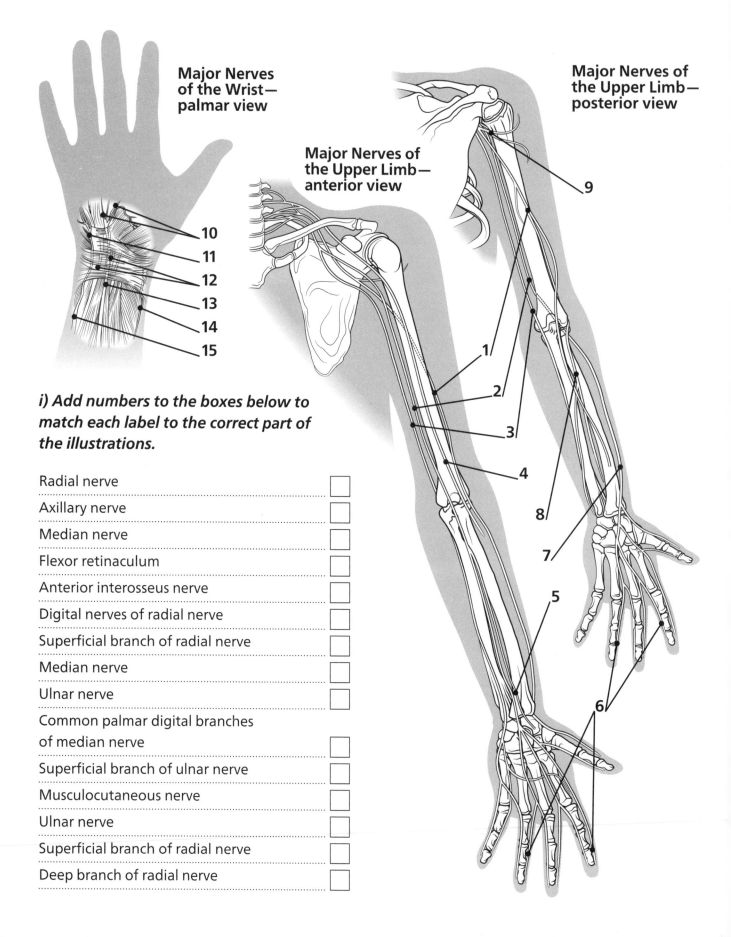

Major Nerves of the Wrist— palmar view

Major Nerves of the Upper Limb— anterior view

Major Nerves of the Upper Limb— posterior view

i) Add numbers to the boxes below to match each label to the correct part of the illustrations.

Radial nerve ☐

Axillary nerve ☐

Median nerve ☐

Flexor retinaculum ☐

Anterior interosseus nerve ☐

Digital nerves of radial nerve ☐

Superficial branch of radial nerve ☐

Median nerve ☐

Ulnar nerve ☐

Common palmar digital branches of median nerve ☐

Superficial branch of ulnar nerve ☐

Musculocutaneous nerve ☐

Ulnar nerve ☐

Superficial branch of radial nerve ☐

Deep branch of radial nerve ☐

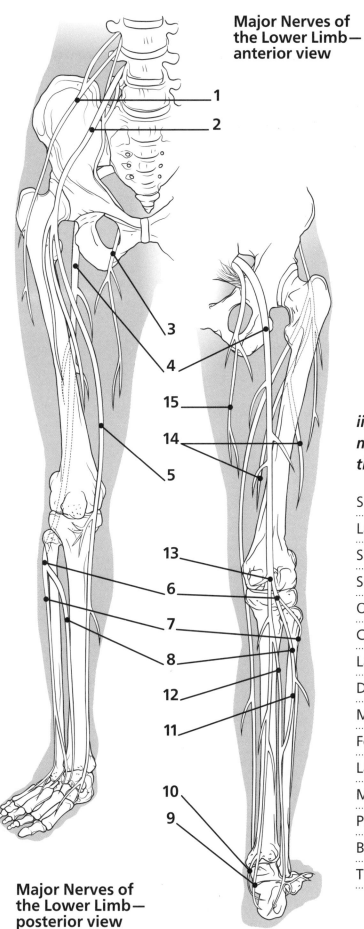

Major Nerves of the Lower Limb— anterior view

1
2
3
4
15
14
5
13
6
7
8
12
11
10
9

Major Nerves of the Lower Limb— posterior view

ii) Add numbers to the boxes below to match each label to the correct part of the illustrations.

Superficial fibular nerve	☐
Lateral plantar nerve	☐
Saphenous nerve	☐
Sciatic nerve	☐
Obturator nerve	☐
Common fibular nerve	☐
Lateral femoral cutaneous nerve	☐
Deep fibular nerve	☐
Medial plantar nerve	☐
Femoral nerve	☐
Lateral sural cutaneous nerve	☐
Medial sural cutaneous nerve	☐
Posterior femoral cutaneous nerve	☐
Branches from femoral nerve	☐
Tibial nerve	☐

The Brain

The brain, together with the spinal cord, makes up the central nervous system. The brain is responsible for processing sensory information about the world around us and the state of the internal organs of the body. It uses that information to make decisions, whether conscious or unconscious, and produces changes to the environment, either through movement of skeletal or voluntary muscles to influence the external environment or by actions on glands and smooth muscle to affect the internal environment.

Key terms:

Brainstem A term generally used to mean the midbrain, pons, and medulla oblongata.

Cerebellum The part of the brain below the cerebrum and posterior to the pons of the brainstem. It has a folded cerebellar cortex and is concerned with motor coordination.

Cerebrum The largest part of the brain, including the cerebral hemispheres and the deep structures (e.g., the thalamus) within the forebrain.

Corpus callosum A large fiber bundle joining the two cerebral hemispheres. It carries over 300 million axons to enable cross-talk between the two sides of the forebrain.

Fornix An arching fiber bundle that passes from the hippocampus and nearby temporal lobe to the septal area and hypothalamus.

Gyrus An elevation of the surface of the cerebral hemisphere. Gyri are separated by sulci.

Hypothalamus The inferior part of the diencephalon. It is concerned with homeostasis (maintaining a constant internal environment), motivation, and control of the autonomic nervous system.

Insula cortex An area of the cerebral cortex hidden in the depths of the lateral fissure.

Pineal gland A gland attached to the superior surface of the diencephalon. It secretes the hormone melatonin and is involved in biological rhythms. It is also known as the pineal body.

Sagittal fissure The deep midline groove separating the two cerebral hemispheres. It runs along the line of the sagittal suture of the skull and is occupied by a sheet of dura mater known as the falx cerebri.

Spinal cord The part of the central nervous system that extends from the foramen magnum at the skull base to the intervertebral disk between lumbar vertebrae 1 and 2.

Sulcus A groove on the surface of the cerebral hemisphere. A deep sulcus may be called a fissure.

Thalamus An egg-shaped region on each side of the third ventricle. It acts as a relay station between the brainstem and the cortex. The thalamus contains nuclei concerned with vision, hearing, and touch, as well as nuclei engaged in feedback motor circuits.

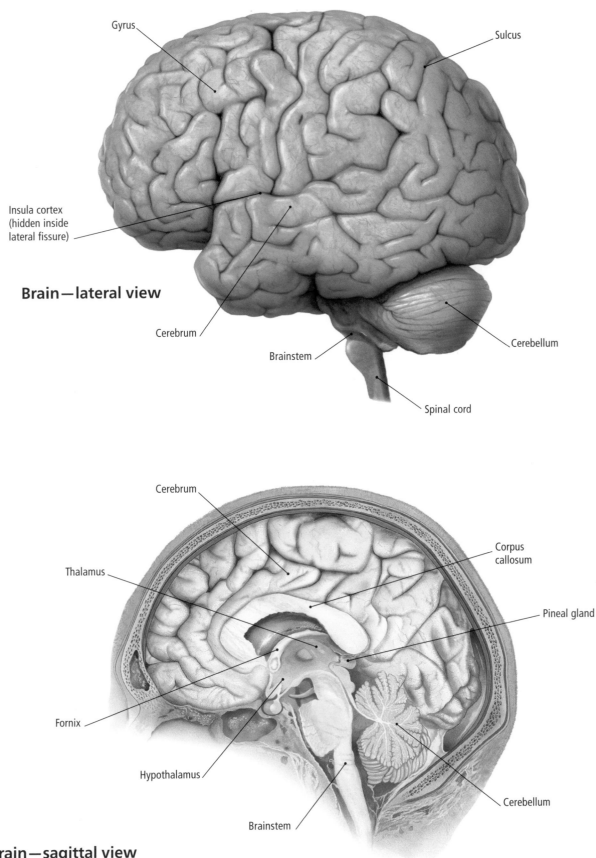

Gyrus

Sulcus

Insula cortex
(hidden inside
lateral fissure)

Brain—lateral view

Cerebrum

Brainstem

Cerebellum

Spinal cord

Cerebrum

Corpus
callosum

Thalamus

Pineal gland

Fornix

Hypothalamus

Brainstem

Cerebellum

**Brain—sagittal view
(cut along sagittal fissure)**

The Brain: Lobes and Functional Areas

Each cerebral hemisphere is divided into lobes. Four of these (frontal, parietal, occipital, temporal) are visible on the external surface of the brain, while the fifth (the insula) lies deep within the lateral fissure. The main sensory areas are the primary somatosensory (touch/pain/temperature) area in the parietal lobe, the primary auditory area (temporal lobe), the primary visual area (occipital lobe), and the primary olfactory (smell) area in the medial temporal lobe. The primary motor cortex is in the frontal lobe.

Key terms:

Auditory association area The area of cortex around the primary auditory cortex. It is concerned with higher processing of auditory information (e.g., pitch, timbre, and sound localization).

Frontal lobe The lobe of the brain deep to the frontal bone. It contains the prefrontal cortex, areas for control of eye movements (frontal eye fields), two types of motor cortex (premotor and primary motor), and Broca's area.

Gyrus An elevation of the surface of the cerebral hemisphere. Gyri are separated by sulci.

Lateral fissure The fissure between the frontal and parietal lobes (superiorly) and the temporal lobe (inferiorly). It conceals the insula.

Motor speech area (Broca's) One of the two traditionally recognized cortical language areas, said to be involved in expressive aspects of speech, but other areas in the insula may also be important. It is located in the inferior frontal gyrus.

Occipital lobe The most posterior lobe of the brain. It contains primary and association visual cortex.

Parietal lobe The lobe of the brain broadly deep to the parietal bone. It contains the association cortex for higher processing of somatosensory, auditory, and visual information.

Postcentral gyrus The gyrus that contains the primary somatosensory cortex. It is organized somatotopically (i.e., different parts of the body are represented in different regions of the cortex).

Precentral gyrus (motor cortex) The gyrus that contains the primary motor cortex. It is organized musculotopically (i.e., different parts—muscles—of the body are controlled by different regions).

Prefrontal cortex The part of the frontal cortex concerned with forethought, planning, and social interactions.

Primary auditory cortex The part of the cortex concerned with processing auditory information sent up from the medial geniculate nucleus, the auditory relay nucleus of the thalamus. It projects to the auditory association cortex.

Primary motor cortex The part of the cortex that directly drives motor neurons in the brainstem and spinal cord through long axon pathways. The pathway to the brainstem is called the corticobulbar tract; that to the spinal cord is called the corticospinal tract.

Primary somatosensory cortex The area of the cortex that receives input from the ventral posterior nucleus, the somatosensory thalamic relay nucleus. It projects in turn to the somatosensory association cortex.

Primary visual cortex The area of the cortex that receives input from the lateral geniculate nucleus, the visual thalamic relay nucleus. It is organized visuotopically (i.e., different parts of the visual world are represented in different regions of the cortex). It projects to the visual association cortex for further visual processing.

Reading comprehension area Also called the visuolexic area. It is concerned with comprehension of written language.

Sensory speech area (Wernicke's) One of the two traditionally recognized cortical language areas. It is said to be concerned with comprehension of spoken language and is usually located on the posterior superior temporal gyrus (planum temporale) but may also be found in the inferior parietal lobe.

Somatosensory association area Cortex concerned with higher processing of somatosensory information. It generates a three-dimensional model of the world around the individual and of shapes held in the hand.

Sulcus A groove on the surface of the cerebral hemisphere. A deep sulcus may be called a fissure.

Temporal lobe The lobe of the brain adjacent to the temporal bone. It contains the primary auditory and association auditory cortical areas, the hippocampus, and the amygdala.

Visual association area The part of the cerebral cortex concerned with higher processing of visual information (color, visual texture, shape, and form).

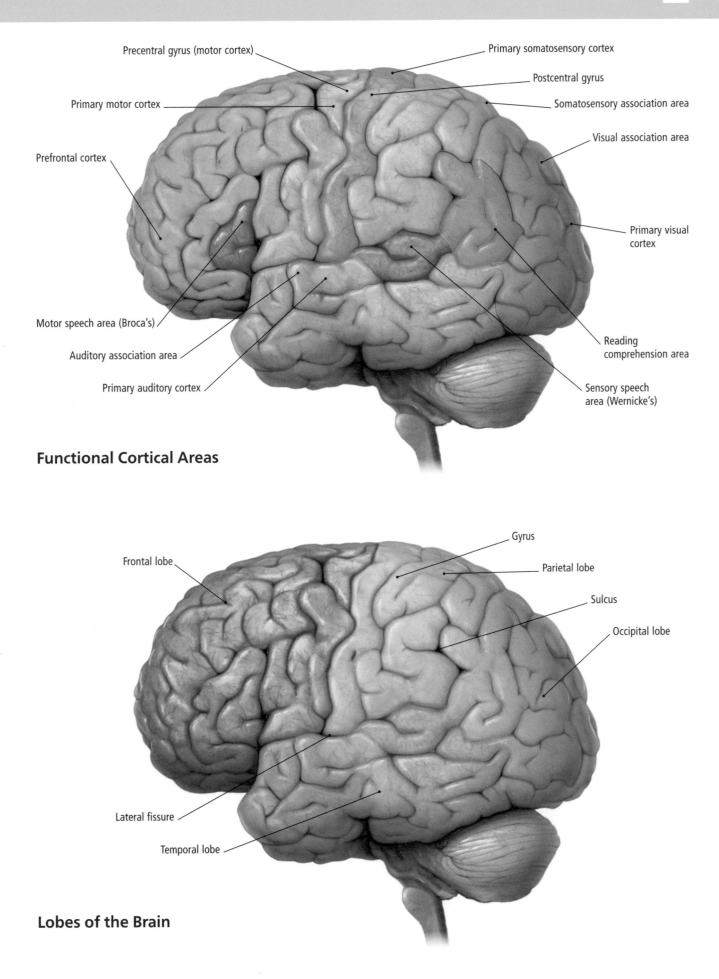

Functional Cortical Areas

Precentral gyrus (motor cortex)

Primary somatosensory cortex

Postcentral gyrus

Primary motor cortex

Somatosensory association area

Prefrontal cortex

Visual association area

Primary visual cortex

Motor speech area (Broca's)

Reading comprehension area

Auditory association area

Primary auditory cortex

Sensory speech area (Wernicke's)

Lobes of the Brain

Frontal lobe

Gyrus

Parietal lobe

Sulcus

Occipital lobe

Lateral fissure

Temporal lobe

True or false?

1   *The cell type in the brain that controls the ionic balance of the space between nerve cells is the astrocyte.*

2   *The main role of the dendrite is to transmit information to other nerve cells.*

3   *The myelin sheath of an axon can increase the speed of impulse conduction by as much as 100 times.*

4   *The major cell type in the cerebral cortex is the bipolar cell.*

5   *The primary motor cortex is located on the postcentral gyrus.*

6   *The auditory cortex is located on the upper surface of the temporal lobe.*

7   *The amygdala lies within the temporal lobe of the brain.*

8   *The corpus callosum is important for the transfer of information between functional areas of the two hemispheres.*

9   *The thalamus is an important way station in the transfer of information from the spinal cord to the cerebral cortex.*

10   *The highest concentration of nerve cells in the brain is in the granule cell layer of the cerebellum.*

11   *The cerebral cortex controls the spinal cord motorneurons by the dorsal column pathways.*

12   *The caudate and putamen play key roles in reward systems and addiction.*

13   *Parkinson's disease is usually due to degeneration of nerve cells in the substantia nigra.*

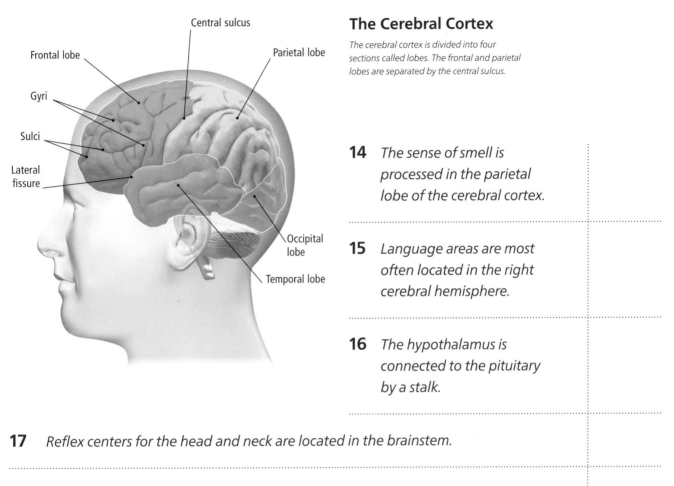

Central sulcus

Frontal lobe

Gyri

Sulci

Lateral
fissure

Parietal lobe

Occipital
lobe

Temporal lobe

The Cerebral Cortex

The cerebral cortex is divided into four sections called lobes. The frontal and parietal lobes are separated by the central sulcus.

14 *The sense of smell is processed in the parietal lobe of the cerebral cortex.*

15 *Language areas are most often located in the right cerebral hemisphere.*

16 *The hypothalamus is connected to the pituitary by a stalk.*

17 *Reflex centers for the head and neck are located in the brainstem.*

18 *The optic nerve carries motor signals for pupillary dilation.*

19 *Damage to the underside of the temporal lobe on both sides causes an inability to recognize faces.*

20 *Damage to the prefrontal cerebral cortex can cause a change in personality.*

Glioma

Mature neurons do not form tumors, but primary brain tumors (gliomas) may arise within the brain from cells that retain the ability to divide—for example, astrocytes and oligodendrocytes and their precursor cells. The most aggressive primary brain tumor is the glioblastoma multiforme, which probably comes from primitive astrocytes. Gliomas cause headaches, vomiting, seizures, focal neurological signs such as paralysis, sensory loss or specific behavioral change, and cranial nerve dysfunction.

Multiple choice

1   *Which cell type in the brain makes myelin?*
(A)   oligodendrocyte
(B)   neuron
(C)   astrocyte
(D)   microglia
(E)   choroid plexus

2   *Which cell type in the brain is responsible for immune surveillance?*
(A)   Purkinje cell
(B)   ganglion cell
(C)   endothelium
(D)   microglia
(E)   choroid plexus

3   *Respiratory centers in the brain are primarily located in the:*
(A)   cerebral cortex
(B)   thalamus
(C)   pons
(D)   medulla oblongata
(E)   both C and D are correct

4   *Which cranial nerve serves the sense of touch from the face?*
(A)   glossopharyngeal nerve
(B)   trigeminal nerve
(C)   facial nerve
(D)   vestibulocochlear nerve
(E)   trochlear nerve

5   *The role of the facial nerve is to:*
(A)   control the muscles of facial expression
(B)   control the parotid salivary gland
(C)   serve sensation from the pharynx
(D)   control sweat glands of the face
(E)   regulate the respiratory rhythm

6   *The vestibulocochlear nerve carries information concerning:*
(A)   auditory function
(B)   taste from the anterior two-thirds of the tongue
(C)   visual function
(D)   balance and orientation of the head
(E)   both A and D are correct

7   *The hunger and satiety centers are located in the:*
(A)   cerebral cortex
(B)   thalamus
(C)   hypothalamus
(D)   cerebellum
(E)   pons

8   *The part of the brain responsible for tagging memories with emotional significance is the:*
(A)   amygdala
(B)   hypothalamus
(C)   hippocampus
(D)   pons
(E)   midbrain

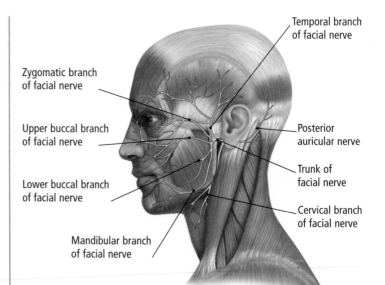

Temporal branch of facial nerve
Zygomatic branch of facial nerve
Upper buccal branch of facial nerve
Lower buccal branch of facial nerve
Posterior auricular nerve
Trunk of facial nerve
Cervical branch of facial nerve
Mandibular branch of facial nerve

Facial Nerve Branches

Note: In this illustration the parotid gland has been removed to reveal the facial nerve branches.

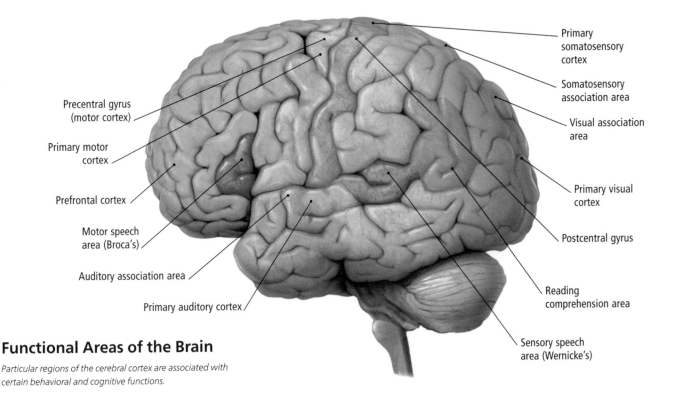

Functional Areas of the Brain

Particular regions of the cerebral cortex are associated with certain behavioral and cognitive functions.

9 *The ability to lay down new memories is greatly impaired by bilateral damage to the:*
(A) hippocampus
(B) amygdala
(C) hypothalamus
(D) thalamus
(E) parietal lobe

10 *The cerebellum receives sensory input from:*
(A) joint receptors
(B) inner ear balance organs
(C) touch receptors on the face
(D) muscle stretch receptors of the limbs
(E) all of the above are correct

11 *The primary somatosensory cortex is located on the:*
(A) precentral gyrus
(B) postcentral gyrus
(C) occipital lobe
(D) orbital cortex
(E) temporal lobe

12 *Damage to the right parietal lobe can cause:*
(A) a problem with expressive aspects of speech
(B) loss of vision in the left visual field
(C) sensory neglect on the left side of the body
(D) a problem with working memory
(E) loss of the sense of smell

13 *Visual information from the left visual field is transmitted to the:*
(A) left superior colliculus
(B) right parietal lobe
(C) right precentral gyrus
(D) right occipital lobe
(E) left pons

14 *Working memory (the ability to hold the steps in a simple task in one's memory for a few minutes) is mainly served by the:*
(A) prefrontal cortex
(B) parietal lobe
(C) cerebellum
(D) hypothalamus
(E) temporal lobe

Color and label

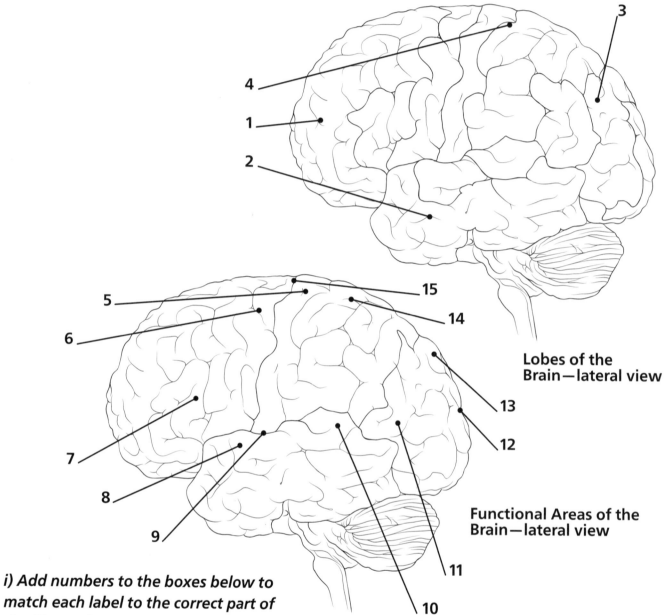

Lobes of the Brain—lateral view

Functional Areas of the Brain—lateral view

i) Add numbers to the boxes below to match each label to the correct part of the illustrations.

Parietal lobe	☐	Auditory cortex	☐
Occipital lobe	☐	Visual cortex	☐
Temporal lobe	☐	Visual association area	☐
Frontal lobe	☐	Wernicke's sensory speech area	☐
Motor speech (Broca's) area	☐	Reading comprehension area	☐
Primary motor cortex	☐	Primary somatosensory cortex	☐
Central sulcus	☐	Somatic sensory association area	☐
Auditory association area	☐		

*ii) Label each structure shown
on the illustration.*

1 ....

2 ....

3 ....

4 ....

5 ....

6 ....

7 ....

8 ....

9 ....

10 ....

11 ....

12 ....

Fill in the blanks

1   The cerebrospinal fluid is mainly made by the _____.

2   The _____ ventricles lie within the cerebral hemispheres.

3   The _____ fluid surrounds the brain and spinal cord.

4   The _____ are veins within the cranial cavity that have been reinforced with dura mater.

5   The _____ is a fiber bundle joining the hippocampus with the hypothalamus.

6   The _____ allows transfer of information between the two cerebral hemispheres.

7   The _____ contains axons connecting the thalamus and cerebral cortex.

8   The _____ is the region where optic nerve axons cross the midline.

9   The _____ nerve supplies the muscles of the tongue.

10   The _____, _____, and _____ nerves supply the skeletal muscles that move the eye.

11   The _____ nucleus is the part of the thalamus that is most concerned with vision.

12   The _____ nucleus is the part of the thalamus that is most concerned with hearing.

13   The _____ peduncle contains axons from the pontine nuclei to the cerebellum.

14   The _____ and _____ nerves are involved in the corneal (blink) reflex.

15   The _____ and _____ nerves are involved in the gag reflex.

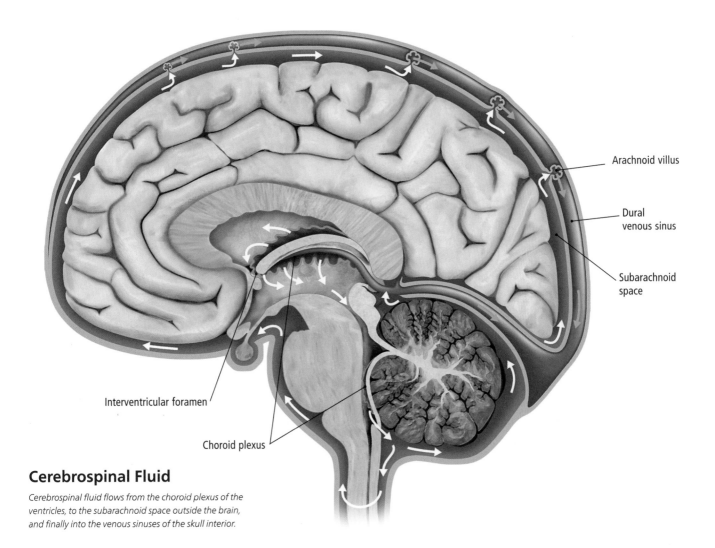

Arachnoid villus

Dural venous sinus

Subarachnoid space

Interventricular foramen

Choroid plexus

Cerebrospinal Fluid

Cerebrospinal fluid flows from the choroid plexus of the ventricles, to the subarachnoid space outside the brain, and finally into the venous sinuses of the skull interior.

16 *The _____sulcus separates the primary motor and somatosensory cortex.*

17 *The primary visual cortex is located around the _____sulcus.*

18 *The _____fissure separates the frontal and temporal lobes.*

19 *The _____ventricle lies between the pons and the cerebellum.*

Match the statement to the reason

1 Information from ganglion cells in the nasal half of the left retina reaches the left visual cortex because…

a fibers cross in the sensory decussation at the level of the medulla oblongata.

2 Information about touch in the right upper limb is represented on the left postcentral gyrus because…

b many descending and ascending axons pass through a narrow space.

3 A hemorrhage in the internal capsule can have profound effects on neurological function because…

c the satiety center is located here.

4 A grand mal epileptic seizure appears to spread gradually across the body because…

d axons cross in the chiasm below the hypothalamus.

5 A tumor in the hypothalamus can cause obesity because…

e the primary motor cortex has an ordered map of the body parts on it.

Epilepsy

Epilepsy is a neurological disturbance characterized by abnormal electrical discharges in brain circuitry. Epilepsy may manifest as sensory disturbances, loss of consciousness, and convulsions. Epilepsy has many variants ranging from the mild absence attacks of petit mal, where the subject transiently loses attention, to tonic/clonic seizures seen in grand mal, where convulsions spread across the body (Jacksonian march) as the abnormal electrical discharge moves across the brain surface. Temporal lobe epilepsy can cause feelings of déjà vu, jamais vu (familiar things seeming new—the opposite of déjà vu), amnesia, olfactory hallucinations, and a sudden sense of fear and anxiety.

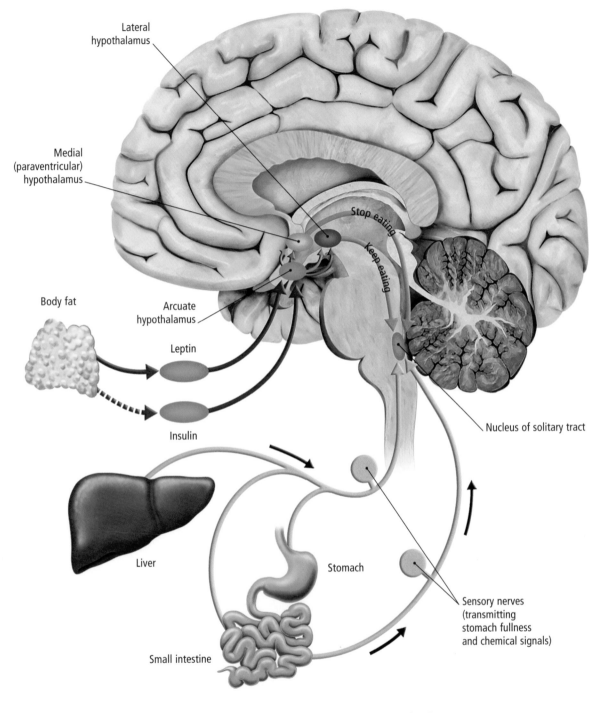

Lateral hypothalamus

Medial (paraventricular) hypothalamus

Body fat

Arcuate hypothalamus

Leptin

Insulin

Stop eating

Keep eating

Nucleus of solitary tract

Liver

Stomach

Sensory nerves (transmitting stomach fullness and chemical signals)

Small intestine

Hypothalamus and Appetite Regulation

Many factors influence the regulation of appetite, and the hypothalamus plays a key role. The amount of body fat influences the hypothalamus through leptin and insulin, hormones that tell the brainstem to stop or continue eating. This interacts with other signals from the liver and gut, via the nucleus of the solitary tract, to influence how hungry we feel.

Color and label

i) Label each structure shown on the illustration.

ii) Use the key to color these structures.

- ■ Atlas
- ▨ Pineal body
- ■ Second cervical nerve

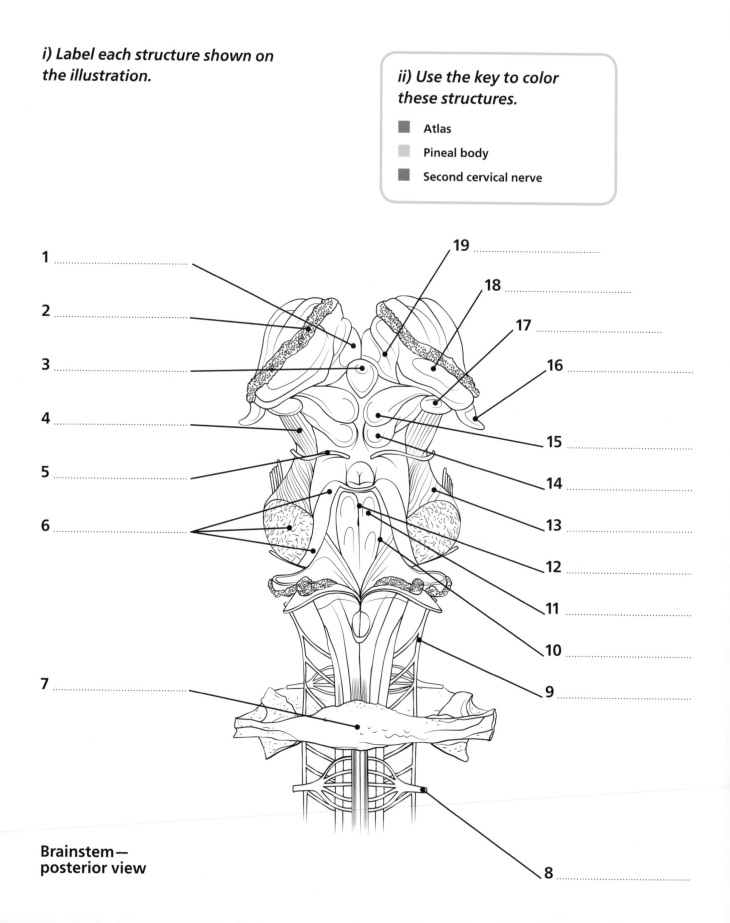

Brainstem— posterior view

iii) Label each structure shown on the illustrations.

Spinal Nerves

1 ....

2 ....

3 ....

4 ....

5 ....

Spinal Cord—anterior view

10 ....

9 ....

8 ....

7 ....

6 ....

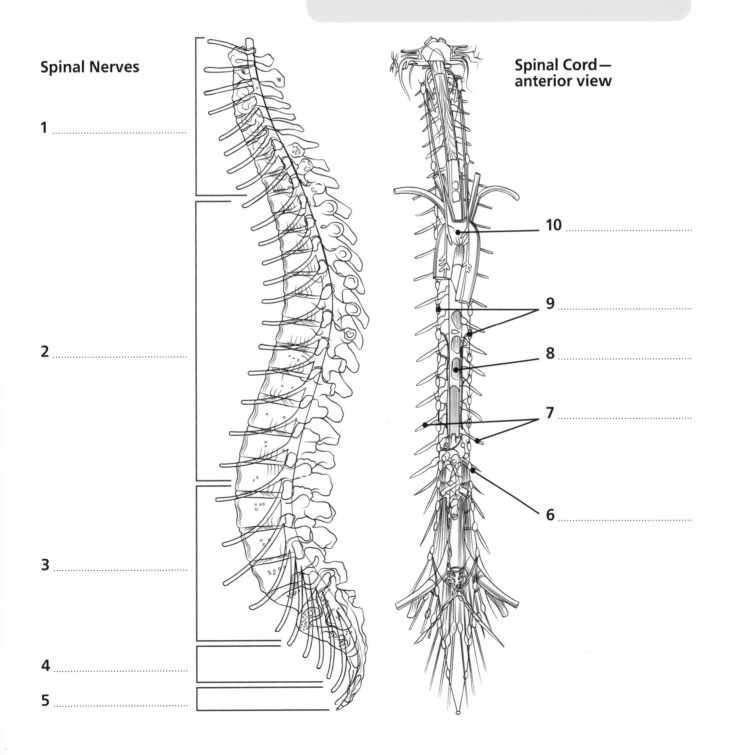

Special Senses: Sight and Hearing

Key terms:

Anterior chamber The fluid chamber between the posterior surface of the cornea and the iris. It is filled with a watery liquid, the aqueous humor.

Choroid The pigmented and vascularized middle coat of the eye. It provides nutrients by diffusion to the outer retinal layers.

Ciliary body The anterior extension of the choroidal layer of the eye. It connects the choroid with the iris and contains the ciliary muscle and ciliary processes.

Ciliary muscle Smooth muscle that adjusts the tension of the suspensory ligaments on the equator of the lens.

Cornea The thin, convex, transparent surface of the eye. It has no blood vessels and is nourished by diffusion from the surrounding tissues. It is responsible for most of the refraction that produces the image on the retina.

Lens A transparent structure of modifiable shape that allows focusing on near or far objects. Its natural shape is close to spherical, but tension along its equator produces a flattened profile.

Optic nerve The nerve that carries retinal ganglion cell axons from the retina to the visual nucleus of the thalamus (lateral geniculate nucleus) and the midbrain (superior colliculus).

Posterior chamber The fluid chamber between the iris in front and the ciliary body, suspensory ligaments, and lens behind. It is filled with a liquid, the aqueous humor.

Retina The neural layer of the eye. It has layers of photo-receptors, bipolar cells, and ganglion cells and is richly supplied with blood.

Suspensory ligaments Ligaments attaching the ciliary body to the equator of the lens. When these are relaxed, the lens returns to a more globular shape.

Vitreous body The posterior transparent cavity of the globe of the eye. It is filled with a liquid, the vitreous humour.

The Eye—lateral view

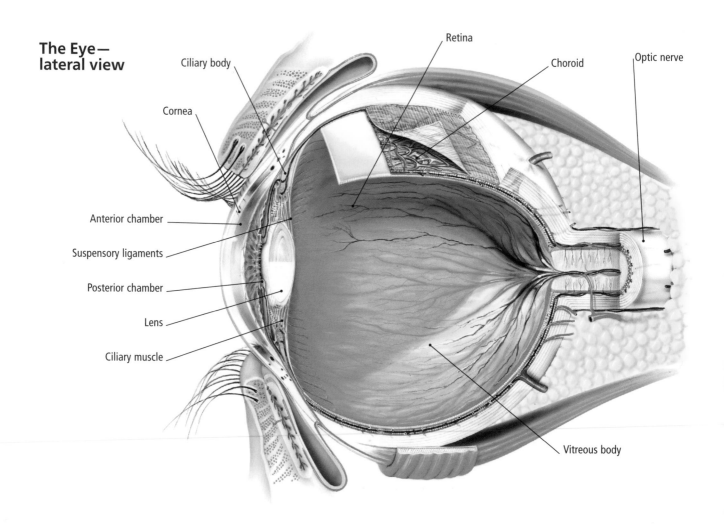

Retina
Choroid
Optic nerve
Ciliary body
Cornea
Anterior chamber
Suspensory ligaments
Posterior chamber
Lens
Ciliary muscle
Vitreous body

Key terms:

Ampullae Enlarged parts of each semicircular duct. Each one houses a crista, an apparatus for detecting head rotation.

Cochlea Spiral tube of the auditory part of the inner ear.

Cochlear duct The spiral tube that houses the organ of Corti. It is flanked by vestibular and tympanic ducts.

Cochlear nerve The aggregated nerve fibers from the organ of Corti that carry auditory information to the brainstem.

Cochlear (round) window A window between the inner and middle ear. It is covered by the epithelium of the middle ear cavity and bulges outward when the oval window is depressed by the stapes footplate.

Eustachian (auditory) tube A tube connecting the middle ear cavity with the nasopharynx. Allows equalization of pressure between the two.

External ear canal (meatus) The tube leading from the external environment to the tympanic membrane.

Helicotrema The point near the apex of the cochlea where the vestibular and tympanic ducts communicate.

Incus The middle bone of the auditory ossicle chain.

Malleus The auditory ossicle in direct contact with the tympanic membrane.

Ossicles Three tiny bones arranged in a chain across the middle ear cavity.

Promontory covering first coil of cochlea Bony projection of the medial wall of the middle ear formed by the cochlea's base.

Saccular macula The sensory region of the saccule. Its hair cells are stimulated during linear acceleration of the head.

Saccule One of the components of the vestibular apparatus. It detects linear acceleration of the head.

Semicircular canals Three bony channels (lateral, posterior, and anterior) that house the semicircular ducts.

Stapes A stirrup-shaped bone of the middle ear.

Stapes footplate covering vestibular (oval) window The base or footplate of the stapes sits on the vestibular window and imparts vibrations from the tympanic membrane to the perilymph (a fluid) of the inner ear.

Tympanic duct The spiral channel (scala tympani) that ascends the cochlea from base to apex.

Tympanic membrane (eardrum) The membrane separating the external from the middle ear.

Utricle One of the components of the vestibular apparatus.

Vestibular duct The spiral channel (scala vestibuli) that ascends the cochlea from the base to the apex.

Vestibular nerve branches Nerve fibers from the components of the vestibular apparatus (maculae of the utricle and saccule and ampullae of the semicircular ducts).

The Ear— lateral view

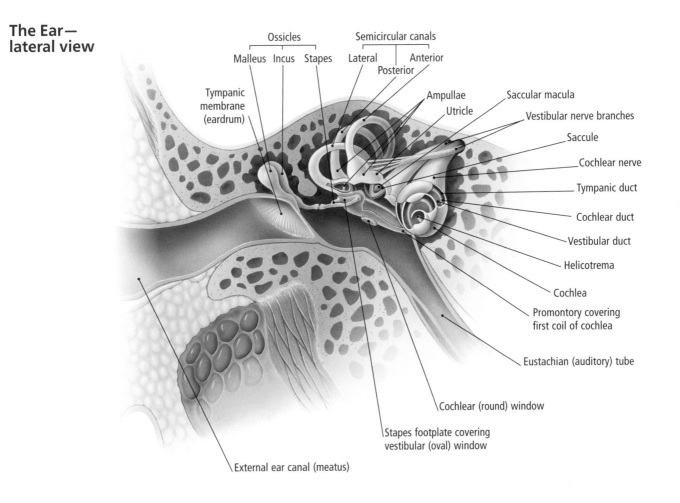

Special Senses: Taste and Smell

Key terms:

Epiglottis A leaf-shaped cartilage that protects the entrance to the larynx. It folds over the laryngeal (airway) entrance during swallowing.

Filiform papillae Conical projections of the dorsum of the tongue. They have no taste buds but grip food to move it around the mouth.

Foliate papillae Ridges on the posterior edge of the tongue. They have taste buds on their surface.

Fungiform papillae Mushroom-shaped elevations of the tongue surface. Taste buds are found on their surface.

Lingual tonsil (lingual nodules) A lymphoid organ on the surface of the posterior third of the tongue. It is part of an immune surveillance system (Waldeyer's ring) for the entrance to the digestive and respiratory tracts.

Median sulcus The midline groove on the dorsum of the tongue.

Palatine tonsil A lymphoid organ lying at the lateral wall of the oropharynx. It lies between the palatoglossal and palatopharyngeal arches. It is part of an immune surveillance system (Waldeyer's ring) at the entrance to the digestive and respiratory tract.

Palatoglossus muscle and arch The arch at the side of the oral cavity that is produced by mucosa over the palatoglossus muscle. The palatoglossus muscle runs from the soft palate to the tongue.

Palatopharyngeus muscle and arch The arch at the side of the oropharynx that is produced by mucosa over the palatopharyngeus muscle. The palatopharyngeus muscle runs from the soft palate to the muscular wall of the pharynx.

Terminal sulcus A V-shaped depression that separates the anterior two-thirds of the tongue from the posterior third.

Vallate papillae Papillae occupying a depression in the tongue surface. They form a V-shaped line anterior to the terminal sulcus of the tongue. Each vallate papilla has an encircling trench where taste receptors (buds) are located.

Vallecula The depression between the posterior third of the tongue and the epiglottis. It is a potential site for chicken or fish bones to become caught.

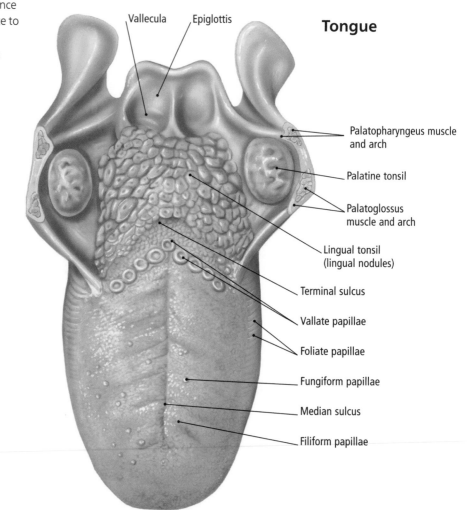

Tongue

Vallecula · Epiglottis · Palatopharyngeus muscle and arch · Palatine tonsil · Palatoglossus muscle and arch · Lingual tonsil (lingual nodules) · Terminal sulcus · Vallate papillae · Foliate papillae · Fungiform papillae · Median sulcus · Filiform papillae

Key terms:

Axon The long process of a nerve cell that transmits the impulse to another part of the brain or body.

Bowman's gland (olfactory gland) Gland that produces a serous fluid in which odorant molecules dissolve so they can be presented to the cilia. The secretions also contain an odorant-binding protein that aids the process.

Cilia Hairlike processes from the apexes of sensory cells in the nasal mucosa. Odorant molecules lock into receptors on the surface of the cilia.

Cribriform plate of ethmoid bone A delicate plate of bone that forms the roof of the nasal cavity. It is perforated by olfactory nerve fibers.

Fila olfactoria Olfactory nerve fibers passing through the cribriform plate of the ethmoid bone to enter the olfactory bulb.

Frontal lobe The lobe of the brain deep to the frontal bone. It contains the prefrontal cortex, areas for control of eye movements, two types of motor cortex, and Broca's area.

Mitral cell The major output neuron of the olfactory bulb. They project information about smell to the olfactory cortex and olfactory tubercle in the brain.

Olfactory bulb The elongated melon-shaped structure above the cribriform plate of the ethmoid bone. It receives fila olfactoria and processes olfactory information before sending it through the olfactory tract to the olfactory cortex in the brain.

Olfactory mucosa The epithelium of the upper nasal cavity that is sensitive to odorants. It contains four types of cells: mature nerve cells, basal proliferative cells, immature nerve cells, and supporting cells.

Olfactory nerve cell Sensory neuron in the olfactory epithelium. Each cell has an apical process with a knoblike ending and 10–20 cilia that sample the inhaled air for odorant molecules. The other end of the cell has an axon that runs to the olfactory bulb.

Olfactory tract The fiber pathway running from the olfactory bulb to the olfactory centers (olfactory cortex and olfactory tubercle) in the brain.

Olfactory Apparatus

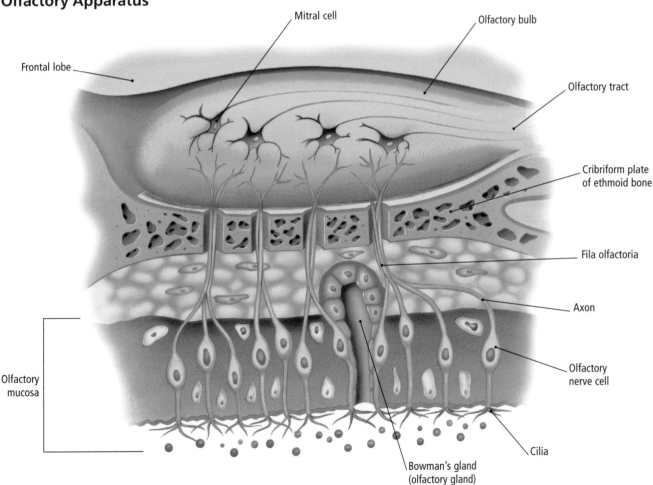

True or false?

**Retina—
cross-sectional view**

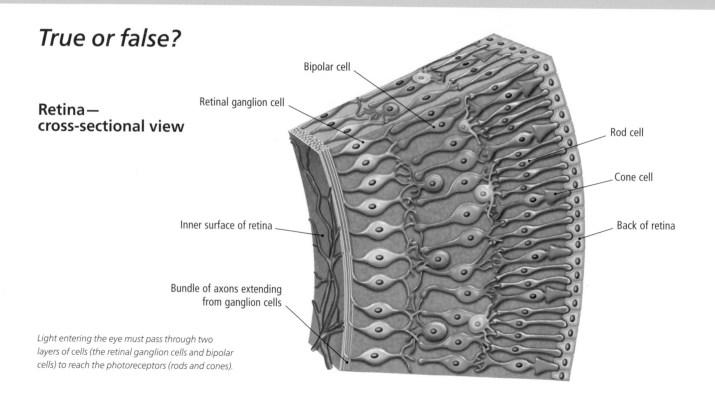

*Light entering the eye must pass through two
layers of cells (the retinal ganglion cells and bipolar
cells) to reach the photoreceptors (rods and cones).*

1   *Olfactory receptor cells send axons through the cribriform plate of the skull.*

2   *The olfactory region of the cerebral cortex is on the medial temporal lobe.*

3   *The olfactory tract carries parasympathetic axons to the olfactory mucosa.*

4   *Olfactory sensory input has a strong influence on emotional memory.*

5   *Circumvallate papillae have taste buds on their exposed surface.*

6   *Sweet, sour, salt, bitter, and umami are the five taste sensations.*

7   *Olfaction plays no role in the tasting of food and drink.*

8   *The lacrimal gland is located in the upper medial part of the orbit.*

9   *The bulbar conjunctiva covers the sclera of the eyeball.*

10   *The ciliary muscle is controlled by the parasympathetic nervous system.*

Cataract

Cataract is a common condition where the lens becomes progressively cloudy or opacified, leading to loss of vision. Patients complain of faded colors, blurring of vision, halos around lights, and difficulty seeing at night. It causes half of the blindness in the world and is most commonly due to aging, although it can be present at birth. Risk factors for early onset include diabetes mellitus and tobacco smoking. The condition can be treated by removal of the old opacified lens core and replacement with a foldable prosthetic lens.

11   The shape of the lens is controlled by contraction of the ciliary muscle.

12   The vitreous humor fills the anterior chamber of the eye.

13   Rod photoreceptors are most useful for vision in low light levels.

14   Cone photoreceptors are concentrated into a fovea located medial to the optic disk.

15   Retinal ganglion cells send their axons into the optic nerve.

16   The middle ear cavity is filled with perilymph.

17   The pharyngotympanic (auditory) tube connects the middle ear with the oropharynx.

18   The footplate of the stapes lies over the oval window.

19   The macula of the utricle primarily detects linear acceleration of the head.

20   Contraction of the stapedius muscle helps to open the pharyngotympanic (auditory) tube.

Multiple choice

1 *Which part of the nasal cavity contains the olfactory sensory region?*
(A) the roof
(B) the medial wall
(C) the lateral wall
(D) the floor
(E) the nasal vestibule

2 *Nerve fibers from the olfactory sensory region of the nasal cavity terminate in the:*
(A) temporal lobe
(B) occipital lobe
(C) amygdala
(D) olfactory bulb
(E) thalamus

3 *Taste information is carried to the brainstem by which nerve(s)?*
(A) trigeminal
(B) facial
(C) glossopharyngeal
(D) vagus
(E) answers B, C, and D are correct

4 *The facial nerve serves taste to the:*
(A) anterior two-thirds of the tongue
(B) soft palate
(C) posterior one-third of the tongue
(D) epiglottis
(E) floor of the mouth

5 *Tear fluid from the conjunctival sac drains to the nasal cavity by the:*
(A) frontonasal duct
(B) ethmoid air cells
(C) nasolacrimal duct
(D) sphenoid sinus
(E) nasociliary canal

6 *The most powerful refractive element of the eye is the:*
(A) aqueous humor
(B) cornea
(C) lens
(D) vitreous humor
(E) choroid

7 *The three layers of the eyeball are, from outside to inside:*
(A) retina, choroid, sclera
(B) retina, sclera, choroid
(C) choroid, sclera, retina
(D) sclera, choroid, retina
(E) choroid, retina, sclera

8 *The size of the pupil is controlled by smooth muscle in the:*
(A) ciliary body
(B) choroid
(C) iris
(D) retina
(E) orbital margin

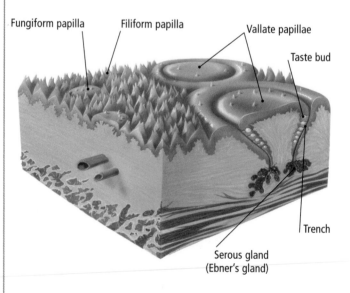

Fungiform papilla   Filiform papilla   Vallate papillae
Taste bud
Trench
Serous gland (Ebner's gland)

Tongue—cross-sectional view

The majority of taste buds are found on the tongue. These taste buds are located in the papillae, which are projections on the upper surface of the tongue.

9 *Which of the following is found in the innermost layer of the retina?*

(A) photoreceptor cell bodies
(B) bipolar cell bodies
(C) amacrine cells
(D) retinal ganglion cell bodies
(E) retinal ganglion cell axons

10 *Which of the following is true concerning the optic disk?*

(A) it is the most light-sensitive part of the retina
(B) it is the site where retinal ganglion cell axons leave the retina
(C) it is the site where the central retinal artery enters the globe of the eye
(D) it is full of rod and cone photoreceptors
(E) both B and C are correct

11 *The boundary between the middle and external ear lies at the:*

(A) oval window
(B) round window
(C) helicotrema
(D) tympanic membrane
(E) external acoustic meatus

12 *The three tiny bones that transmit sound across the middle ear are, from lateral to medial:*

(A) incus, stapes, and malleus
(B) malleus, incus, and stapes
(C) malleus, stapes, and incus
(D) stapes, incus, and malleus
(E) stapes, malleus, and incus

13 *The main function of the outer hair cells of the organ of Corti is to:*

(A) transduct fluid vibrations in the inner ear to nerve impulses
(B) keep the tectorial membrane clear of debris
(C) move fluid from the scala vestibuli to the cochlear duct
(D) act as a cochlear amplifier to improve auditory sensitivity
(E) produce the fluid of the tunnel of Corti

14 *The sensory organ primarily responsible for detection of rotation of the head in a horizontal plane is the:*

(A) macula of the utricle
(B) macula of the saccule
(C) ampulla of the superior semicircular duct
(D) ampulla of the horizontal semicircular duct
(E) cochlea

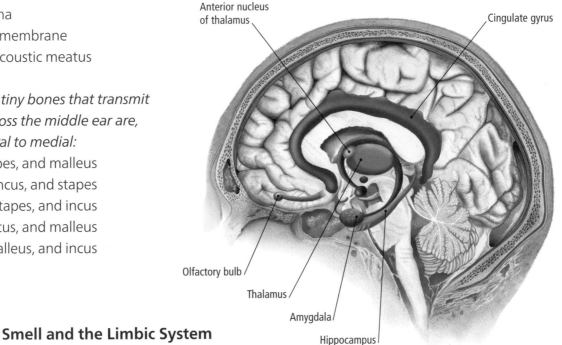

Anterior nucleus of thalamus

Cingulate gyrus

Olfactory bulb

Thalamus

Amygdala

Hippocampus

Smell and the Limbic System

The olfactory bulb is directly connected to the hippocampus and amygdala in the limbic system, which is important for memory and emotion. This is why smells are evocative of past places and feelings and can trigger responses such as fear and pleasure.

Color and label

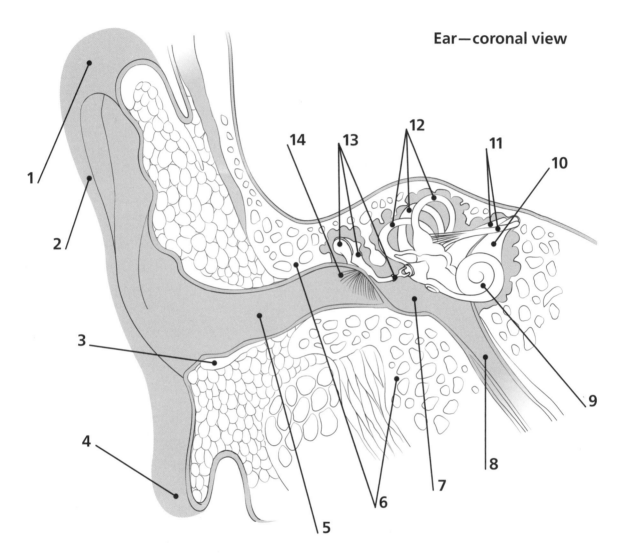

Ear—coronal view

i) Add numbers to the boxes below to match each label to the correct part of the illustration.

Lobule ☐

Pinna ☐

Cartilage ☐

Helix ☐

Middle ear (tympanic cavity) ☐

Temporal bone ☐

External auditory canal (meatus) ☐

Tympanic membrane (eardrum) ☐

Semicircular canals ☐

Cochlear nerve branch ☐

Vestibular nerve branches ☐

Cochlea ☐

Ossicles (malleus, incus, and stapes) ☐

Eustachian (auditory) tube ☐

ii) Label each structure shown on the illustrations.

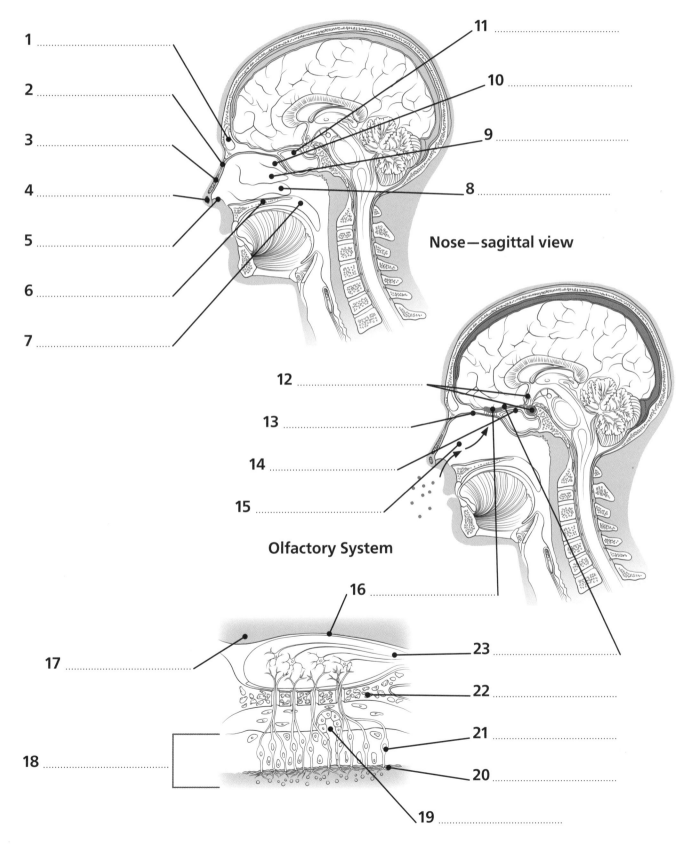

1 ....

2 ....

3 ....

4 ....

5 ....

6 ....

7 ....

11 ....

10 ....

9 ....

8 ....

Nose—sagittal view

12 ....

13 ....

14 ....

15 ....

Olfactory System

16 ....

17 ....

18 ....

23 ....

22 ....

21 ....

20 ....

19 ....

Fill in the blanks

1   The sphincter pupillae and dilator pupillae muscles are under control of the _____ and _____ nervous systems, respectively.

2   The highly colored part of the eye is the _____.

3   The ciliary muscle is under the control of the _____ nervous system and regulates the shape of the lens in a process called_____.

4   Much of the oxygen supply to the photoreceptors comes by diffusion from the _____.

5   The _____ photoreceptors are specialized for monochromatic vision in low light levels, whereas the _____ photoreceptors serve color vision under bright light conditions.

6   Bipolar cells receive input from _____ and pass information to the _____.

7   Cone photoreceptors are concentrated in a central region of the retina called the _____, which is specialized for high acuity color vision.

8   The utricle and saccule lie within _____ of the _____.

9   The _____ of the organ of Corti are the auditory sensory neurons, whereas the _____ act as cochlear amplifiers to enhance perception of key frequencies.

10   The fluid within the scala media of the cochlea is called the _____.

11   Blockage of the _____ can cause a middle ear infection known as _____.

12   The endolymph of the inner ear is manufactured by the _____.

13 The _____ ducts detect rotation of the head by the flow of fluid past the sensory _____ of the ducts.

14 The expanded basal part of the cochlea serves _____ sounds.

15 The _____ division of the _____ nerve arises from utricle, saccule, and semicircular duct ampullae of the inner ear.

Ménière's disease

Ménière's disease is a condition affecting the inner ear, where the patient experiences dizziness, tinnitus (ringing in the ear), hearing loss, and a feeling of fullness in the ear. The episodes of dizziness tend to last between 20 minutes and an hour. Causes are hereditary and environmental (vascular, viral, or autoimmune). The fundamental problem is the buildup of fluid in the inner ear. There is no cure currently, but a low-salt diet, diuretics, and anti-inflammatory medication have been tried.

16 The organ of Corti sits on a _____ membrane, the width of which determines the frequency sensitivity of that part of the cochlea.

17 The olfactory receptor neurons have apical processes with _____ and 10 to 20 _____ that sample inhaled air for odorant molecules.

18 Epilepsy in the _____ lobe of the brain can cause olfactory hallucinations.

19 Central regions for taste include the _____ of the brainstem and the _____ of the cerebral cortex.

20 The _____ branch of the _____ nerve conveys taste information from the anterior two-thirds of the tongue.

Organ of Corti

The highly sensitive and receptive hairlike cells in the organ of Corti enable us to hear even very faint sounds. The cells are triggered by movement in the cochlear fluid and, when activated, send nerve impulses to the auditory cortex of the brain.

Inner hair cell

Nerve fibers

Tectorial membrane

Outer hair cell

Phalangeal cell

Basilar membrane

Pillar cell

Match the statement to the reason

1   The human retina receives oxygen and nutrients from two arterial supplies (choroid and central retinal artery branches) because…

a   the axons from retinal ganglion cells that "see" the opposite visual fields pass to the same side of the brain (i.e., the left cortex receives visual information from right visual field halves for each eye).

2   Tumors of the pituitary gland can cut off vision in the temporal fields of each eye because…

b   the delicate axons of the olfactory receptor neurons pass through tiny holes in the bony cribriform plate and can be severed by a skull base fracture.

3   Damage to the primary visual cortex of one side of the brain causes blindness in the paired visual fields toward the opposite side of the body—i.e., blindness in the temporal field for the right eye and nasal field for the left eye from a left visual cortex lesion—because…

c   the thickness of the human retina is too great for supply of the inner retinal layers from the choroid circulation alone.

4   Progressive deafness with age for high auditory tones results from loss of cochlear hair cells from the basal turns of the cochlea because…

d   axons from the nasal retina (which "sees" the temporal visual field due to the image inversion of the eye) cross in the midline directly superior to the anterior lobe of the pituitary.

5   Fractures of the skull base from a blow to the nasal bridge can cause loss of the sense of smell (anosmia) because…

e   inner hair cells in the basal parts of the cochlea spiral are specifically tuned to perceive sounds of high frequency and are most readily damaged by prolonged exposure to loud noise.

1 Middle ear infections are more common in children because…

a hair cells in particular parts of the cochlear spiral are tuned to specific frequencies, and an implanted electrode can stimulate the remaining nerve pathways to allow perception of sound.

2 The glands of the external ear produce waxy cerumen because…

b a fluid environment is necessary to present taste molecules (tastants) to the receptors.

3 Cochlear implants can alleviate deafness by threading an electrode around the turns of the cochlear because…

c fatty acids are toxic to many pathogenic microorganisms.

4 Damage to the olfactory mucosa and nerves leads to loss of enjoyment of food because…

d olfaction plays a key role in the "taste" of food.

5 Food cannot be tasted with a dry mouth because…

e the juvenile auditory tube is more horizontal and drains poorly.

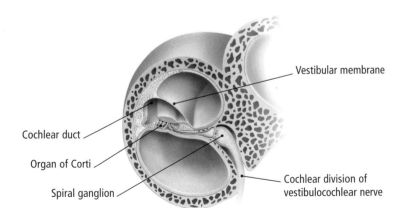

Cochlear duct

Organ of Corti

Spiral ganglion

Vestibular membrane

Cochlear division of vestibulocochlear nerve

Cochlea

The cochlea is a spiral structure of the inner ear containing three fluid-filled channels. The central channel contains the organ of Corti (see p. 117).

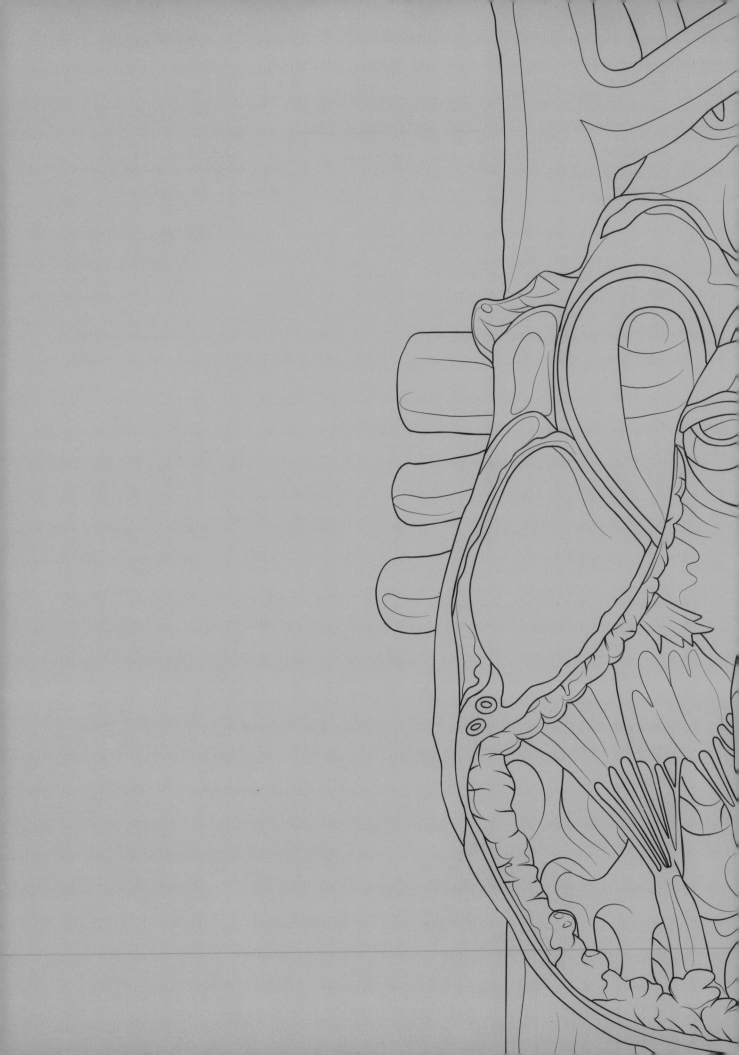

CHAPTER 4:
THE CIRCULATORY
SYSTEM

The Heart

The heart is a muscular pump that circulates blood around both the systemic and pulmonary circulations. The systemic circulation carries oxygenated blood from the left side of the heart to the capillaries of all organs except the peripheral lung and returns the deoxygenated blood to the right side of the heart. The pulmonary circulation carries deoxygenated blood from the right side of the heart to the capillaries of the lungs and returns it as oxygenated blood to be pumped through the systemic circulation again.

Key terms:

Aortic arch The most superior part of the aorta. It gives rise to the brachiocephalic artery (trunk), left common carotid artery, and left subclavian artery.

Aortic valve The semilunar valve at the beginning of the aorta. It prevents regurgitation of aortic blood into the left ventricle during ventricular diastole.

Ascending aorta The initial part of the aorta, from the aortic valve to the origin of the brachiocephalic artery (trunk). It gives rise to the coronary arteries.

Brachiocephalic artery (trunk) The brachiocephalic artery is the first branch of the aortic arch. It gives off the right subclavian and right common carotid arteries.

Chordae tendineae Fibrous bands ("heartstrings") that anchor the edges of atrioventricular valve leaflets to the apices of papillary muscles.

Descending aorta The part of the aorta distal to the arch. It supplies the posterior chest wall, spinal cord, abdomen, pelvis, and lower limbs.

Inferior vena cava The largest vein of the abdomen. It carries blood from the lower limb, pelvis, kidneys, and posterior abdominal wall.

Leaflet/cusp of mitral valve One of two leaflets of the mitral valve. The edge of each is anchored by chordae tendineae and a papillary muscle to the left ventricle wall.

Leaflet/cusp of tricuspid valve One of three leaflets of the tricuspid valve. The edge of each is anchored by chordae tendineae and a papillary muscle to the right ventricle wall.

Left atrium The chamber that receives relatively oxygenated blood from the four pulmonary veins and pumps it to the left ventricle.

Left brachiocephalic vein The vein draining systemic venous blood from the left head and neck and left upper limb.

Left common carotid artery The branch of the arch of the aorta that supplies the left head and neck.

Left inferior pulmonary vein One of four veins returning relatively oxygenated blood to the left atrium.

Left pulmonary artery The branch of the pulmonary artery that supplies relatively deoxygenated blood to the left lung.

Left subclavian artery The branch of the arch of the aorta that supplies the left upper limb.

Left superior pulmonary vein One of four veins returning relatively oxygenated blood to the left atrium.

Left ventricle The heart chamber that develops the highest pressure (120 mm Hg) and therefore has the thickest muscle wall. Its interior cavity pumps blood to the ascending aorta.

Opening of coronary sinus The opening of a large vein that drains blood from most of the heart muscle. It enters the right atrium above the opening of the tricuspid valve.

Papillary muscle A small muscular projection that anchors the chordae tendineae and hence atrioventricular valve leaflets to the heart wall. The papillary muscles contract during ventricular systole to keep the valve leaflets closed.

Pericardium A multilayered sac that encloses the heart. The outermost fibrous layer anchors the heart to adjacent structures. The inner, serous double layer provides a fluid-filled space to reduce friction during beating of the heart.

Pulmonary valve The semilunar valve at the ouflow from the right ventricle. It prevents regurgitation of blood from the pulmonary trunk into the right ventricle during diastole.

Right atrium The heart chamber that receives systemic venous blood from the superior and inferior vena cava and the veins draining blood from the heart itself.

Right brachiocephalic vein The vein draining systemic venous blood from the right head and neck and right upper limb.

Right inferior pulmonary vein One of four veins returning relatively oxygenated blood to the left atrium.

Right pulmonary artery The branch of the pulmonary artery that supplies relatively deoxygenated blood to the right lung.

Right superior pulmonary vein One of four veins returning relatively oxygenated blood to the left atrium.

Right ventricle The heart chamber that pumps blood into the pulmonary circulation. It has a thinner wall than the left ventricle because it develops lower pressures (25 mm Hg).

Superior vena cava The large vein that drains blood from the head, neck, and upper limbs into the right atrium of the heart.

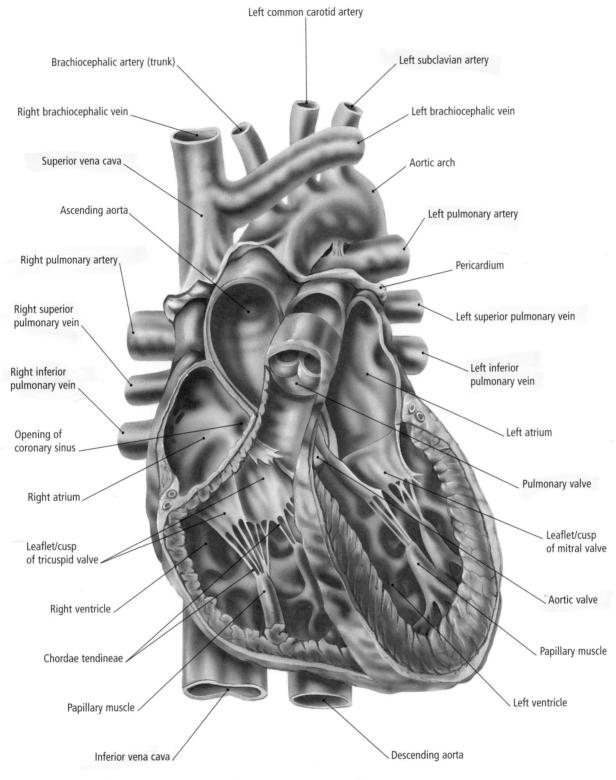

Left common carotid artery

Brachiocephalic artery (trunk)

Left subclavian artery

Right brachiocephalic vein

Left brachiocephalic vein

Superior vena cava

Aortic arch

Ascending aorta

Left pulmonary artery

Right pulmonary artery

Pericardium

Right superior
pulmonary vein

Left superior pulmonary vein

Right inferior
pulmonary vein

Left inferior
pulmonary vein

Opening of
coronary sinus

Left atrium

Right atrium

Pulmonary valve

Leaflet/cusp
of tricuspid valve

Leaflet/cusp
of mitral valve

Right ventricle

Aortic valve

Chordae tendineae

Papillary muscle

Papillary muscle

Left ventricle

Inferior vena cava

Descending aorta

Heart—cross-section

Heart Valves

There are four valves that control the flow of blood through the heart. The two atrioventricular valves (tricuspid or right atrioventricular valve, and mitral or left atrioventricular valve) prevent backflow of blood from the ventricles to the atria at the start of ventricular contraction. The two semilunar valves (pulmonary at the right ventricular outflow, and aortic at the left ventricular outflow) prevent backflow of blood from the pulmonary trunk and ascending aorta into the ventricles at the end of contraction.

Key terms:

Aortic valve See pp. 122–123.

Heart valves Heart valves are found between the atria and ventricles (atrioventricular valves such as the tricuspid and mitral) or at the outflow tracts from the left and right ventricles (aortic and pulmonary semilunar valves, respectively).

Leaflet/cusp of mitral valve See pp. 122–123.

Leaflet/cusp of tricuspid valve See pp. 122–123.

Left atrium See pp. 122–123.

Left ventricle See pp. 122–123.

Mitral valve A two-cusp valve between the left atrium and left ventricle. It is also called the bicuspid valve and is open during ventricular diastole.

Pulmonary valve See pp. 122–123.

Right atrium See pp. 122–123.

Right ventricle See pp. 122–123.

Tricuspid valve A three-cusp valve located between the right atrium and right ventricle. It is also called the right atrioventricular valve.

Ventricular diastole The period of the cardiac cycle when the ventricles are relaxed, and blood flows through the open tricuspid and mitral valves into the ventricles.

Ventricular systole The period of the cardiac cycle when the ventricles contract, expelling blood through the pulmonary and aortic valves.

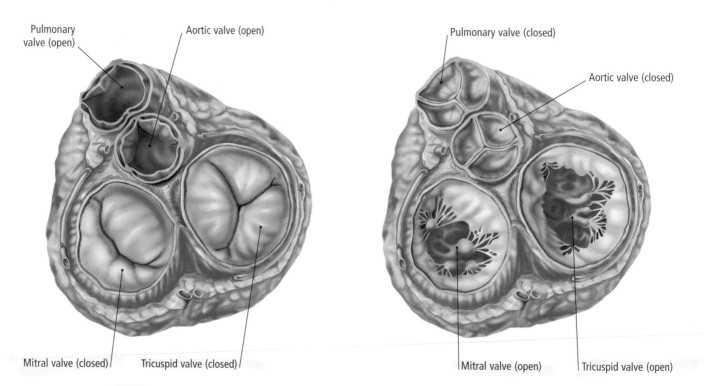

Pulmonary valve (open)

Aortic valve (open)

Mitral valve (closed)

Tricuspid valve (closed)

Pulmonary valve (closed)

Aortic valve (closed)

Mitral valve (open)

Tricuspid valve (open)

Ventricular Systole

Ventricular Diastole

Tricuspid Valve

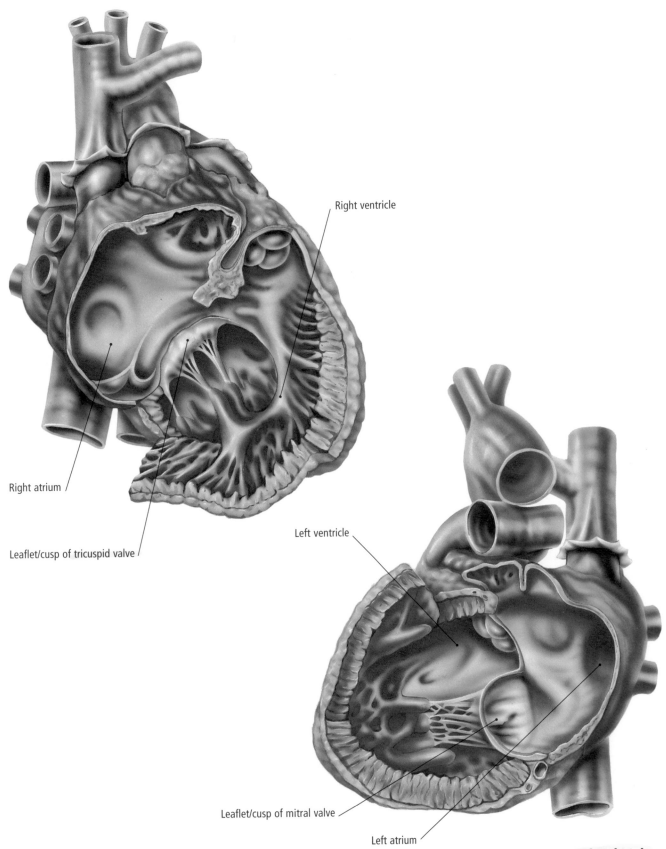

Right ventricle

Right atrium

Leaflet/cusp of tricuspid valve

Left ventricle

Leaflet/cusp of mitral valve

Left atrium

Mitral Valve

True or false?

1   The heart is mainly found behind the right side of the chest wall.

2   The heart is surrounded by an inner fibrous and outer serous pericardium.

3   The heart is divided into two atria and two ventricles.

4   The right side of the heart receives blood from the pulmonary veins.

5   The right atrium receives blood from the superior and inferior vena cava.

6   The aorta arises from the left ventricle.

7   There is one pulmonary vein on each side.

8   The muscle layer of the left ventricular wall is thinner than that of the left atrium.

9   The left atrium receives venous blood from the pulmonary circulation.

10   The septum between the left and right atria is the thickest wall in the heart.

11   The two cardiac semilunar valves are the aortic and pulmonary.

12   Highly oxygenated blood flows through the pulmonary artery.

13   The mitral valve is between the left and right atria.

14   The trabeculae carneae are the muscular ridges and bridges on the interior of the ventricles.

15   The heart chambers are lined with pericardium.

16   The right atrium has a completely smooth interior.

17   *Both atria have anterior ear-shaped appendages called the auricles.*

18   *The bridge from the ventricular septum to the base of the anterior papillary muscle carries part of the right bundle branch of the conducting/pacemaking system.*

19   *The heart chamber at the posterior base of the heart is the right atrium.*

20   *A prominent interventricular sulcus marks the boundary between the two ventricles.*

Myocardial infarction

Myocardial infarction is the death (infarction) of heart muscle (myocardium) due to inadequate blood supply. This is usually due to obstruction, by atherosclerosis, of the coronary arteries supplying the heart muscle. When a partially obstructed artery is completely blocked by a thrombus (blood coagulation), the patient experiences the sudden onset of a crushing pain behind the sternum, which cannot be relieved by anti-angina medication. Complications of infarction include arrhythmias (problems with heart rhythm), which may be the often-fatal ventricular fibrillation, and heart rupture due to weakening of the heart wall.

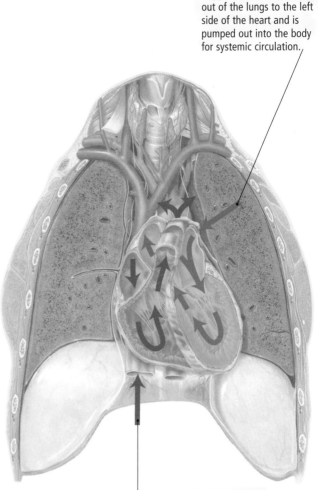

Oxygenated blood flows out of the lungs to the left side of the heart and is pumped out into the body for systemic circulation.

Oxygen-depleted blood enters the right ventricle of the heart and is pumped into the lungs to be oxygenated by the alveoli.

Pulmonary Circulation

The pulmonary system carries deoxygenated blood to the lungs, where gas exchange occurs. Carbon dioxide—picked up by the blood from cells and tissues in exchange for oxygen—is exchanged in the lungs for fresh supplies of oxygen. That carbon dioxide is then exhaled, while the now oxygen-rich blood is returned to the heart for another circuit of the body.

Multiple choice

1 *Which of the following structures has the thickest wall?*

- (A) left atrium
- (B) right atrium
- (C) left ventricle
- (D) right ventricle
- (E) aorta

2 *At which of the following sites would the oxygen concentration be highest?*

- (A) left atrium
- (B) pulmonary artery
- (C) superior vena cava
- (D) right atrium
- (E) right ventricle

3 *Auricular appendages are attached to which heart chambers?*

- (A) left atrium and left ventricle
- (B) left and right atria
- (C) left and right ventricles
- (D) right ventricle only
- (E) right atria and right ventricle

Rheumatic heart disease

The lining of the heart's valves shares some antigens (molecules capable of inducing an immune response) with those on the cell wall of group A streptococcal bacteria. When the body's immune system mounts a response to those bacteria following throat infection, damage may also occur to the heart valves. This can cause serious damage, leading to fibrosis and scarring of the valve leaflets. Valve leaflets may stick together, causing valvular stenosis (narrowing), or stick to the vessel wall, causing valvular incompetence. Both of these have long-term adverse effects on heart function.

4 *Which chamber or structure forms the base of the heart?*

- (A) left ventricle
- (B) right ventricle
- (C) left atrium
- (D) right atrium
- (E) arch of aorta

5 *Which chamber or structure forms the apex of the heart?*

- (A) left atrium
- (B) right atrium
- (C) arch of aorta
- (D) right ventricle
- (E) left ventricle

6 *Which of the following structures attaches directly to the edges of the atrioventricular valve cusps?*

- (A) musculi pectinati
- (B) papillary muscles
- (C) trabeculae carneae
- (D) chordae tendineae
- (E) interventricular septum

7 *Which of the following structures prevents regurgitation of blood into the left atrium?*

- (A) tricuspid valve
- (B) aortic valve
- (C) pulmonary valve
- (D) mitral valve
- (E) interatrial septum

8 *How thick is the left ventricle wall compared to the right ventricle wall?*

- (A) the same thickness
- (B) half as thick as the right ventricle wall
- (C) one quarter as thick as the right ventricle wall
- (D) twice as thick as the right ventricle wall
- (E) almost three times as thick as the right ventricle wall

9 *What is the source of oxygenated blood to the heart muscle?*
- Ⓐ coronary sinus
- Ⓑ inferior vena cava
- Ⓒ coronary arteries
- Ⓓ pulmonary trunk
- Ⓔ pulmonary veins

10 *Which vessel usually supplies the largest volume of heart muscle?*
- Ⓐ pulmonary trunk
- Ⓑ right coronary artery
- Ⓒ phrenic artery
- Ⓓ left coronary artery
- Ⓔ coronary sinus

11 *Where is the sinoatrial node located?*
- Ⓐ left atrium
- Ⓑ right atrium
- Ⓒ left ventricle
- Ⓓ right ventricle
- Ⓔ superior vena cava

12 *Where is the atrioventricular node located?*
- Ⓐ right atrium
- Ⓑ left atrium
- Ⓒ left ventricle
- Ⓓ right ventricle
- Ⓔ superior vena cava

13 *Which structure drains most venous blood from the heart?*
- Ⓐ right coronary vein
- Ⓑ superior vena cava
- Ⓒ azygos vein
- Ⓓ coronary sinus
- Ⓔ left coronary vein

14 *The first heart sound is due to the:*
- Ⓐ closure of the aortic valve
- Ⓑ closure of the pulmonary valve
- Ⓒ opening of the aortic valve
- Ⓓ opening of the tricuspid and mitral valves
- Ⓔ closure of the tricuspid and mitral valves

15 *The second heart sound is due to the:*
- Ⓐ closure of the aortic and pulmonary valves
- Ⓑ closure of the mitral valve only
- Ⓒ opening of the aortic valve
- Ⓓ opening of the tricuspid and mitral valves
- Ⓔ closure of the tricuspid and mitral valves

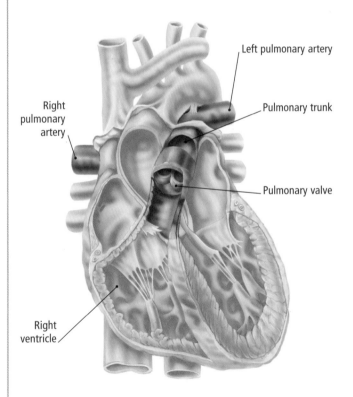

Right pulmonary artery

Left pulmonary artery

Pulmonary trunk

Pulmonary valve

Right ventricle

Pulmonary Artery

Deoxygenated blood is pumped from the right side of the heart to the lungs through the pulmonary artery and its branches.

Color and label

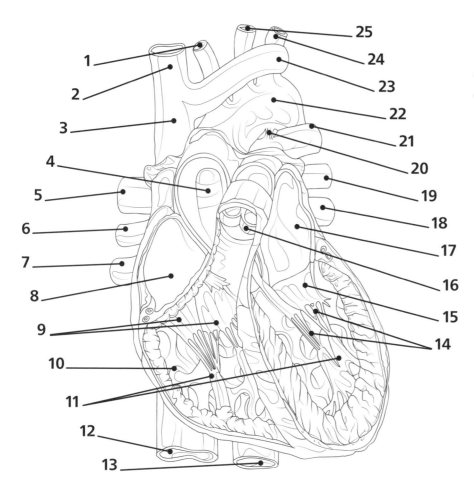

Heart—cross-sectional view

i) Color the arteries in red and the veins in blue.

ii) Add numbers to the boxes below to match each label to the correct part of the illustration.

Left pulmonary artery	☐
Left superior pulmonary vein	☐
Cusp of mitral valve	☐
Right atrium	☐
Aortic arch	☐
Descending thoracic aorta	☐
Left subclavian artery	☐
Left common carotid artery	☐
Cusp of tricuspid valve	☐
Right ventricle	☐
Papillary muscles	☐

Inferior vena cava	☐
Left atrium	☐
Superior vena cava	☐
Ascending aorta	☐
Right pulmonary artery	☐
Chordae tendineae	☐
Right inferior pulmonary vein	☐
Right superior pulmonary vein	☐
Right brachiocephalic vein	☐
Left brachiocephalic vein	☐
Left inferior pulmonary vein	☐
Ligamentum arteriosum	☐
Brachiocephalic artery	☐
Pulmonary valve	☐

iii) Label each structure shown on the illustration.

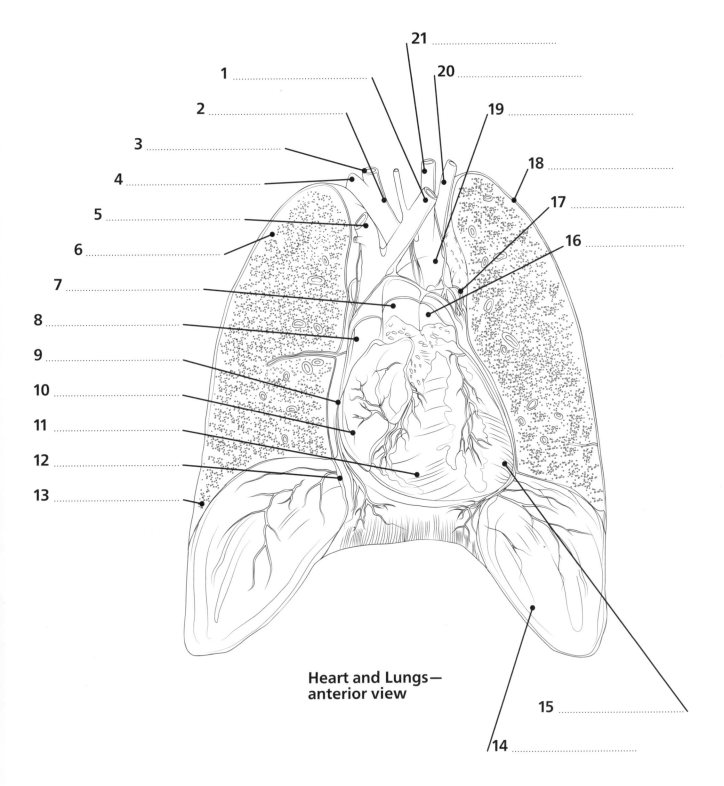

Heart and Lungs— anterior view

Fill in the blanks

1   Venous blood from the heart muscle mainly drains into the right atrium through the
_____ .

2   The coronary arteries to the heart muscle branch from the _____ .

3   The _____ of the heart points downward and to the left.

4   The interventricular septum has both _____ and _____ parts.

5   The_____ valve is between the right atrium and right ventricle.

6   The _____ carries deoxygenated blood to the lungs.

7   Deoxygenated blood enters the heart at the
_____ .

8   The right auricle curls around the
_____ .

9   The muscular ridges on the interior of the
ventricles are called _____ .

10   The muscular ridges on the interior of the
atria are called _____ .

11   The _____ is the source of
blood for the pulmonary artery.

12   The _____ forms the left or
pulmonary side of the heart.

Position of
sinoatrial
node

Electrical
pathways
carry the
impulse
throughout
the heart

Heartbeat

*The rhythmic beating of the heart is controlled by the sinoatrial node, which
transmits electrical impulses to initiate contraction of the heart muscle.*

Match the statement to the reason

1   *The coronary arteries arise from the aorta because…*

a   *it must develop very high pressure during contraction (over 120 mm Hg) and needs to be very strong.*

2   *The cardiac veins drain into the right atrium because…*

b   *the atrial muscle must be activated as a single electrical unit before the ventricles are activated, so that atrial contraction precedes ventricular contraction.*

3   *The muscular wall of the left ventricle is very thick because…*

c   *they must supply oxygenated blood to cardiac muscle, and it is best that this comes from the outflow of oxygenated blood from the left ventricle.*

4   *A papillary muscle would contract along with the rest of the ventricular muscle because…*

d   *the atrioventricular valve leaflets must be pulled shut to prevent regurgitation of blood into the atria during ventricular contraction.*

5   *There is an insulating layer of connective tissue between the atrial and ventricular muscles because…*

e   *they carry deoxygenated blood from the cardiac muscle, and this is logically best returned to the chamber of the heart where deoxygenated blood is received.*

Arteries

Arteries carry blood away from the heart and are high-pressure vessels. The largest artery in the body is the aorta, which gives off branches to the head, neck, and upper limb (brachiocephalic trunk, common carotid, subclavian) before descending to supply the chest wall (by the intercostal arteries); gut organs (by the celiac and mesenteric arteries); kidneys (by the renal arteries); and pelvic organs, buttock, and lower limb (by the internal iliac, external iliac, and femoral arteries).

Key terms:

Anterior tibial artery A branch of the popliteal artery that supplies the anterior compartment of the leg.

Aorta The largest artery of the body. It consists of an initial arch arising from the heart, followed by thoracic and abdominal segments.

Aortic arch The most superior part of the aorta. It gives rise to the brachiocephalic artery (trunk), left common carotid artery, and left subclavian artery.

Axillary artery The continuation of the subclavian artery. It gives off branches to the adjacent muscles, a collateral circulation around the glenohumeral joint, and becomes the brachial artery at the lower border of teres minor.

Brachial artery The largest artery of the arm. It is a continuation of the axillary artery and gives off the profunda brachii, radial, and ulnar branches, as well as collateral branches around the elbow joint.

Common carotid artery Supplies the head and neck. Each common carotid divides into an external carotid artery (for the face, scalp, and throat) and an internal carotid artery (for the brain, pituitary gland, and eye).

Common iliac artery The two terminal branches of the abdominal aorta. They divide into external and internal iliac arteries.

Descending aorta The part of the aorta distal to the arch. It supplies the posterior chest wall, spinal cord, abdomen, pelvis, and lower limbs. It divides at the level of lumbar vertebra 4 into two common iliac arteries.

Dorsal arch The dorsal arterial arch of the foot is supplied by the anterior tibial artery and its superficial continuation, the dorsalis pedis artery.

External iliac artery The branch of the common iliac artery that becomes the femoral artery at the inguinal ligament.

Facial artery A branch of the external carotid artery that supplies the face, throat, and anterior neck.

Femoral artery The continuation of the external iliac artery after it passes under the inguinal ligament. It passes through the femoral triangle and gives off the profunda femoris artery before passing through the adductor opening to become the popliteal artery.

Fibular artery A branch of the posterior tibial artery that supplies the posterior compartment of the leg.

Heart A four-chambered muscular pump in the middle mediastinum of the chest.

Intercostal arteries Branches of the aorta or highest intercostal artery that run in the costal groove beneath each rib. They supply the lateral chest wall and anastomose with branches of the internal thoracic artery toward the front of the chest.

Internal iliac artery The branch of the common iliac artery that supplies the gluteal region, pelvic organs, medial pelvic wall, and perineum (space between the thighs).

Obturator artery A branch of the internal iliac (or sometimes the external iliac) artery that supplies the medial pelvic wall and medial upper thigh.

Palmar arterial arches Superficial and deep palmar arches receive blood from the ulnar and radial arteries and give off branches to the palm and digits.

Plantar arch An arterial arch on the plantar side of the distal foot, fed by the posterior tibial artery.

Popliteal artery The continuation of the femoral artery that passes through the popliteal fossa behind the knee. It gives off posterior tibial and fibular arteries.

Posterior auricular artery A branch of the external carotid artery that runs through the parotid gland to divide into auricular and occipital branches near the mastoid process.

Posterior tibial artery A branch of the large artery behind the knee (popliteal artery).

Radial artery One of two arteries of the forearm. It is palpable on the lateral anterior surface of the distal wrist and is used to assess the arterial pulse.

Renal artery The artery supplying the kidney. It also gives branches to the suprarenal (adrenal) gland and the ureter. It arises from the abdominal aorta at the level of the disk between the first and second lumbar vertebrae.

Subclavian artery Arises from brachiocephalic trunk (on right) or the aortic arch (on left). It crosses first rib posterior to the scalene tubercle to become the axillary artery.

Ulnar artery The branch of the brachial artery that supplies the medial forearm.

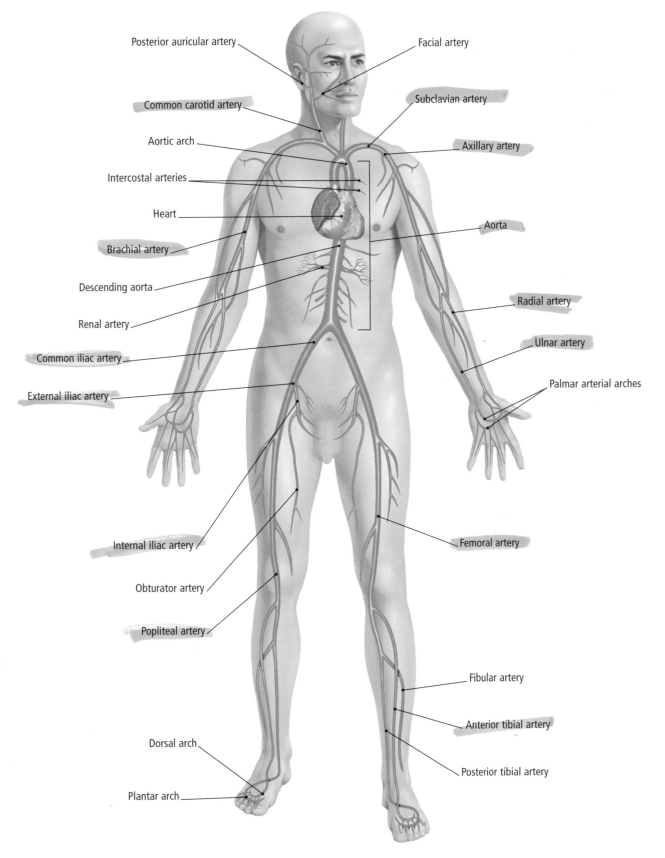

Posterior auricular artery

Facial artery

Common carotid artery

Subclavian artery

Aortic arch

Axillary artery

Intercostal arteries

Heart

Aorta

Brachial artery

Descending aorta

Radial artery

Renal artery

Ulnar artery

Common iliac artery

Palmar arterial arches

External iliac artery

Internal iliac artery

Femoral artery

Obturator artery

Popliteal artery

Fibular artery

Anterior tibial artery

Dorsal arch

Posterior tibial artery

Plantar arch

**Major Arteries of the
Body—anterior view**

Arteries of the Brain

The brain is supplied by four arteries, two internal carotid arteries and two vertebral arteries, although the contribution by the vertebral arteries is probably only 10 percent. These vessels feed an arterial circle (the circle of Willis), which helps to evenly distribute arterial flow to all cerebral artery branches. The blood from the internal carotid is distributed to the front and middle forebrain, whereas blood from the vertebral system flows to the brainstem, cerebellum, and posterior cerebral hemisphere.

Key terms:

Anterior cerebral artery The anterior branch of the internal carotid artery at the circle of Willis. It supplies the orbital cortex, medial frontal lobe, and corpus callosum.

Anterior communicating artery The small artery joining the anterior cerebral arteries to complete the circle of Willis.

Anterior inferior cerebellar artery A branch of the basilar artery. Supplies the superior pons and anterior cerebellum.

Basilar artery A large midline artery formed from the junction of the vertebral arteries. Supplies the pons and divides to give the posterior cerebral arteries.

Calcarine branch The branch of the posterior cerebral artery that supplies the central part of the primary visual cortex.

Callosomarginal artery A branch of the anterior cerebral artery that ascends into the callosomarginal sulcus.

Circle of Willis The vascular ring at the base of the brain that receives the internal carotid and vertebral arteries and gives off branches to the forebrain and superior brainstem. It helps to maintain blood supply to the brain when one branch suffers partial or temporary obstruction.

Dorsal branch to corpus callosum One of the terminal branches of the anterior cerebral artery to the corpus callosum.

Internal carotid artery The artery supplying the anterior and lateral parts of the cerebral cortex and much of the deep white matter. It is a branch of the common carotid artery.

Labyrinthine artery A basilar artery branch to the inner ear. It supplies the cochlear and vestibular apparatus.

Medial frontal branches Branches (anterior, intermediate, and posterior) of the anterior cerebral artery to the supplementary motor cortex.

Medial frontobasal artery An arterial branch to the orbital surface of the frontal lobe.

Medial occipital artery (branch of posterior cerebral artery) An artery supplying the medial primary and association visual cortex.

Medial striate artery An arterial branch of the circle of Willis to the deep structures of the medial forebrain.

Middle cerebral artery The largest branch of the internal carotid artery. Runs in the lateral fissure to supply the lateral parts of the frontal, parietal, and temporal cortex, and the insula. Its region of supply includes most of the motor and primary somatosensory cortex, primary auditory cortex, and language areas. It also supplies the deep white and gray matter structures of the forebrain.

Paracentral artery A branch of the anterior cerebral artery to the paracentral lobule (lower limb region of the primary motor and somatosensory cortex).

Parietooccipital branch A branch of the posterior cerebral artery that runs in the parietooccipital fissure to supply the visual association cortex.

Pericallosal artery A branch of the anterior cerebral artery that courses around the corpus callosum.

Polar frontal artery A branch of the anterior cerebral artery to the frontal pole of the cerebrum.

Posterior cerebral artery The terminal branch of the basilar artery. It supplies the posterior occipital lobe, in particular the primary and association visual cortex.

Posterior communicating artery Artery joining the carotid- and basilar-derived parts of the circle of Willis. It runs between the internal carotid and posterior cerebral arteries.

Posterior inferior cerebellar artery A branch of the vertebral artery that supplies the lateral parts of the medulla and the posterior inferior cerebellum.

Precuneal artery A branch of the anterior cerebral artery that supplies the part of the medial cerebral surface between the primary somatosensory cortex and the visual association cortex.

Right anterior cerebral artery An anterior branch of the circle of Willis that supplies the medial anterior forebrain, including the supplementary motor cortex and the medial parts of the primary motor and somatosensory cortex.

Superior cerebellar artery A branch of the basilar artery that supplies the superior parts of the cerebellum.

Vertebral artery A branch of the subclavian artery that ascends the neck through the transverse foramina of the upper six cervical vertebrae. It enters the cranial cavity through the foramen magnum to supply the brainstem and occipital lobe.

Arteries of the Base of the Brain

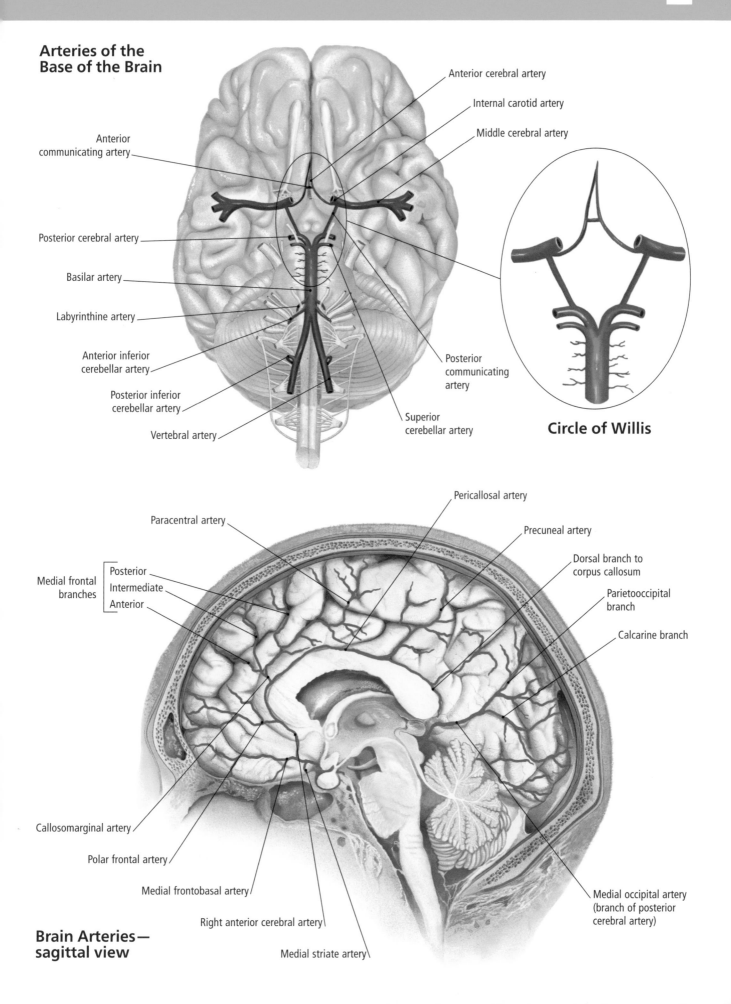

Anterior cerebral artery

Internal carotid artery

Middle cerebral artery

Anterior communicating artery

Posterior cerebral artery

Basilar artery

Labyrinthine artery

Anterior inferior cerebellar artery

Posterior inferior cerebellar artery

Vertebral artery

Posterior communicating artery

Superior cerebellar artery

Circle of Willis

Pericallosal artery

Paracentral artery

Precuneal artery

Dorsal branch to corpus callosum

Parietooccipital branch

Calcarine branch

Medial frontal branches

Posterior

Intermediate

Anterior

Callosomarginal artery

Polar frontal artery

Medial frontobasal artery

Right anterior cerebral artery

Medial striate artery

Medial occipital artery (branch of posterior cerebral artery)

Brain Arteries— sagittal view

True or false?

1   Oxygenated blood to the brain passes mainly through the subclavian artery.

2   The brachial artery passes on the medial side of the humerus.

3   The common iliac artery is a branch of the aorta.

4   The internal iliac artery supplies the lower limb.

5   Arterial blood to the intestines is from the superior and inferior mesenteric arteries.

6   The celiac artery usually supplies the jejunum and ileum.

7   The radial artery is often used to assess the pulse during clinical examination.

8   The ulnar artery is palpable on the dorsum of the wrist.

9   The external iliac artery becomes the popliteal artery at the base of the thigh.

10   The femoral artery accompanies the sciatic nerve.

11   The brachial artery is often used to test arterial blood pressure.

12   To slow blood loss from the forearm, the brachial artery can be compressed against the medial surface of the humerus.

13   The subclavian artery could be compressed against the first rib to control hemorrhage in the arm.

14   There are two arterial arches in the palm of the hand.

15   Renal arteries branch off the aorta above the muscular diaphragm.

16 *The pelvic organs are supplied by the internal iliac artery.*

17 *The femoral artery passes behind the knee joint.*

18 *The penis is supplied by branches of the internal pudendal arteries.*

19 *The pulse in the posterior tibial artery can be felt behind the medial malleolus.*

20 *The posterior tibial artery is palpable on the dorsum of the foot.*

Cerebral aneurysms

An aneurysm is a localized dilation of a vessel wall. When these form on cerebral arteries (usually at junction points), they are called cerebral aneurysms. Cerebral aneurysms may press against nerves and cause neurological deficits, or rupture, allowing blood under pressure to force its way through brain tissue or flood the subarachnoid space (subarachnoid hemorrhage). Hemorrhage into the brain (a type of stroke) can have catastrophic effects and lead to sudden death or permanent disability.

Anterior branch of superficial temporal artery

Posterior branch of superficial temporal artery

Supraorbital artery

Supratrochlear artery

Transverse facial artery

Occipital artery

Superficial temporal artery

Facial artery

Transverse cervical artery

External carotid artery

Surface Arteries of the Head and Neck—lateral view

The carotid arteries are the main arteries of the neck, receiving their supply from the aorta. They supply blood to the neck and its structures, as well as the brain, face, and head via the branches of the carotid arteries.

Multiple choice

1 *Which of the following arteries lies along the thumb side of the upper limb?*

(A) ulnar artery
(B) brachial artery
(C) subclavian artery
(D) radial artery
(E) axillary artery

2 *Which of the following is a branch of the abdominal aorta?*

(A) pulmonary artery
(B) renal artery
(C) right coronary artery
(D) brachial artery
(E) femoral artery

3 *Which artery can be felt behind the medial malleolus of the ankle?*

(A) femoral artery
(B) popliteal artery
(C) posteror tibial artery
(D) dorsalis pedis artery
(E) anterior tibial artery

4 *Which artery could be lacerated in a fracture of the middle of the humerus?*

(A) brachial artery
(B) radial artery
(C) ulnar artery
(D) subclavian artery
(E) axillary artery

5 *Which artery is most important for the supply of the organs of the pelvis?*

(A) external iliac artery
(B) obturator artery
(C) femoral artery
(D) renal artery
(E) internal iliac artery

6 *Which of the following is a direct branch of the thoracic aorta?*

(A) axillary artery
(B) left subclavian artery
(C) right subclavian artery
(D) right common carotid artery
(E) internal carotid artery

7 *Which of the following is the first branch of the aorta after the coronary arteries?*

(A) left subclavian artery
(B) first intercostal artery
(C) left common carotid artery
(D) brachiocephalic trunk
(E) pulmonary artery

8 *Which of the following arteries supplies the stomach and liver?*

(A) superior mesenteric artery
(B) celiac trunk
(C) inferior mesenteric artery
(D) renal artery
(E) common iliac artery

9 *Which vessel would be penetrated by a knife stabbing from the umbilicus through to the vertebral column?*

(A) abdominal aorta
(B) thoracic aorta
(C) internal iliac artery
(D) common iliac artery
(E) external iliac artery

10 *Which artery is the main supply to the lower anterior abdominal wall?*

(A) subclavian artery
(B) external iliac artery
(C) internal iliac artery
(D) femoral artery
(E) lumbar artery

11 *Which vessel is palpable behind the knee?*
(A) femoral artery
(B) iliac artery
(C) tibial artery
(D) sciatic artery
(E) popliteal artery

12 *Which artery could be lacerated in a fracture of the lower shaft of the femur just above the knee?*
(A) popliteal artery
(B) femoral artery
(C) posterior tibial artery
(D) patellar artery
(E) dorsalis pedis artery

13 *How does the femoral artery reach the popliteal fossa?*
(A) by passing through the biceps femoris
(B) by passing through the adductor longus
(C) by penetrating the iliotibial tract
(D) by penetrating the adductor magnus
(E) none of the above is correct

14 *Which artery can be felt on the top of the foot above the first metatarsal?*
(A) femoral artery
(B) popliteal artery
(C) posteror tibial artery
(D) dorsalis pedis artery
(E) anterior tibial artery

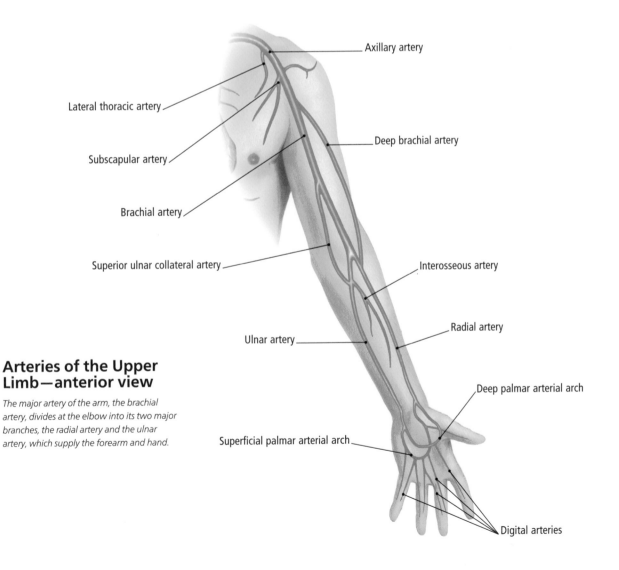

Arteries of the Upper Limb—anterior view

The major artery of the arm, the brachial artery, divides at the elbow into its two major branches, the radial artery and the ulnar artery, which supply the forearm and hand.

Color and label

i) Label each structure shown on the illustrations.

1 ....

2 ....

3 ....

4 ....

5 ....

6 ....

7 ....

8 ....

9 ....

10 ....

19 ....

18 ....

17 ....

16 ....

15 ....

14 ....

13 ....

12 ....

11 ....

Portal System— anterior view

ii) Use the key to color these structures.

☐ Liver

⬛ Colon

⬛ Small intestine

20 ....

21 ....

22 ....

23 ....

24 ....

36 ....

35 ....

34 ....

33 ....

32 ....

31 ....

30 ....

29 ....

28 ....

27 ....

26 ....

25 ....

Arterial System of Abdomen— anterior view

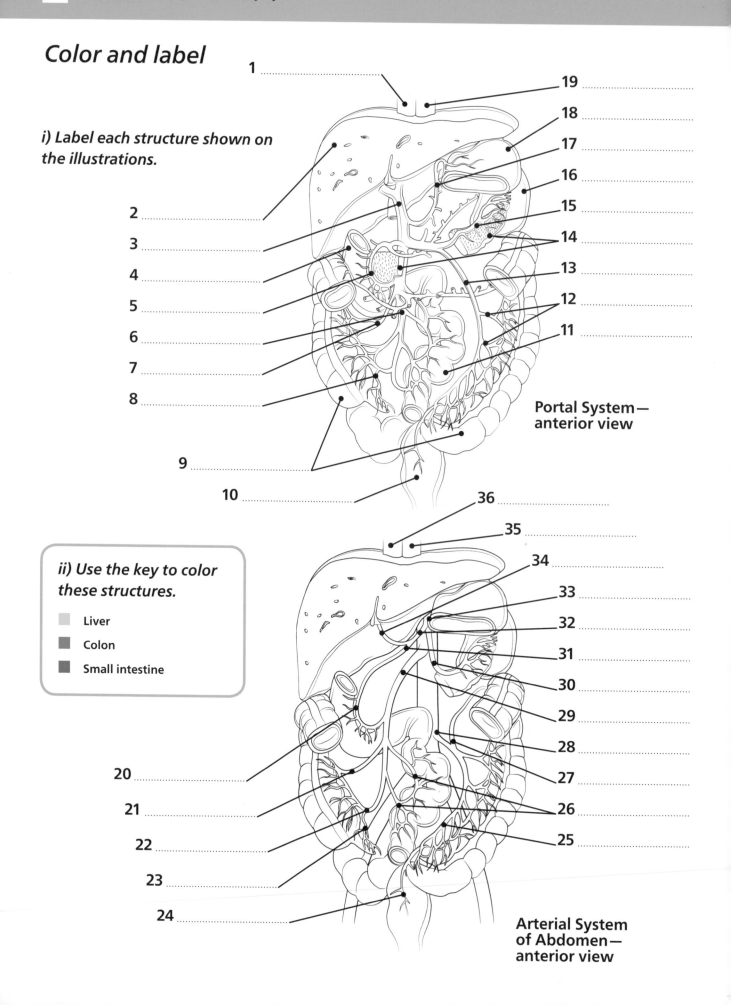

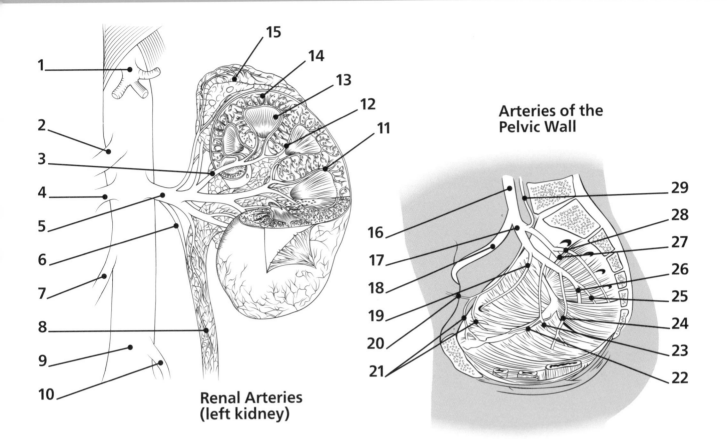

Renal Arteries (left kidney)

Arteries of the Pelvic Wall

iii) Add numbers to the boxes below to match each label to the correct part of the illustrations.

Segmental artery	☐	Obturator artery	☐
Ureter	☐	Internal iliac artery	☐
Celiac trunk	☐	Middle rectal artery	☐
Cortex	☐	Common iliac artery	☐
Left renal artery	☐	Obliterated umbilical artery	☐
Inferior mesenteric artery	☐	Superior vesicle artery	☐
Right gonadal artery	☐	Iliolumbar artery	☐
Superior mesenteric artery	☐	Vaginal artery	☐
Interlobar artery	☐	External iliac artery	☐
Left gonadal artery	☐	Internal pudendal artery	☐
Arcuate artery	☐	Superior gluteal artery	☐
Abdominal aorta	☐	Inferior gluteal artery	☐
Renal pyramid (medulla)	☐	Lateral sacral artery	☐
Right renal artery	☐	Uterine artery	☐
Left adrenal gland	☐		

Fill in the blanks

1   The _____ is the largest artery in the body.

2   The _____ and _____ arteries supply the brain.

3   The face and throat are supplied by the _____ artery.

4   The right common carotid and right subclavian arteries are branches of the _____.

5   The chest wall is supplied by a series of _____ arteries.

6   The posterior abdominal wall is supplied by a series of _____ arteries.

7   The arterial supply to the pelvic organs is primarily from the _____ arteries.

8   Arterial supply to the leg and foot is from the _____ artery.

9   The _____ and _____ arteries supply the hand.

10 The gut is supplied by _____ , _____ , and _____ arteries.

Arteries of the Lower Limb—anterior view

Blood supply to the leg is provided mainly by the external iliac artery, which becomes the femoral artery as it enters the leg.

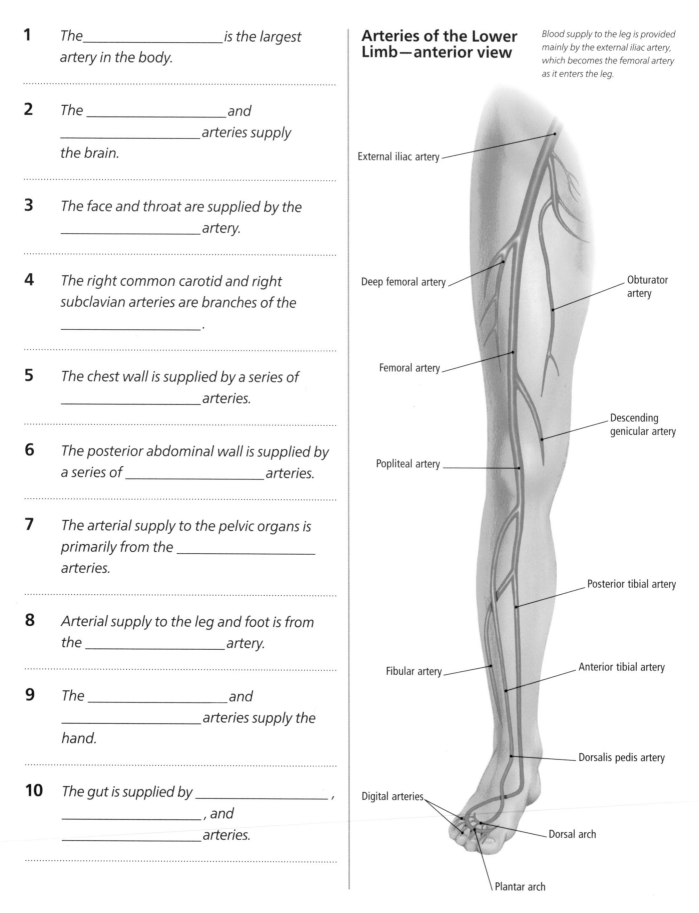

External iliac artery

Deep femoral artery

Femoral artery

Popliteal artery

Fibular artery

Digital arteries

Obturator artery

Descending genicular artery

Posterior tibial artery

Anterior tibial artery

Dorsalis pedis artery

Dorsal arch

Plantar arch

Match the statement to the reason

1   *The pulse can be taken at the groin because…*

a   *both the radial and ulnar arteries supply the forearm and hand.*

2   *Atherosclerosis and blockage of coronary arteries are a prime cause of heart attack (myocardial infarction) because…*

b   *the axillary artery passes close to the humerus in the axilla.*

3   *Obstruction of the radial artery doesn't necessarily cause gangrene of the hand because…*

c   *these arteries supply the brainstem and cerebellum.*

4   *Obstruction of the vertebral arteries can cause dizziness and collapse because…*

d   *the coronary arteries are the sole supply of oxygenated blood to heart muscle*

5   *A fracture of the upper end of the humerus could cause blood loss because…*

e   *the femoral artery passes under the inguinal ligament here.*

Atherosclerosis

Atherosclerosis is a condition of arteries where there is an accumulation of fatty, fibrous, and sometimes calcific material in the vessel wall to form plaques. The material primarily accumulates in the tunica intima, narrowing the lumen of the vessel and increasing the risk of thrombosis on the vessel wall. Thrombosis can cause sudden constriction of the vessel with consequent death of the tissue, e.g., heart muscle or brain tissue. Risk factors include smoking, drinking alcohol, elevated blood pressure, diabetes mellitus, and elevated blood lipids.

Veins

Veins are blood vessels that carry blood back to the heart in both the systemic and pulmonary circulations. Systemic veins have valves to facilitate the flow of blood to the heart, and these are particularly prominent in the limbs. Veins are low-pressure vessels and have much thinner walls than arteries, although they also provide an important reservoir function by storing blood volume in distensible vessels.

Key terms:

Axillary vein The vein of the axilla. It is a continuation of the brachial vein after it has been joined by the basilic vein. It becomes the subclavian vein at the lateral border of the first rib.

Azygos vein Literally the unpaired vein (i.e., there is only one in the body) of the chest. It receives blood from the posterior chest and abdominal wall and drains into the superior vena cava.

Basilic vein One of the superficial veins of the arm. It runs along the medial side of the forearm and arm to join the deep system at the brachial vein.

Brachial vein The deep vein of the arm. It receives the radial and ulnar veins and becomes the axillary vein at the lower border of the teres minor.

Brachiocephalic vein Formed from the junction of the subclavian and jugular veins in the superior mediastinum of the chest. The two brachiocephalic veins join to form the superior vena cava.

Cephalic vein One of the superficial veins of the upper limb. It runs along the lateral aspect of the forearm and arm to join the deep system at the axillary vein.

Common iliac vein The vein formed from the junction of the external and internal iliac veins. The two common iliac veins join to form the inferior vena cava.

Dorsal venous arch The venous arch on the dorsal surface of the foot. It communicates with the plantar venous arch and drains into the great saphenous vein.

External iliac vein The continuation of the femoral vein after it passes under the inguinal ligament. The external iliac vein joins the internal iliac vein to form the common iliac vein.

External jugular vein The large vein draining the scalp, maxilla, throat, and face. It passes external to the sternocleidomastoid, where it may be visible.

Femoral vein A deep vein that is the continuation of the popliteal vein. It runs through the femoral triangle, receives the great saphenous vein, and passes under the inguinal ligament to become the external iliac vein.

Great saphenous vein A long superficial vein arising on the medial dorsum of the foot. It passes anterior to the medial malleolus to run up the leg and thigh before entering the femoral vein at the groin.

Inferior vena cava The largest vein of the abdomen. It carries blood from the lower limb, pelvis, kidneys, and posterior abdominal wall. It passes through the diaphragm and enters the right atrium of the heart.

Internal iliac vein The vein that drains the organs of the pelvis and the buttock region. It joins with the external iliac vein to form the common iliac vein.

Internal jugular vein The large vein draining the inside of the skull. It passes deep to the sternocleidomastoid muscle alongside the common carotid artery and vagus nerve.

Median antebrachial vein A superficial vein running up the midline of the forearm.

Palmar venous arch A venous arch draining the digits and palm of the hand.

Plantar venous arch A venous arch draining the digits and sole of the foot.

Renal vein The vein draining the kidney. The left renal vein is much longer than the right and receives the gonadal (testicular or ovarian) and suprarenal veins.

Small saphenous vein A superficial vein draining the skin over the posterior calf. It enters the deep system by joining the popliteal vein behind the knee.

Subclavian vein The vein draining blood from the upper limb. It is the continuation of the axillary vein and joins with the external and internal jugular veins to form the brachiocephalic vein.

Superior vena cava The large vein that drains blood from the head, neck, and upper limbs into the right atrium of the heart.

Vein A low-pressure vessel returning blood to the heart. Veins have a relatively thin tunica media.

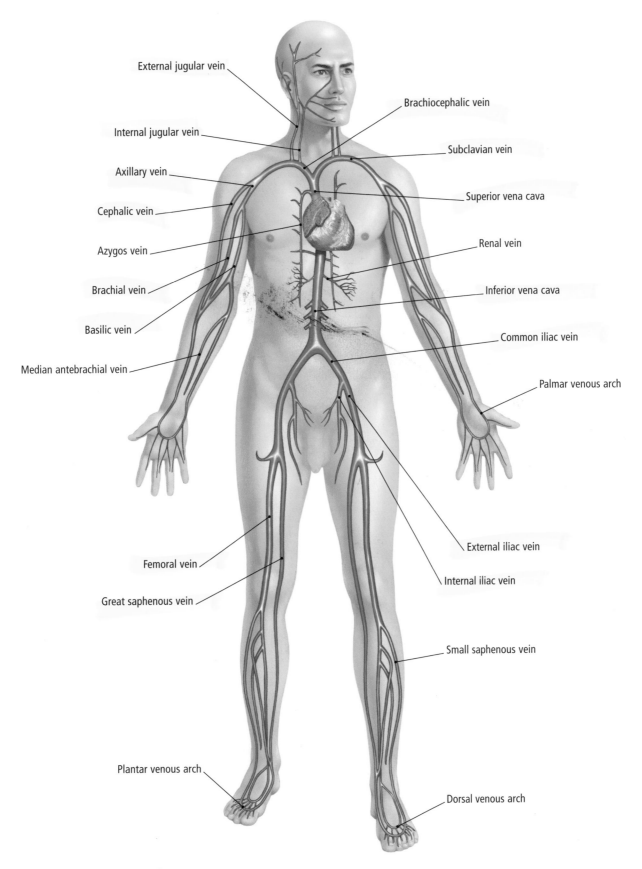

External jugular vein

Internal jugular vein

Axillary vein

Cephalic vein

Azygos vein

Brachial vein

Basilic vein

Median antebrachial vein

Brachiocephalic vein

Subclavian vein

Superior vena cava

Renal vein

Inferior vena cava

Common iliac vein

Palmar venous arch

Femoral vein

Great saphenous vein

Plantar venous arch

External iliac vein

Internal iliac vein

Small saphenous vein

Dorsal venous arch

**Major Veins of the Body—
anterior view**

True or false?

1 The radial vein drains blood from the thumb side of the forearm.

2 The pulmonary veins drain oxygenated blood from the lungs.

3 The great saphenous vein is the longest vein in the body.

4 The short or small saphenous vein runs up the inner side of the thigh.

5 The superior vena cava drains the abdomen.

6 The right testicular vein drains into the inferior vena cava.

7 The portal vein drains blood from the stomach.

8 The left renal vein is shorter than the right.

9 The subclavian vein passes over the first rib.

10 Blood from the brain drains into the internal jugular vein.

Veins of the Lower Limb—anterior view

The venous system of the lower limb includes the great saphenous vein.

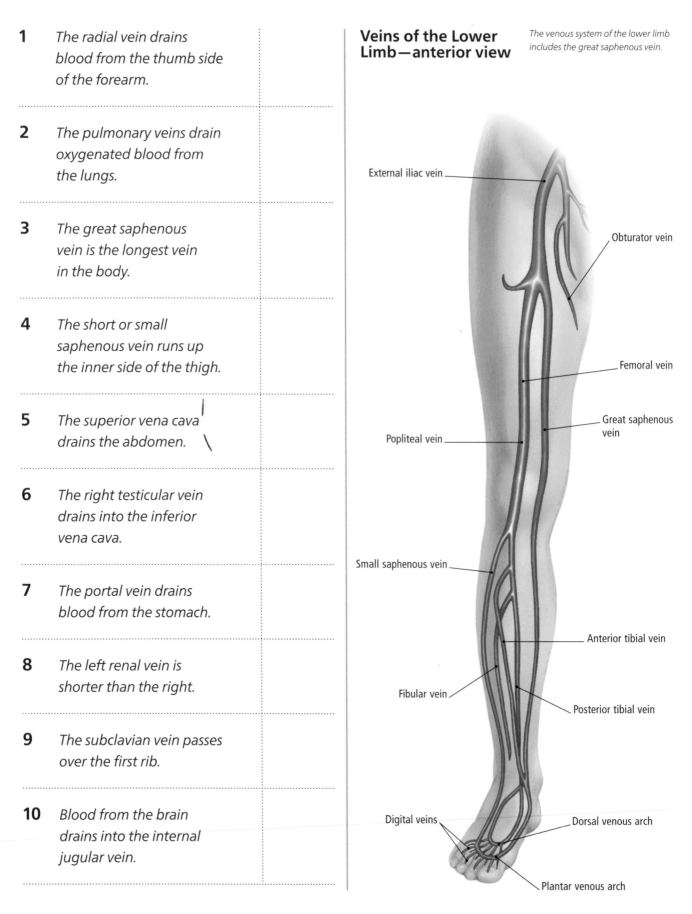

External iliac vein

Obturator vein

Femoral vein

Great saphenous vein

Popliteal vein

Small saphenous vein

Anterior tibial vein

Fibular vein

Posterior tibial vein

Digital veins

Dorsal venous arch

Plantar venous arch

Multiple choice

1 *Which vein drains blood from both the head and upper limb?*

Ⓐ inferior vena cava
Ⓑ superior vena cava
Ⓒ internal jugular vein
Ⓓ external jugular vein
Ⓔ pulmonary vein

2 *Which of the following drains directly into the inferior vena cava?*

Ⓐ celiac vein
Ⓑ portal vein
Ⓒ renal vein
Ⓓ inguinal vein
Ⓔ popliteal vein

3 *Which organ receives the portal ve...*

Ⓐ liver
Ⓑ kidney
Ⓒ stomach
Ⓓ testis
Ⓔ urinary bladder

4 *Which of the following is a superficial ve... of the lower limb?*

Ⓐ popliteal vein
Ⓑ femoral vein
Ⓒ posterior tibial vein
Ⓓ obturator vein
Ⓔ great saphenous vein

5 *Which of the following is a superficial vein of the upper limb?*

Ⓐ brachial vein
Ⓑ radial vein
Ⓒ cephalic vein
Ⓓ ulnar vein
Ⓔ subclavian vein

6 *With respect to the veins of the lower limb:*

Ⓐ venous v...
Ⓑ ...rficial to deep veins
Ⓒ ...pressure when standing is only a few
...of blood
Ⓓ ...of venous ... present
Ⓔ ...above is correct

Which ...drain blood from the liver to the ...

portal ...
...superficial...
...
hepato...
...

Deep vein thrombosis

Deep vein thrombosis is the formation of a thrombus (coagulated blood) within the lumen of the deep veins of the lower limb. This may occur as a result of venous stasis (slow venous blood flow), for example, when sitting for long periods in an airplane or automobile. The patient may experience pain, swelling, redness, and localized warmth over the thrombosis. The hazard is that the thrombus may move (embolize) through the veins and obstruct venous return to the heart and lungs. This can cause sudden collapse and death.

Fill in the blanks

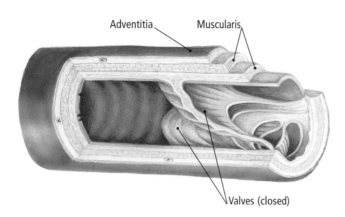

Adventitia   Muscularis

Valves (closed)

Structure of a Normal Vein

Varicose veins

Varicose veins occur when pressure in the veins causes them to dilate and become tortuous. Dilation of veins separates the leaflets of their internal valves so blood flow can no longer be controlled by the internal valve system. Varicose veins are most common in the lower parts of the body where hydrostatic pressure is highest, i.e., lower limb and pelvis. The effect of elevated pressure is made worse by failure of the valves that channel blood from the superficial unsupported veins of the lower limb to the deep muscle-embedded veins. Once these valves fail, so much blood may pool in the leg veins that the circulation of blood through capillaries of the foot and leg is impaired. Chronic venous stasis ulcers may then develop on the thin skin of the shins.

1   The _____ and _____ drain into the right atrium.

2   Venous blood from the head drains through the _____ and _____ .

3   The two largest superficial veins in the upper limb are the _____ and _____ veins.

4   Venous drainage from the kidneys is through the _____ veins.

5   The _____ and _____ veins are the two largest superficial veins of the lower limb.

6   Almost all venous blood from the lower limb passes through the _____ vein.

7   The radial and ulnar veins join to form the _____ vein in the arm.

8   The main deep veins draining the leg are the _____ and _____ .

9   Blood from the gut passes through the _____ vein to the liver.

10   Venous blood may be readily sampled from the _____ vein in front of the elbow.

Match the statement to the reason

1 The level of pressure in the systemic venous compartment can be assessed in the neck because…

a the external jugular vein is in a position superficial to the sternocleidomastoid muscle.

2 Venous return to the heart is increased by breathing in deeply because…

b contraction of the diaphragm lowers the pressure in the right atrium relative to the inferior vena cava.

3 Blood in the lower limb is channeled from the superficial to deep veins because…

c a large uterus can compress the common iliac veins.

4 Blood from the gut drains to the liver via the portal vein because…

d nutrients and toxins from the intestine must be processed before they reach the systemic circulation.

5 Varicose veins of the lower limb may develop during pregnancy because…

e there is a series of valved perforating veins that join the two systems.

6 The thickness of the wall of the pulmonary artery is less than the aorta because…

f pressure in cranial venous channels is below atmospheric when standing upright.

7 Venous sinuses in the skull could collapse without the reinforcement of tough dural membranes because…

g pressure in the pulmonary artery is only 20 percent of that in the aorta.

Color and label

i) Label each structure shown on the illustration.

Circulatory System—anterior view

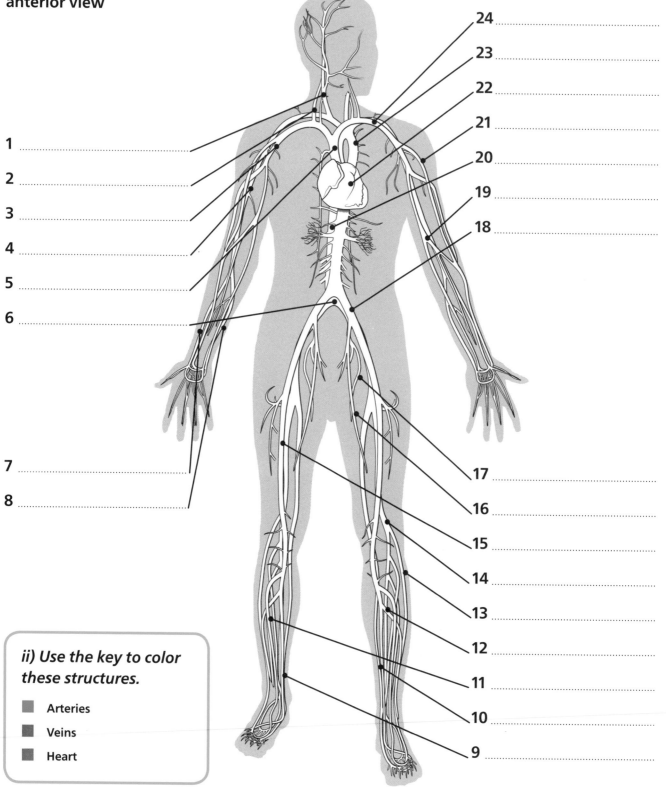

1 ....

2 ....

3 ....

4 ....

5 ....

6 ....

7 ....

8 ....

24 ....

23 ....

22 ....

21 ....

20 ....

19 ....

18 ....

17 ....

16 ....

15 ....

14 ....

13 ....

12 ....

11 ....

10 ....

9 ....

ii) Use the key to color these structures.

■ Arteries

■ Veins

■ Heart

Superficial Arteries of the Head and Neck— lateral view

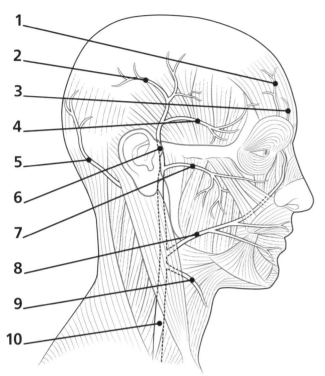

Superficial Veins of the Head and Neck— lateral view

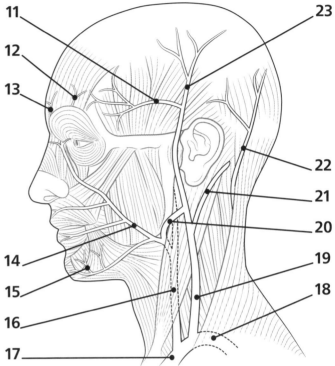

iii) Add numbers to the boxes below to match each label to the correct part of the illustrations.

Occipital artery ☐

Posterior branch of superficial temporal artery ☐

Facial artery ☐

Anterior branch of superficial temporal artery ☐

Supraorbital artery ☐

Superficial temporal artery ☐

External carotid artery ☐

Supratrochlear artery ☐

Transverse facial artery ☐

Transverse cervical artery ☐

Anterior branch of superficial temporal vein ☐

Posterior auricular vein ☐

Supraorbital vein ☐

Brachiocephalic vein ☐

Occipital vein ☐

Internal jugular vein ☐

Subclavian vein ☐

Facial vein ☐

External jugular vein ☐

Retromandibular vein ☐

Supratrochlear vein ☐

Submental vein ☐

Posterior branch of superficial temporal vein ☐

True or false?

1   *Veins have the thickest walls of all vessels.*

2   *Thick-walled arteries often have accompanying vessels called vasa vasorum.*

3   *Arteries have internal valves.*

4   *All capillaries have fenestrations (tiny openings in their walls).*

5   *High-pressure arteries like the aorta have a high content of elastic fibers in their walls.*

6   *The pressure in the femoral vein is higher than in the femoral artery.*

7   *The primary role of systemic capillaries is exchange of nutrients and waste products with body tissues.*

8   *The main role of systemic arterioles is to regulate the flow of blood to capillary beds.*

9   *The tunica intima of arteries has abundant smooth muscle.*

10   *The endothelium is the inner lining of blood vessels.*

Subclavian vein

Axillary vein

Cephalic vein

Brachial vein

Basilic vein

Median cubital vein

Median antebrachial vein

Ulnar vein

Radial vein

Palmar venous arch

Digital veins

Veins of the Upper Limb—anterior view

The venous system of the arm includes the digital veins of the hands, the cephalic and median veins of the forearm, the basilic vein, and the brachial vein.

Multiple choice

1 Which vessels provide a capacitance or reservoir function, storing excess blood volume?

- (A) systemic arteries
- (B) pulmonary arteries
- (C) systemic capillaries
- (D) systemic veins
- (E) pulmonary veins

2 Which of the following would NOT be found in the tunica adventitia of an artery?

- (A) nerves
- (B) companion arteries
- (C) companion veins
- (D) lymphatics
- (E) mucous glands

3 The disease atherosclerosis is characterized by fatty and fibrous deposits mainly in the:

- (A) tunica intima
- (B) tunica media
- (C) tunica adventitia
- (D) submucosa
- (E) serosa

4 Which layer of the typical arterial wall has the most smooth muscle?

- (A) endothelium
- (B) submucosa
- (C) tunica intima
- (D) tunica media
- (E) tunica adventitia

5 The main type of cell forming a capillary is the:

- (A) smooth muscle cell
- (B) endothelial cell
- (C) fibroblast
- (D) mucous gland
- (E) nerve cell

6 Fenestrated capillaries are found in the:

- (A) brain
- (B) skeletal muscle
- (C) skin
- (D) lungs
- (E) intestines

7 The thickness of most capillary walls is about:

- (A) 0.1 μm
- (B) 1 μm
- (C) 10 μm
- (D) 100 μm
- (E) 1,000 μm

Venous stasis ulcers

Venous stasis (or venous insufficiency) ulcers are areas where the epidermis of the skin is lost due to poor venous drainage. They are commonly found on the shin and ankle of the elderly and arise when the poor drainage of the skin in this region causes inadequate supply of oxygen and nutrients to the tissue. Minor injuries of the skin fail to heal, and these wounds may become infected, forming long-standing ulcers that are difficult to treat.

Fill in the blanks

1   The _____ cells line the interior of all vessels.

2   The thickest layer of an artery is the tunica _____.

3   The valves of veins are extensions of the tunica _____.

4   The elastic fibers of large arteries are mainly found in the tunica _____.

5   The endothelium of capillaries is surrounded by undifferentiated cells called _____.

6   Companion vessels and nerves of large vessels are concentrated in the tunica _____.

7   Capillaries in the intestines have _____ in the endothelium to improve diffusion.

8   The interior of a vessel is called the _____.

9   The elastic layer between the intima and media is called the _____.

10   The elastic layer between the media and adventitia is called the _____.

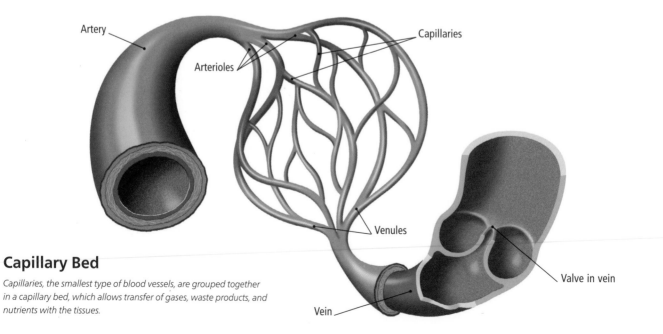

Capillary Bed

Capillaries, the smallest type of blood vessels, are grouped together in a capillary bed, which allows transfer of gases, waste products, and nutrients with the tissues.

Match the statement to the reason

1 The large systemic arteries of the body (aorta, brachiocephalic trunk, common iliac arteries) contain abundant elastic fibers because…

a they need to change diameter to regulate blood flow through systemic capillaries.

2 The systemic arterioles contain smooth muscle sphincters because…

b they need to allow diffusion of nutrients and gases between blood and tissues.

3 The veins of the lower limb contain valves because…

c they are very low-pressure vessels.

4 Capillaries contain only a thin layer of cells because…

d they need to expand when the heart pumps out blood and recoil between beats.

5 Venules have very little smooth muscle in their walls because…

e they need to direct blood back to the heart against gravity.

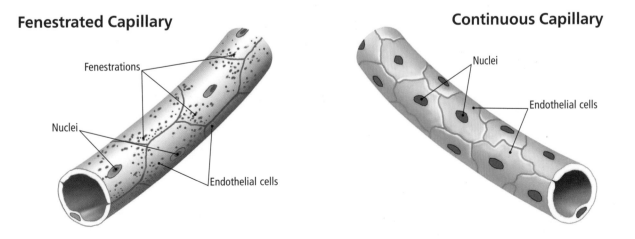

Fenestrated Capillary

Fenestrations

Nuclei

Endothelial cells

Continuous Capillary

Nuclei

Endothelial cells

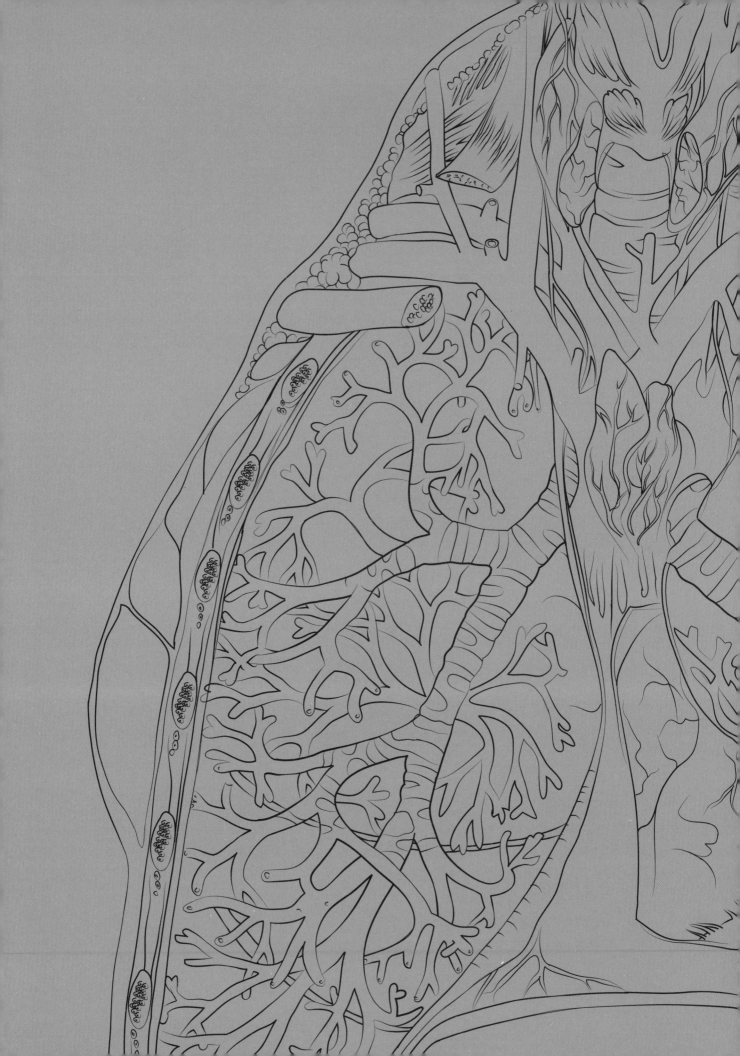

CHAPTER 5: THE RESPIRATORY SYSTEM

The Respiratory System

The respiratory system is concerned primarily with gas exchange between the blood stream and the external air, although it also serves additional functions such as communication, immune surveillance, acid/base regulation, and thermoregulation. The respiratory tract is usually divided into an upper part (nose, paranasal sinuses, and larynx) and a lower part (trachea, bronchi, lungs, and pleura). Other critically important structures are the respiratory muscles and the thoracic cage that those muscles move.

Key terms:

Diaphragm A sheet of muscle and tendon separating the thoracic and abdominal cavities. It is perforated by the aorta, inferior vena cava, and esophagus.

Larynx The part of the airway where phonation (sound production) occurs. The upper larynx (epiglottis and aryepiglottic folds) also protects the airway entrance.

Left primary bronchus The bronchus for the left lung. It gives off an upper (superior) lobar bronchus and a lower (inferior) lobar bronchus.

Lower lobar bronchus The bronchus for the lower (inferior) lobe of the left (or right) lung.

Middle lobar bronchus The bronchus for the middle lobe of the right lung. It branches from the right primary bronchus.

Nasal cavity The internal nose is divided into two cavities by a bony and cartilaginous septum. Each cavity has three bony elevations (conchae or turbinates) on its lateral wall. The superior part of each cavity has an olfactory region.

Pharynx The vertical tube behind the nasal, oral, and laryngeal cavities. It provides passage for both the airway and gastrointestinal tract.

Right primary bronchus The bronchus of the right lung. It is wider and more vertical than the left primary bronchus and is therefore more likely to receive inhaled foreign bodies.

Superior lobar bronchus The bronchus to the upper (superior) lobe of the lung.

Trachea The airway between the larynx and the tracheal bifurcation into right and left primary bronchi. The trachea consists of 16–20 U-shaped cartilages, with the trachealis as its posterior wall.

Upper lobar bronchus The bronchus to the upper (superior) lobe of the lung.

Respiratory System— anterior view

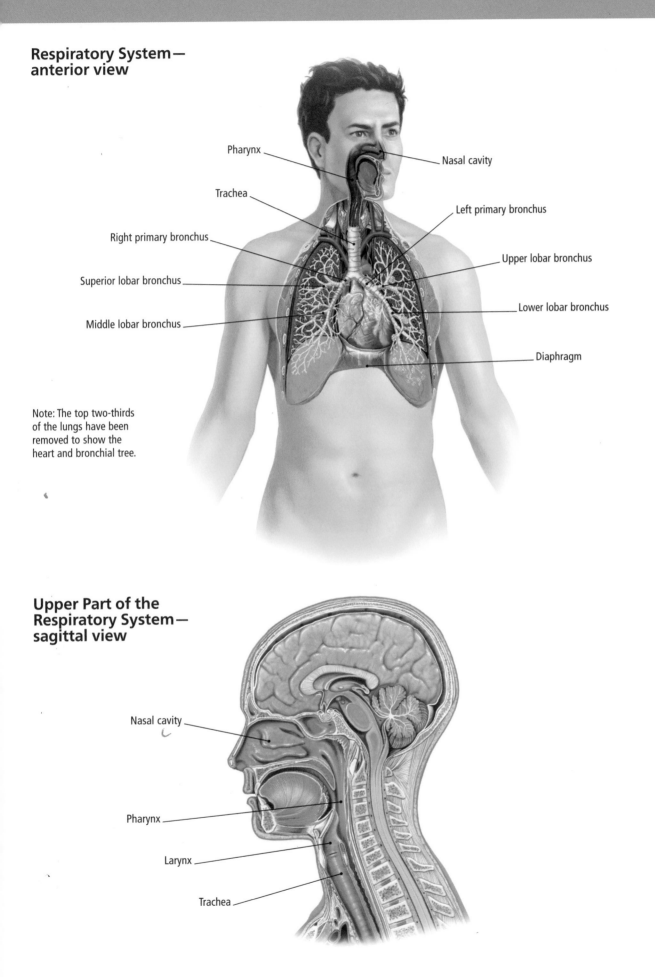

Pharynx

Nasal cavity

Trachea

Left primary bronchus

Right primary bronchus

Superior lobar bronchus

Upper lobar bronchus

Middle lobar bronchus

Lower lobar bronchus

Diaphragm

Note: The top two-thirds of the lungs have been removed to show the heart and bronchial tree.

Upper Part of the Respiratory System— sagittal view

Nasal cavity

Pharynx

Larynx

Trachea

True or false?

1   *There are usually four nasal conchae or turbinates on each side of the nose.*

2   *The initial part of the nasal cavity is the vestibule.*

3   *The role of the nasal conchae or turbinates is to warm and moisten air and assist with the removal of dust and pathogens.*

4   *The sphenoid sinus opens into the floor of the nasal cavity.*

5   *The respiratory part of the nose includes all of the mucosa of the middle and nasal conchae, but only part of the surface of the superior concha.*

6   *The olfactory region of the nasal cavity lies immediately inferior to the cribriform plate of the ethmoid bone.*

7   *The typical epithelium of the respiratory part of the nasal cavity is nonkeratinized stratified squamous epithelium.*

8   *The maxillary sinus opens beneath the inferior concha.*

9   *The lateral wall of the nasal cavity contains openings from the ethmoidal, maxillary, and frontal paranasal sinuses.*

10   *The larynx is built around a framework of bones.*

11   *The laryngeal entrance (aditus) is located in a region called the laryngopharynx.*

Carcinoma of the larynx

Cancer (carcinoma) of the larynx is usually the result of cigarette or pipe smoking. Chronic exposure of the respiratory epithelium to carcinogens will result in mutations that allow epithelial cells to divide out of control. The cancer will produce a growth, which may ulcerate, leading to coughing of blood (hemoptysis), or invade surrounding tissues, causing nerve damage, and spread to lymph nodes and distant sites. Radical surgery is often necessary to remove the tumor.

12   The rim of the laryngeal entrance is strengthened by the hyoid bone.

13   During swallowing, the larynx rises and the epiglottis is pulled inferiorly to close the laryngeal entrance.

14   The vocal fold is the most superior of the internal folds of the larynx.

15   The sound of the voice (phonation) is produced by vibration of the vestibular folds of the larynx.

16   The vocal folds are moved by rotation of the cuneiform cartilages.

17   The vocal folds are lubricated during speech by fluid that drips from the oropharynx.

18   The inferior part of the larynx is encircled by the epiglottis.

19   The airway passes through the center of a ring-like cartilage called the cricoid to reach the trachea.

20   The muscles of the larynx are all supplied by the hypoglossal nerve.

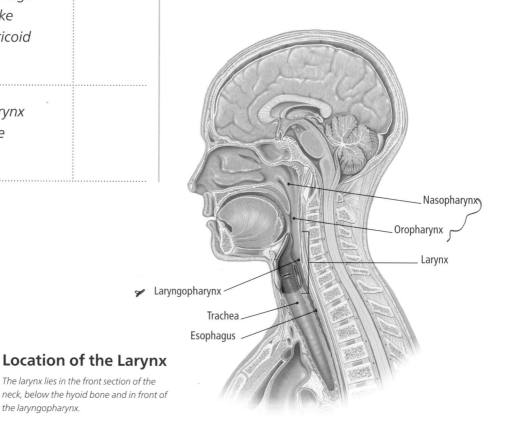

Nasopharynx

Oropharynx

Larynx

Laryngopharynx

Trachea

Esophagus

Location of the Larynx

The larynx lies in the front section of the neck, below the hyoid bone and in front of the laryngopharynx.

Multiple choice

1   *Which of the following would be found in or on the epithelium of the respiratory region of the nasal cavity?*

Ⓐ   cilia

Ⓑ   pseudostratified columnar cells

Ⓒ   goblet cells

Ⓓ   mucus

Ⓔ   all of the above

2   *Which of the following opens into the meatus inferior to the inferior concha?*

Ⓐ   anterior ethmoidal air cells (paranasal sinuses)

Ⓑ   sphenoidal paranasal sinus

Ⓒ   frontal paranasal sinus

Ⓓ   nasolacrimal duct

Ⓔ   maxillary paranasal sinus

3   *Which of the following structures make up most of the nasal septum?*

Ⓐ   maxillary crest and plate and frontal bones

Ⓑ   perpendicular plate of palatine and inferior concha

Ⓒ   vomer bone and perpendicular plate of ethmoid bone

Ⓓ   maxillary crest and perpendicular plate of palatine bone

Ⓔ   vertical plates of sphenoid and frontal bones

4   *Which of the following make up the floor of the nasal cavity?*

Ⓐ   maxilla and palatine horizontal plate

Ⓑ   sphenoid and maxilla bones

Ⓒ   ethmoid and palatine bones

Ⓓ   frontal and nasal bones

Ⓔ   occipital and temporal bones

5   *Inhaled nasal air is under direct surveillance from which part of the immune system?*

Ⓐ   lingual tonsil

Ⓑ   palatine tonsil

Ⓒ   nasopharyngeal tonsil

Ⓓ   paranasal sinuses

Ⓔ   thymus gland

6   *Which of the following laryngeal cartilages is most mobile during breathing and speech (phonation)?*

Ⓐ   cricoid

Ⓑ   arytenoid

Ⓒ   thyroid

Ⓓ   corniculate

Ⓔ   epiglottis

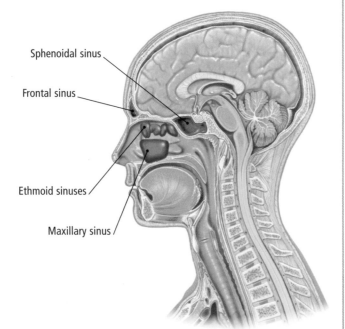

Sphenoidal sinus

Frontal sinus

Ethmoid sinuses

Maxillary sinus

Paranasal Sinuses—sagittal view

The paranasal sinuses are membrane-lined, air-filled cavities in the bones of the skull, which are connected to the nose by passageways.

7   *Which of the following is/are found within the aryepiglottic fold?*

(A)   epiglottis
(B)   corniculate cartilage
(C)   arytenoid cartilage
(D)   both A and B are correct
(E)   both A and C are correct

8   *Which of the following structures contributes to the "Adam's apple"?*

(A)   cricoid cartilage
(B)   hyoid bone
(C)   arytenoid cartilages
(D)   epiglottis
(E)   thyroid cartilage

9   *Which of the following is associated with the deeper male voice?*

(A)   shorter vocal folds
(B)   longer vestibular folds
(C)   shorter vestibular folds
(D)   longer vocal ligaments
(E)   shorter vocal ligaments

10   *Which of the following structures makes up the core of the vocal fold (vocal cord) of the larynx?*

(A)   epiglottis
(B)   vocal ligament
(C)   cricoid cartilage
(D)   thyroid cartilage
(E)   hyoid bone

11   *Which structures are most likely to be affected by spread of tumor from a carcinoma (cancer) of the larynx?*

(A)   cervical spinal cord
(B)   postvertebral muscles
(C)   cervical lymph nodes
(D)   soft palate
(E)   trachea

Sinusitis

Sinusitis is inflammation of the paranasal sinuses. This usually occurs following an upper respiratory tract infection, when a large amount of mucus is produced by the respiratory epithelium of the paranasal sinuses and nasal cavity. If this mucus thickens, drainage from the sinuses is impaired and the pressure alone can cause pain. If there is an additional bacterial infection, then the pain may become intense. Pain will be felt over the forehead (frontal sinus), the cheeks (maxillary sinus), or centrally in the head (ethmoid and sphenoidal sinuses). Treatment is by nasal decongestants and antibiotics if there is bacterial infection.

12   *Which type of epithelium is found in the larynx?*

(A)   pseudostratified ciliated columnar epithelium with goblet cells
(B)   keratinized stratified squamous epithelium with glands
(C)   nonkeratinized stratified squamous epithelium with goblet cells
(D)   simple columnar epithelium with submucosal glands
(E)   transitional uroepithelium with goblet cells and mucus

13   *Which of the following describes what happens during whispering?*

(A)   both vocal folds are wide apart
(B)   both vocal folds are completely together
(C)   one vocal fold is in the midline and the other to the side
(D)   anterior vocal folds are separated, but posterior folds are together in the midline
(E)   posterior vocal folds are separated, but anterior folds are together in the midline

Color and label

i) Label each structure shown on the illustrations.

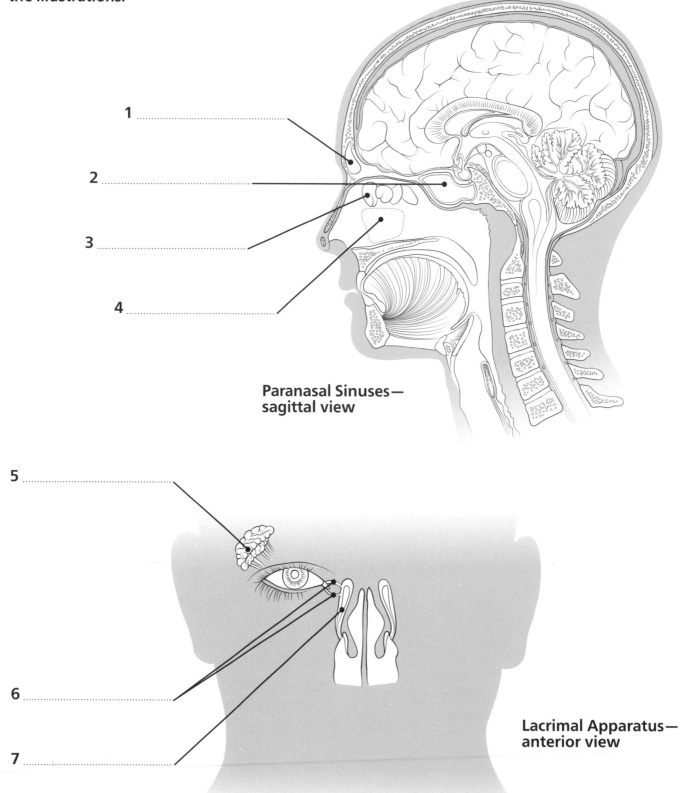

1 ....

2 ....

3 ....

4 ....

**Paranasal Sinuses—
sagittal view**

5 ....

6 ....

7 ....

**Lacrimal Apparatus—
anterior view**

Pharynx—posterior view

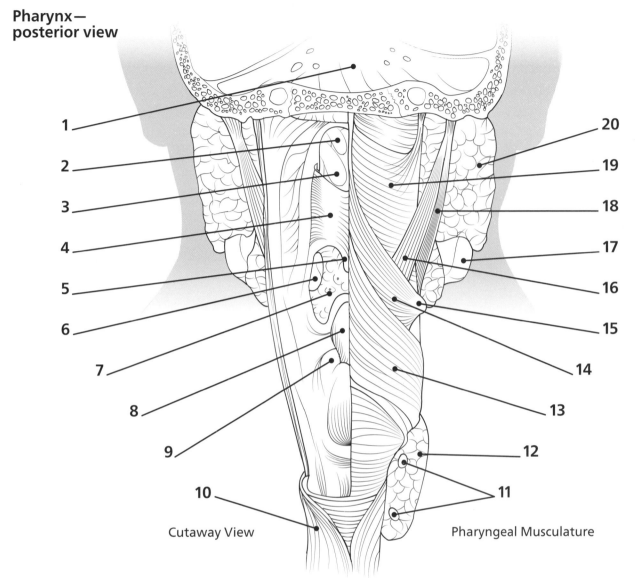

1
2
3
4
5
6
7
8
9
10

20
19
18
17
16
15
14
13
12
11

Cutaway View

Pharyngeal Musculature

ii) Add numbers to the boxes below to match each label to the correct part of the illustration.

Uvula ☐	Parathyroid glands ☐
Inferior nasal concha ☐	Thyroid gland (lateral lobe) ☐
Middle nasal concha ☐	Dorsum of tongue ☐
Soft palate ☐	Epiglottis ☐
Base of skull ☐	End of greater horn of hyoid bone ☐
Palatine tonsil ☐	Stylohyoid muscle ☐
Angle of mandible ☐	Inferior constrictor muscle ☐
Aryepiglottic fold ☐	Stylopharyngeus muscle ☐
Parotid gland ☐	Superior constrictor muscle ☐
Esophagus ☐	Middle constrictor muscle ☐

Fill in the blanks

1   The nasal lining is classified as _____ tissue because it contains numerous vascular spaces that can become filled with blood under appropriate stimulation.

2   The frontal paranasal sinus opens into the anterior part of the _____.

3   Neurosurgeons operating on the pituitary can reach it through the _____ paranasal sinus.

4   The maxillary paranasal sinus opens into the nasal cavity inferior to the _____.

5   The part of the nasal cavity where hair grows during middle and old age is the nasal _____.

6   The upper part of the laryngeal cavity (superior to the vestibular folds) is known as the laryngeal _____.

7   Loss of the voice during laryngitis is due to inflammation of the _____.

8   The two key functions of the larynx are _____ and _____.

9   The three cartilages that strengthen the rim of the laryngeal entrance are the _____.

10   The _____ bone protects the upper airway by preventing collapse during deep inhalation.

Greater horn of hyoid bone

Cricothyroid muscle

Tracheal cartilages

Epiglottis

Thyrohyoid membrane

Thyroid cartilage

Cricoid cartilage

Trachea

Larynx—anterior view

The larynx is composed of cartilages that are joined by ligaments and held in place and controlled by skeletal muscles.

Match the statement to the reason

1 *Blowing the nose draws mucus from the paranasal sinuses because…*

a *mucus builds up in the frontal and maxillary paranasal sinuses, raising pressure on the sensitive sinus mucosa and providing a rich culture medium for bacterial infection.*

2 *Nasal mucosa is prone to bleeding because…*

b *the pathways for food, fluids, and air pass through the upper part of the laryngopharynx posterior to the laryngeal entrance.*

3 *Excess mucus generated during an upper respiratory tract infection can cause pain in the medial forehead and cheek because…*

c *it contains a plexus of large, delicately walled veins.*

4 *Humans are prone to choking during eating and drinking because…*

d *rapid flow of air through the narrow parts of the nose lowers the pressure at the sinus openings by the Venturi effect.*

5 *Phonation (voice production) always occurs during breathing out (expiration) because…*

e *expired air must be forced through a gap between the vocal folds to produce vibration.*

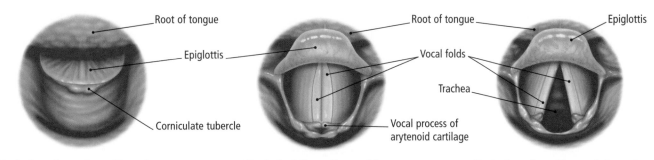

Epiglottis—Swallowing   **Epiglottis—Speaking**   **Epiglottis—Breathing In**

The Lungs

The trachea leads from the larynx to the tracheal bifurcation, where the two main bronchi branch to each lung. Altogether there are about 23 generations of branching of the airways until the alveoli are reached. These are the main sites of gas exchange, where oxygen from inhaled air is taken up by the blood, and carbon dioxide is released to be exhaled. The pleural sacs surround the lungs and allow free movement of the lungs during ventilation.

Key terms:

Aortic arch See pp. 122–123.

Brachial plexus A network of nerves formed from spinal nerves cervical 5 to thoracic 1 to supply the upper limb. Major nerves arising from the brachial plexus are the radial, ulnar, median, axillary (circumflex), and musculocutaneous.

Cardiac branch of vagus nerve A branch of the vagus nerve that runs to the cardiac plexus. It reduces the spontaneous rate of the sinuatrial node to lower heart rate.

Common carotid artery See pp. 134–135.

Costodiaphragmatic recess A recess in the pleural sac between the costal and diaphragmatic parietal pleura. It accommodates the inferior border of the lung during deep inspiration.

Cricoid cartilage One of the laryngeal cartilages. It forms a ring around the airway and provides a foundation on which the arytenoid cartilages sit.

Cricothyroid muscle A muscle that tilts the thyroid cartilage down, or the cricoid cartilage up, depending on which is fixed by other muscles. It tenses the vocal ligaments and folds.

Diaphragm See pp. 160–161.

External jugular vein See pp. 146–147.

First rib The first rib is an atypical rib. It has only a single facet on its head. It has impressions on its upper surface for the subclavian vein and artery and a tubercle for attachment of the scalenus anterior muscle.

Inferior thyroid vein One of the veins draining the thyroid gland. It is quite variable in its course but often enters the brachiocephalic vein.

Internal jugular vein The large vein draining the inside of the skull. It passes deep to the sternocleidomastoid muscle alongside the common carotid artery and vagus nerve.

Internal thoracic vein A vein draining the anterior chest wall. It runs alongside the internal thoracic artery to enter the brachiocephalic or subclavian vein.

Left brachiocephalic vein See pp. 122–123.

Lower lobe (left lung) Separated from the upper lobe by the oblique fissure, this lobe occupies the lung base and back. Its inferior border sits within the costodiaphragmatic recess.

Lower lobe (right lung) The lower lobe of the right lung lies immediately over the diaphragm and the liver. It is bordered above by the oblique fissure.

Lungs Paired organs for gas exchange between the blood and the external environment. They lie within the pleural sacs of the thoracic cavity.

Middle lobe (right lung) Lying against the front of the chest, this lobe is separated from the upper and lower lobes by the horizontal and oblique fissures, respectively.

Pectoralis major A muscle arising from the medial clavicle and upper six costal cartilages to insert into the crest of the greater tubercle of the humerus. It adducts, medially rotates, and flexes the arm.

Pericardium A multilayered sac that encloses the heart. The outermost fibrous layer anchors the heart to adjacent structures. The inner, serous double layer provides a fluid-filled space to reduce friction during beating of the heart.

Right atrium See pp. 122–123.

Right brachiocephalic vein See pp. 122–123.

Right ventricle See pp. 122–123.

Scalenus anterior Muscle arising from the transverse processes of the lower cervical vertebrae (3–6); inserts into the first rib.

Subclavian artery and vein See pp. 134–135, and pp. 146–147.

Superior vena cava See pp. 122–123.

Thymus A lymphoid organ in the anterior mediastinum of the chest. It is active before puberty to produce T lymphocytes but becomes a fatty-fibrous remnant in adult life.

Thyroid gland A gland that concentrates iodine from the blood to produce the hormones thyroxine and tri-iodothyronine for regulation of the body's metabolic rate.

Trachea See pp. 160–161.

Upper lobe (left lung) The upper lobe of the left lung extends from the left apex to the oblique fissure of the left lung. Sounds from it can be heard over the left anterior chest.

Upper lobe (right lung) The upper lobe of the right lung is in contact with the anterior chest wall. It is bordered below by the horizontal or transverse fissure and behind by the oblique fissure.

Lungs—anterior view

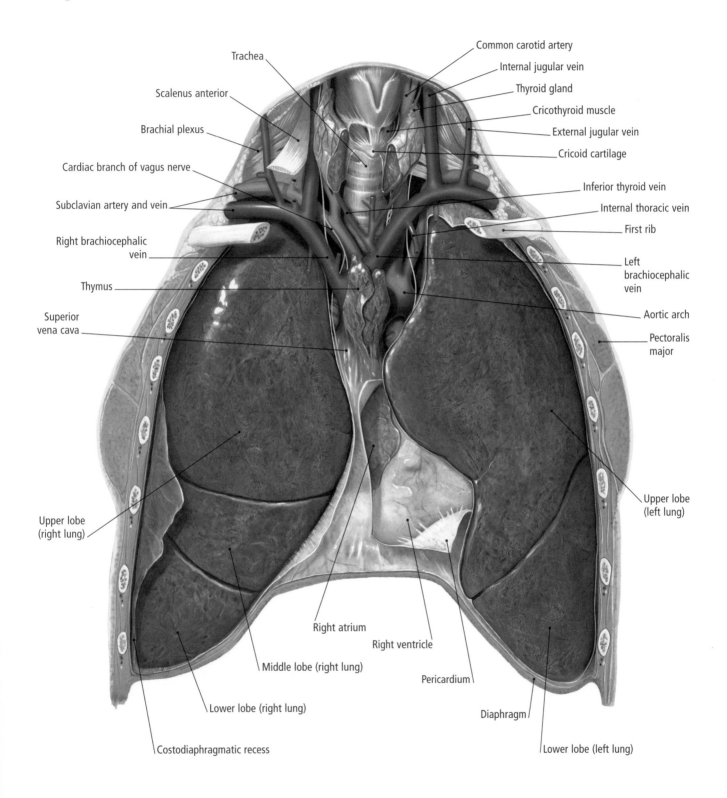

Trachea

Common carotid artery

Internal jugular vein

Scalenus anterior

Thyroid gland

Cricothyroid muscle

Brachial plexus

External jugular vein

Cricoid cartilage

Cardiac branch of vagus nerve

Inferior thyroid vein

Subclavian artery and vein

Internal thoracic vein

First rib

Right brachiocephalic vein

Left brachiocephalic vein

Thymus

Aortic arch

Superior vena cava

Pectoralis major

Upper lobe (left lung)

Upper lobe (right lung)

Right atrium

Right ventricle

Middle lobe (right lung)

Pericardium

Lower lobe (right lung)

Diaphragm

Costodiaphragmatic recess

Lower lobe (left lung)

True or false?

1   The trachea consists of 16 to 20 continuous cartilage rings.

2   The posterior surface of the trachea is made up of dense connective tissue.

3   The tracheal bifurcation has an internal ridge (the carina) that is highly sensitive to contact with inhaled foreign bodies and elicits a cough reflex when touched.

4   The right main bronchus is usually more vertical than the left.

5   The surface of each lung is covered by a smooth, glistening layer called the parietal pleura.

6   The division of the trachea into bronchi is at about the level of the manubriosternal joint.

7   The pleural cavity is filled with a fluid film that allows free movement of the lung within the chest cavity.

8   The normal volume of pleural fluid is about 250 ml.

9   The right lung is usually divided into three lobes by horizontal (transverse) and oblique fissures.

10   The number of lobes is a good indicator of which side the lung belongs on.

11   The two lobes of the left lung are usually separated by a horizontal (transverse) fissure.

Submucosal gland

Cartilage

Trachealis muscle

Trachea—cross-section

The muscular fibroelastic tube of the trachea is reinforced with cartilage. This flexible tube forms a passageway between the larynx and the bronchi of the lungs.

12 *The horizontal fissure is usually aligned with the sternal angle.*

13 *The surface marking of the oblique fissure of the right lung is given by the course of the sixth rib.*

14 *The inferior border of the lung sits in the costodiaphragmatic recess.*

15 *The anterior border of the right lung is interrupted by a notch called the lingula.*

16 *There are usually 10 bronchopulmonary segments in each lung, each served by a segmental bronchus and separated from its neighbor by a connective tissue wall.*

17 *The dome of the diaphragm in a standing subject at the end of quiet expiration reaches the height of the 5th costal cartilage.*

18 *The posterior border of the lung lies alongside the vertebral column.*

19 *The apices of both lungs rise above the anterior end of the first rib.*

20 *During deep inspiration, the lungs may pass between the kidneys and the chest wall.*

Pneumonia

Pneumonia is an inflammatory condition of the lungs that mainly involves accumulation of fluid and cellular debris in the alveolar spaces. Pneumonia is usually due to infection with viruses or bacteria. Symptoms and signs include cough with or without sputum, chest pain, fever, and difficulty breathing. Risk factors include smoking, asthma, chronic obstructive pulmonary disease, and diabetes mellitus. Vaccines are now available against some causative microorganisms, but the condition remains common and serious, particularly in the elderly. Treatment may include support of respiratory function (oxygen), physiotherapy, and antibiotics.

Multiple choice

1   *Structures in close proximity to the trachea during its passage through the chest include all of the following EXCEPT the:*

(A)   left ventricle

(B)   esophagus

(C)   arch of aorta

(D)   recurrent laryngeal nerves

(E)   lymphatic channels

2   *What type of epithelium lines the trachea?*

(A)   keratinized simple squamous epithelium

(B)   pseudostratified ciliated columnar epithelium

(C)   simple columnar epithelium

(D)   mesothelium

(E)   simple cuboidal epithelium

3   *Which of the following contribute(s) to clearing the trachea of inhaled debris and microorganisms?*

(A)   beating of cilia

(B)   goblet cell secretion

(C)   tracheal cartilage

(D)   seromucous tracheal glands

(E)   A, B, and D are all correct

4   *Which of the following structures crosses over the left main bronchus?*

(A)   arch of aorta

(B)   esophagus

(C)   left vagus nerve

(D)   left atrium of heart

(E)   none of the above

5   *Structures in contact with the right lung include all of the following EXCEPT:*

(A)   right atrium

(B)   ribs

(C)   right ventricle

(D)   diaphragm

(E)   arch of aorta

6   *The number of times that the airways divide to reach the alveoli is approximately:*

(A)   10 to 14

(B)   14 to 17

(C)   17 to 19

(D)   20 to 23

(E)   24 to 27

7   *What is the approximate partial pressure of oxygen in the lung alveoli?*

(A)   10 mm Hg

(B)   40 mm Hg

(C)   100 mm Hg

(D)   150 mm Hg

(E)   200 mm Hg

8   *Which structure of the respiratory tract is NOT part of dead space?*

(A)   alveoli

(B)   trachea

(C)   main bronchi

(D)   lobar bronchi

(E)   segmental bronchi

9   *The terminal air passages are the:*

(A)   alveolar ducts

(B)   alveolar sacs

(C)   alveoli

(D)   respiratory bronchioles

(E)   terminal bronchioles

10 *The main role of type 2 alveolar cells is to:*
(A) produce antibodies against invading viruses, bacteria, and fungi
(B) produce pulmonary surfactant to lower the surface tension of alveolar fluid
(C) engulf bacteria and inhaled debris and move these to the lung hilum
(D) participate in gas exchange between pulmonary capillaries and the alveolar air
(E) produce oxygen for the body

11 *Which cell type in the alveoli is primarily responsible for engulfing inhaled foreign material and microorganisms?*
(A) alveolar macrophage
(B) type 1 pneumocyte
(C) alveolar epithelial cell
(D) type 2 pneumocyte
(E) goblet cell

12 *Alveolar macrophages that have ingested soot may migrate to the:*
(A) esophagus
(B) hilar lymph nodes
(C) larynx
(D) parietal pleura
(E) heart

Pneumothorax

Pressure within the pleural sac is normally below atmospheric. This allows chest wall movement to draw out and inflate the lungs without any direct physical attachment between the lungs and chest wall. The pleural cavity also contains a thin film of fluid, which lubricates lung movement. If the chest or lung covering is punctured, air can enter the pleural sac, and the pleural pressure equilibrates with the external atmosphere (pneumothorax). The lung will then collapse, and effective ventilation is lost. If a flap valve develops, pressure in the pleural sac may even exceed the atmosphere (a tension pneumothorax), and the patient may die quickly from loss of lung ventilation and impaired venous return to the heart.

13 *Hemothorax is the clinical term for which of the following?*
(A) air in the pleural sac
(B) blood in the segmental bronchi
(C) pus in the pleural sac
(D) blood in the trachea
(E) blood in the pleural sac

14 *What is the approximate distance across which alveolar oxygen must diffuse to reach the capillary blood?*
(A) less than 0.1 μm
(B) 1 to 2 μm
(C) 5 to 10 μm
(D) 10 to 100 μm
(E) 1 mm

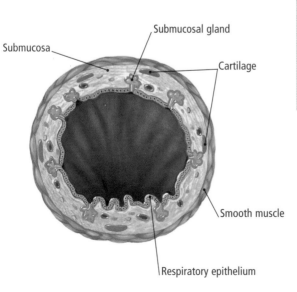

Submucosal gland

Submucosa

Cartilage

Smooth muscle

Respiratory epithelium

Bronchus—cross-section

The bronchi carry air from the trachea to the lungs. The two main bronchi divide into lobar bronchi, which then subdivide into smaller bronchi and bronchioles, which finally terminate at the tiny air sacs of the alveoli.

Color and label

i) Color the trachea red and the inferior lobar bronchi blue.

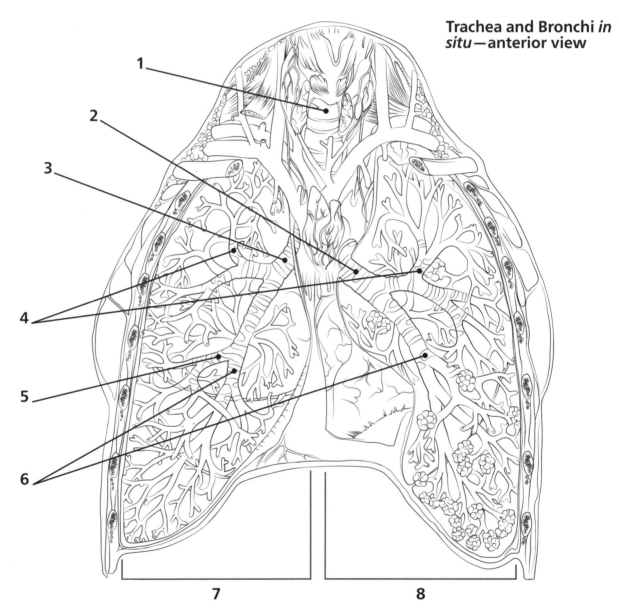

Trachea and Bronchi *in situ*—anterior view

ii) Add numbers to the boxes below to match each label to the correct part of the illustration.

Right primary bronchus	☐	Middle lobar bronchus	☐
Left primary bronchus	☐	Right lung	☐
Trachea	☐	Inferior lobar bronchi	☐
Left lung	☐	Superior lobar bronchi	☐

*iii) Label each structure shown
on the illustrations.*

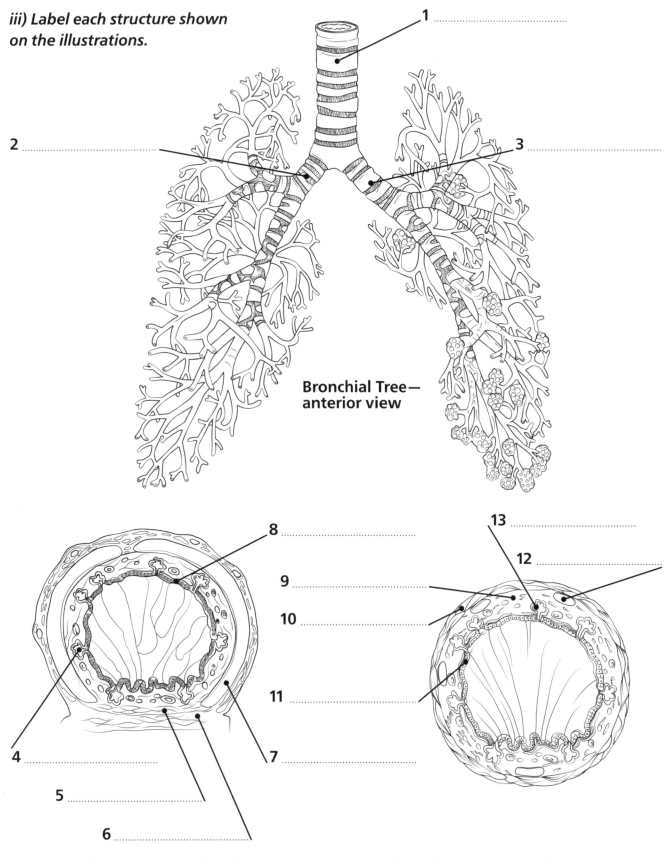

1 ....

2 ....

3 ....

**Bronchial Tree—
anterior view**

8 ....

13 ....

12 ....

9 ....

10 ....

11 ....

4 ....

5 ....

6 ....

7 ....

Trachea—cross-sectional view   **Bronchus—cross-sectional view**

Fill in the blanks

1   The _____ muscle is the sheet of smooth muscle that forms the posterior wall of the trachea.

2   The _____ lies immediately posterior to the trachea.

3   The _____ is an internal ridge located at the bifurcation of the trachea.

4   The two nerves ascending alongside the trachea are the _____ nerves.

5   _____ is a common medical condition where airway smooth muscle contracts, causing airway narrowing and shortness of breath.

6   Most lung cancers arise from the _____ epithelium.

7   Parietal pleura can be divided into _____, _____, _____, and _____ parts.

8   A stab wound to the right posterior chest wall is most likely to penetrate the _____ lobe.

9   The _____ is the tongue-shaped extension of the left upper lobe that passes between the heart and the anterior chest wall.

10 The _____ of each lung sits on the dome of the diaphragm.

11 The _____ is a feature of the anterior border of the left lung due to the extension of the heart toward the anterior chest wall.

12 The _____ is the part of the parietal pleura directly superior to the apex of the lung.

13 The surface marking for the right transverse fissure is a horizontal line that passes through the _____ costal cartilage.

14   The part of the heart in contact with the right lung is the _____.

15   During deep inspiration, the inferior borders of the lungs descend into recesses between the _____ and _____ pleura.

16   The _____ are the last generation of the conducting airways.

17   Oxygen in the alveoli must diffuse across the _____, _____, and _____ to reach the blood.

18   The _____ bronchioles have some alveoli along their walls.

19   Lymph nodes of the lower respiratory tract are divided into _____, _____, _____, and _____ groups.

20   The most abundant cell type in the alveolus is the _____.

Right common carotid artery
Right subclavian artery
Brachiocephalic artery (trunk)
Right brachiocephalic vein
Upper lobe (right lung)
Ascending aorta
Superior vena cava
Horizontal fissure
Pericardium
Right atrium
Visceral pleura
Lower lobe (right lung)

Left common carotid artery
Left subclavian artery
Left brachiocephalic vein
Aortic arch
Left lung
Left pulmonary artery
Pulmonary trunk

Right ventricle
Left ventricle

Lungs and Heart— anterior view

The heart lies in the mediastinum, the region between the two lungs. The heart and lungs work together in the pulmonary circulation, with the two organs connected by the pulmonary vessels.

Match the statement to the reason

1   The posterior wall of the trachea is soft and deformable because…

a   the pulmonary arterial blood is too deoxygenated after having passed through the body tissues, and the large airway walls are too thick to allow direct diffusion of oxygen from the airway to the wall tissues.

2   Inhaled foreign bodies are more likely to become lodged in the right as opposed to the left main bronchus because…

b   most pathogenic microorganisms (viruses, bacteria, fungi) reach the lungs by inhaled air.

3   The larger airways have their own arterial supply from the aorta (bronchial arteries) because…

c   the right main bronchus is wider and more vertical than the left main bronchus.

4   Lymph nodes are located along the major airways because…

d   equilibration of pleural pressure with that in the external atmosphere will cause lung collapse.

5   Entry of air into the pleural cavity can have serious consequences because…

e   space must be allowed for expansion of the esophagus inside the bony ring of the thoracic inlet (first rib and first thoracic vertebra) during swallowing.

1   The costodiaphragmatic recesses are potential spaces that are necessary because…

a   lung expansion requires a space for the inferior lung borders to enter.

2   Lung markings are more obvious in the lower lobes in an erect chest X-ray because…

b   the lung apex passes anterior to the lower roots of the brachial plexus.

3   Lung cancer is most often first seen near the hilum because…

c   no alveoli are present in this zone of the airway.

4   No significant gas exchange occurs in the so-called anatomical dead space because…

d   the larger airways have a large surface area in which cancer may be induced.

5   Tumors of the lung apex can cause nerve lesions of the upper limb because…

e   blood pools in pulmonary veins when the subject is standing upright.

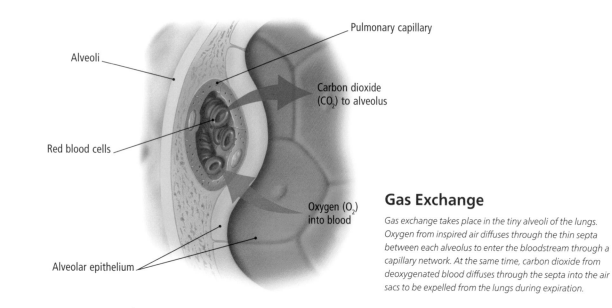

Alveoli

Pulmonary capillary

Carbon dioxide (CO_2) to alveolus

Red blood cells

Oxygen (O_2) into blood

Alveolar epithelium

Gas Exchange

Gas exchange takes place in the tiny alveoli of the lungs. Oxygen from inspired air diffuses through the thin septa between each alveolus to enter the bloodstream through a capillary network. At the same time, carbon dioxide from deoxygenated blood diffuses through the septa into the air sacs to be expelled from the lungs during expiration.

Color and label

**Respiratory System—
anterior view**

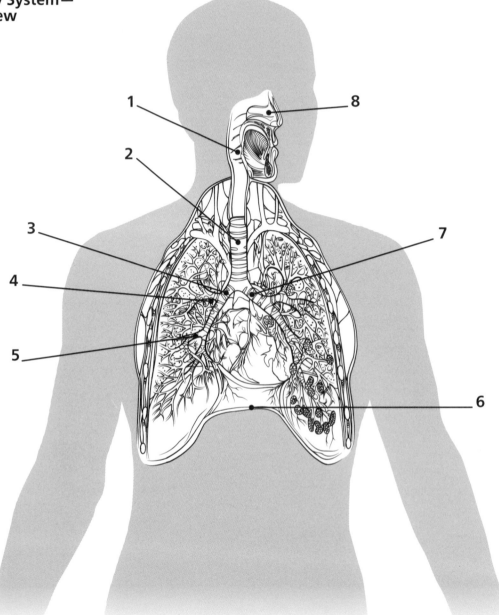

**i) Add numbers to the boxes below to
match each label to the correct part of
the illustration.**

Right primary bronchus	☐	Middle lobar bronchus	☐
Trachea	☐	Superior lobar bronchus	☐
Pharynx	☐	Nasal cavity	☐
Diaphragm	☐	Left primary bronchus	☐

ii) Label each structure shown on the illustrations.

iii) Color the rectus abdominus muscle in red.

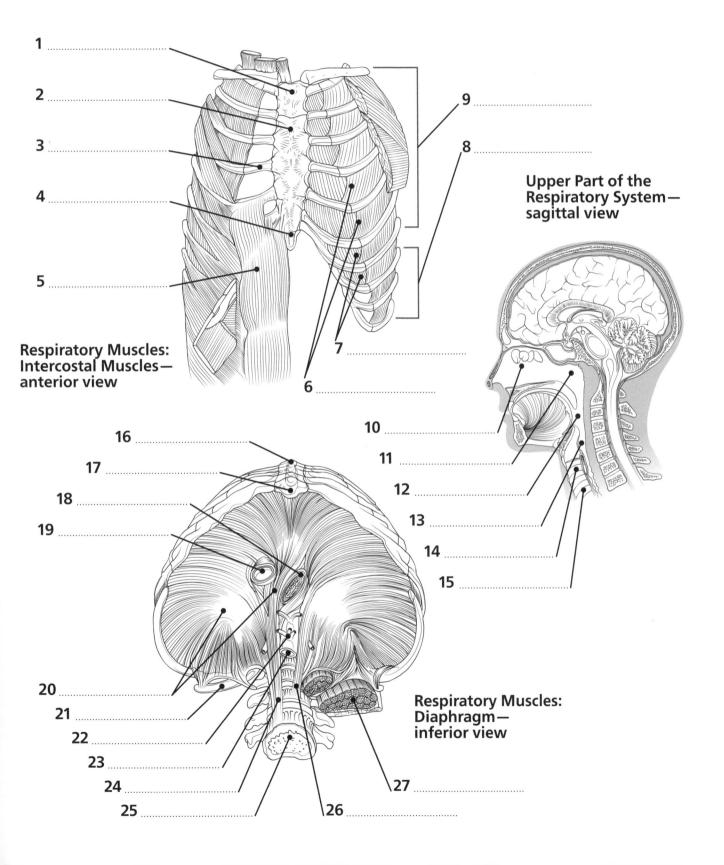

1 ....

2 ....

3 ....

4 ....

5 ....

**Respiratory Muscles:
Intercostal Muscles—
anterior view**

6 ....

7 ....

8 ....

9 ....

**Upper Part of the
Respiratory System—
sagittal view**

10 ....

11 ....

12 ....

13 ....

14 ....

15 ....

16 ....

17 ....

18 ....

19 ....

20 ....

21 ....

22 ....

23 ....

24 ....

25 ....

26 ....

27 ....

**Respiratory Muscles:
Diaphragm—
inferior view**

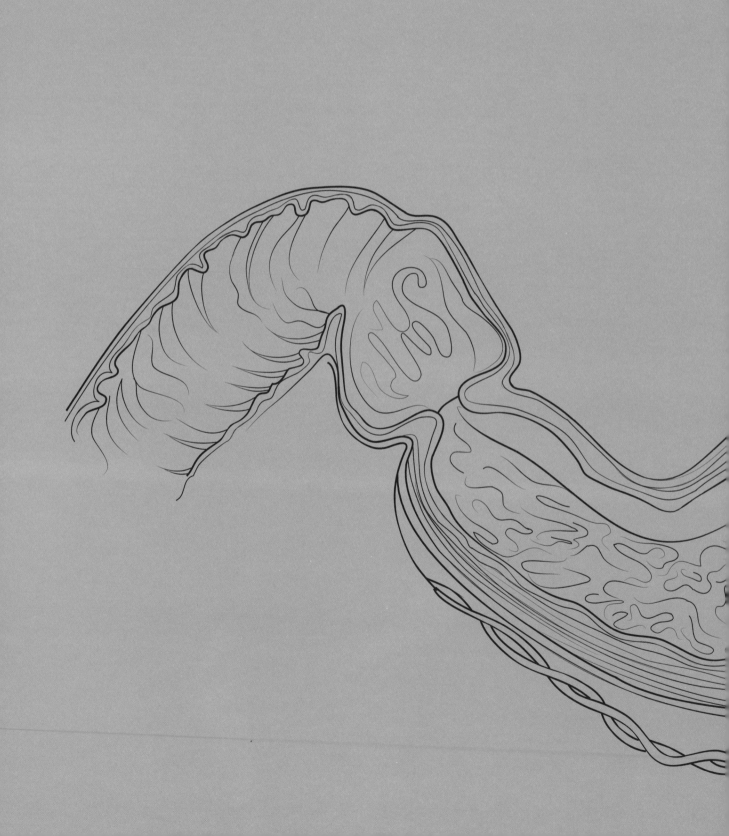

CHAPTER 6:
THE DIGESTIVE SYSTEM

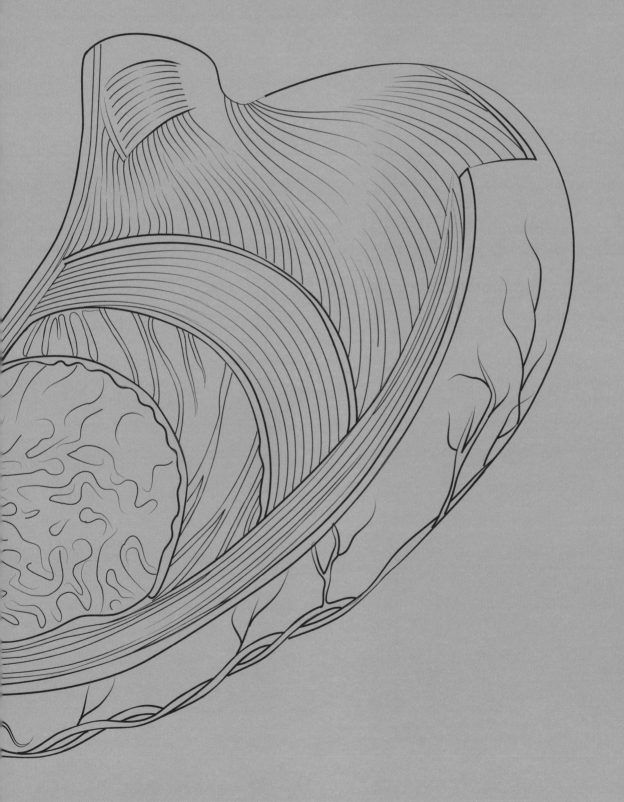

The Digestive System

The digestive system is concerned with the ingestion of food, the mechanical and chemical breakdown of complex molecules in the food, the absorption of simple food chemicals (simple sugars, fats, and amino acids), and the excretion of waste products. It extends from the oral cavity to the anus and can be divided into a gut tube (esophagus, stomach, small intestine, and large intestine) and accessory glandular structures (salivary glands, liver, and pancreas) secreting into the gut tube.

Key terms:

Appendix The vermiform appendix is a blind tube that is 4 inches (10 cm) long and opens into the cecum. Inflammation and obstruction of the venous drainage of the appendix leads to appendicitis.

Ascending colon The part of the large intestine between the cecum and the hepatic (right colic) flexure.

Cecum The blind initial segment of the large intestine. It receives the terminal ileum and the orifice of the appendix and leads on to the ascending colon.

Duodenum A C-shaped part of the small intestine that encircles the head of the pancreas. Four parts are recognized, and the second part receives the bile duct and main and accessory pancreatic ducts.

Esophagus The muscular tube that connects the throat to the stomach.

Gallbladder This saclike structure lies in its fossa on the visceral surface of the liver. It stores and concentrates bile and ejects it through the cystic duct to the bile duct.

Gastroesophageal (cardioesophageal) junction The junction of the abdominal esophagus with the cardia of the stomach. Unlike the pylorus, it does not have a muscular sphincter.

Ileum The last part of the small intestine. Its wall is not as thick or vascular as the jejunum, but it is an important site of absorption of vitamin B12 and has abundant lymphoid tissue (Peyer's patches).

Jejunum The middle part of the small intestine. It is the major site of absorption of nutrients and has abundant vasculature and a highly folded mucosa.

Liver The largest gland of the body. It produces plasma proteins and clotting factors, processes nutrients and toxins from the gut, stores glycogen, and produces bile salts for the emulsification of fat in the gut.

Pancreas A mixed endocrine and exocrine organ. The islets of Langerhans, which make insulin, glucagon, and somatostatin, are the endocrine component. The exocrine component produces proteases, amylase, and lipase for digestion.

Pylorus The part of the stomach that is lined by mucosa containing pyloric glands. It is divided into an antrum to the left and the pyloric canal containing the sphincter to the right.

Rectum The part of the large intestine that is 6 inches (15 cm) long and is between the sigmoid colon and the anal canal. It contains an upper part with transverse rectal folds and a lower dilated part called the ampulla.

Sigmoid colon The S-shaped part of the large intestine running from the pelvic brim to the midpiece of the sacrum. It has a peritoneal attachment called the sigmoid mesocolon.

Spleen A lymphoid organ in the upper left abdomen. It has a diaphragmatic surface in contact with the diaphragm and a visceral surface in contact with the stomach, tail of pancreas, left kidney, and left colic flexure. It recycles old and damaged red blood cells and clears the blood of debris.

Stomach An organ that homogenizes and chemically processes food. It has a thick muscular wall to churn the food, secretes acid to sterilize and chemically digest food, and secretes enzymes to break proteins down into smaller peptides.

Tongue A mobile muscular organ that moves food around the mouth during mastication and is essential for speech.

Transverse colon The segment of large intestine between the right and left colic flexures. It is attached to the back of the greater omentum and has a mesentery called the transverse mesocolon.

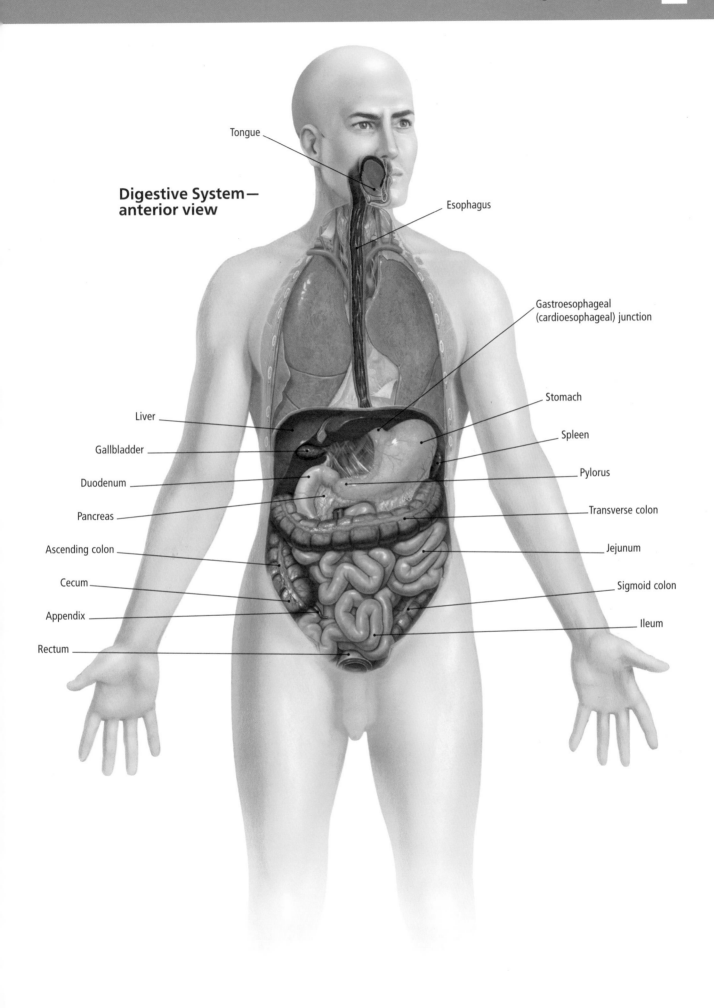

**Digestive System—
anterior view**

Tongue

Esophagus

Gastroesophageal
(cardioesophageal) junction

Stomach

Spleen

Pylorus

Transverse colon

Jejunum

Sigmoid colon

Ileum

Liver

Gallbladder

Duodenum

Pancreas

Ascending colon

Cecum

Appendix

Rectum

True or false?

1   *All major salivary glands open into the floor of the mouth.*

2   *When enlarged, the submandibular gland can be palpated medial and anterior to the angle of the jaw.*

3   *The human adult dentition consists of eight teeth in each quadrant of the mouth.*

4   *The crown of each tooth consists of superficial dentine surrounding a pulp cavity.*

5   *The incisors are chisel-shaped and adapted for cutting food.*

6   *The wisdom teeth (third molars) erupt at about 14 years of age.*

7   *The pulp cavity of the tooth contains nerves and blood vessels.*

8   *Teeth are anchored in the mandible and maxilla by a fibrous joint called a gomphosis.*

9   *The lingual tonsil covers the anterior one-third of the tongue.*

10   *The tongue can be elevated by contraction of the hyoglossus muscle.*

11   *The palatoglossus muscle brings the tongue and soft tissue together during swallowing.*

12   *The muscle fibers of the tongue are arranged in three directions at right angles to each other (vertical, transverse, and longitudinal).*

Salivary gland calculi (sialolithiasis)

Calculi (stones) may form in the major salivary glands when salts precipitate out of solution, in a condition known as sialolithiasis. Calculi are most likely to occur in the duct of the submandibular gland (also called Wharton's duct) and, when present, can be palpated in the floor of the mouth. The usual symptoms are pain and swelling of the affected gland. These get worse when the gland is stimulated because of the thought, sight, or smell of food. Treatment includes hydration, nonsteroidal anti-inflammatory agents, massage, fragmentation of the calculus with ultrasound (lithotripsy), or surgical removal.

13 *Transverse muscle fibers of the tongue will make the tongue narrower but longer and thicker.*

14 *Circumvallate papillae have a shallow moat around a central elevated portion.*

15 *Filiform papillae are shaped like tiny mushrooms.*

16 *The floor of the mouth contains the parotid salivary gland.*

17 *The esophagus may expand during swallowing by compressing the soft posterior wall of the trachea (trachealis muscle).*

18 *The esophagus contains striated (voluntary) muscle down its entire length.*

19 *The esophagus is directly posterior to the left ventricle.*

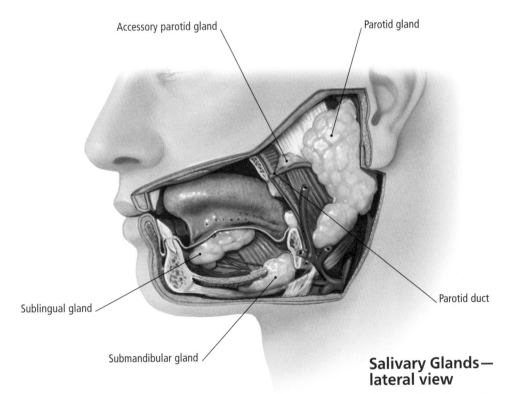

Accessory parotid gland

Parotid gland

Parotid duct

Sublingual gland

Submandibular gland

Salivary Glands— lateral view

The major salivary glands are the parotid glands, which are anterior and inferior to the external ear, and the sublingual and submandibular glands, which are in the floor of the mouth and below the angle of the mandible, respectively.

Multiple choice

1   *Which of the following is true concerning the deciduous teeth?*

(A)   they are usually present at birth
(B)   they include premolars
(C)   molars are the first to erupt
(D)   there are usually 20 in total
(E)   the last molar is called the wisdom tooth

2   *Which type of human tooth is best adapted for crushing food?*

(A)   incisor
(B)   canine
(C)   carnassial
(D)   premolar
(E)   molar

3   *Which type of human tooth may have three roots?*

(A)   maxillary molars
(B)   first premolars
(C)   canines
(D)   lateral incisors
(E)   deciduous molars

4   *Which is the hardest part of a human tooth?*

(A)   dentine
(B)   pseudodentine
(C)   enamel
(D)   pulp
(E)   periodontal ligament

5   *Which of the following is not a function of saliva?*

(A)   provide a fluid phase for tastant molecules
(B)   assist in the formation of a bolus for swallowing
(C)   lubricate the mouth for speech
(D)   initiate digestion of proteins
(E)   assist in forming a tight seal around straws

6   *Which of the following opens on the floor of the mouth at the sublingual papilla?*

(A)   submandibular salivary gland duct
(B)   parotid salivary gland duct
(C)   sublingual salivary gland duct
(D)   nasolacrimal duct
(E)   palatine tonsillar duct

7   *Which statement is true concerning the sublingual gland?*

(A)   it lies anterior to the external ear
(B)   it usually has only a single duct
(C)   it lies deep to the sublingual fold
(D)   it has a duct opening opposite the second upper molar
(E)   it is supplied by the hypoglossal nerve

8   *Which statement is true concerning the parotid gland?*

(A)   it is traversed by the facial nerve
(B)   it is located medial to the angle of the mandible
(C)   the internal carotid artery is embedded within it
(D)   its duct pierces the masseter muscle
(E)   it may be enlarged in chickenpox

9   *Which statement is true concerning the submandibular gland?*

(A)   it has multiple ducts opening at the cheek
(B)   it has superficial and deep parts
(C)   it is divided by the genioglossus muscle
(D)   its duct opens near the second upper molar
(E)   it is purely mucus-secreting

10 Which of the following muscles protrudes the tongue?

(A) hyoglossus

(B) mylohyoid

(C) styloglossus

(D) palatoglossus

(E) genioglossus

11 Which intrinsic muscle fibers allow the tongue tip to touch the roof of the mouth?

(A) transverse intrinsic

(B) hyoglossus

(C) styloglossus

(D) superior longitudinal

(E) inferior longitudinal

12 Which structure is found between the palatoglossal and palatopharyngeal arches?

(A) lingual tonsil

(B) palatine tonsil

(C) nasopharyngeal tonsil

(D) sublingual salivary gland

(E) epiglottis

13 What is the name of the space between the tongue and the epiglottis?

(A) vallecula

(B) lingual fossa

(C) tonsillar fossa

(D) laryngeal inlet

(E) Eustachian tube

14 Which type of lingual papilla is located immediately anterior to the sulcus terminalis?

(A) foliate

(B) circumvallate

(C) fungiform

(D) filiform

(E) palatine

15 What is the embryological significance of the foramen cecum of the tongue?

(A) it is the site of origin of the parotid gland

(B) it is the site of origin of the cervical sinus

(C) it is the site of origin of the thyroid gland

(D) it marks the position of the vomeronasal organ

(E) it is the site where all taste receptors converge

16 During swallowing, which of the following protects the airway?

(A) uvula

(B) nasopharyngeal fold

(C) sublingual fold

(D) palatoglossal fold

(E) epiglottis

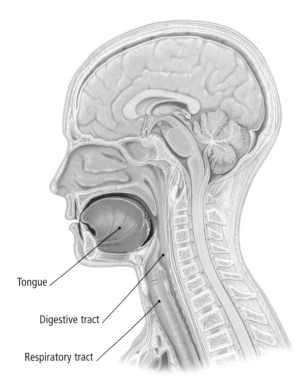

Tongue

Digestive tract

Respiratory tract

Mouth—sagittal view

The mouth has the soft and hard palates as its roof, the tongue in its floor, and extends from the oral fissure (lips) anteriorly to the oropharynx posteriorly.

Fill in the blanks

1   The _____ teeth are adapted for gripping food.

2   The roots of the teeth are anchored in the maxilla or mandible by _____.

3   The _____ teeth have two cusps.

4   The teeth of the upper jaw are supplied by the _____ division of the trigeminal nerve.

5   The teeth of the lower jaw are supplied by the _____ division of the trigeminal nerve.

6   The _____ parotid gland often lies along the superior border of the parotid duct.

7   The _____ muscle divides the submandibular gland into superficial and deep portions.

8   The _____ fibers of the _____ nerve stimulate copious secretion of enzyme-rich fluid by the submandibular and sublingual salivary glands.

9   The _____ separates the anterior two-thirds and posterior one-third of the tongue.

10 The _____ muscle retracts the tongue.

11 The _____ arch joins the soft palate and the tongue.

12 The palatine tonsillar fossa lies between the _____ and _____ arches.

13 The three muscles that compress the contents of the pharynx are called the _____.

14 The _____ nerve supplies the muscle of the esophagus.

Labels: Central incisor, Lateral incisor, Palatopharyngeus arch, Palatine tonsil, Palatoglossus arch, Posterior wall of pharynx, Median sulcus, Soft palate, Canine, Uvula, Wisdom tooth, Second molar, First molar, Second premolar, First premolar

Mouth—anterior view

The mouth consists of a vestibule outside the dental arcades and a true oral cavity within the teeth. The soft palate has a uvula that hangs from its midline.

Match the statement to the reason

1 Dental caries may lead to abscesses in the maxilla or mandible because...

2 An abscess in the palatine tonsil can cause obstruction of the back of the mouth because...

3 Loss of teeth can lead to erosion of the alveolar margin of the mandible because...

4 A carcinoma in the parotid gland can cause facial muscle paralysis because...

5 Paralysis of the tongue on one side causes deviation of the protruded tongue to the affected side because...

6 Saliva contains the enzyme amylase because...

7 A stone in the parotid duct can be palpated in the cheek because...

a the oropharyngeal isthmus is a relatively narrow part of the oral cavity.

b absence of the tooth crown leads to increased pressure on the thin margin of the alveolar bone.

c digestion and absorption of starch carbohydrate is optimal if some breakdown begins in the mouth.

d the pulp cavity of each tooth communicates with the alveolar canals.

e the duct pierces the buccinator muscle to open opposite the crown of the second upper molar.

f the genioglossus of the normal side pushes the tongue across the midline.

g the facial nerve is vulnerable as it passes through the parotid gland.

Color and label

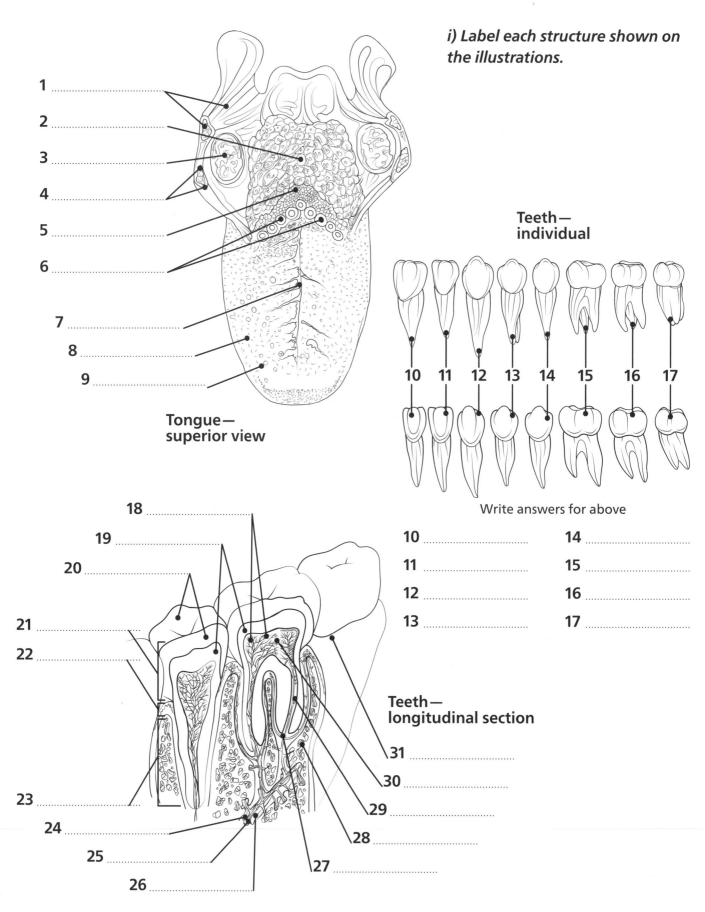

i) Label each structure shown on the illustrations.

1 ....

2 ....

3 ....

4 ....

5 ....

6 ....

7 ....

8 ....

9 ....

**Tongue—
superior view**

**Teeth—
individual**

10   11   12   13   14   15   16   17

Write answers for above

10 ....   14 ....

11 ....   15 ....

12 ....   16 ....

13 ....   17 ....

18 ....

19 ....

20 ....

21 ....

22 ....

23 ....

24 ....

25 ....

26 ....

27 ....

**Teeth—
longitudinal section**

31 ....

30 ....

29 ....

28 ....

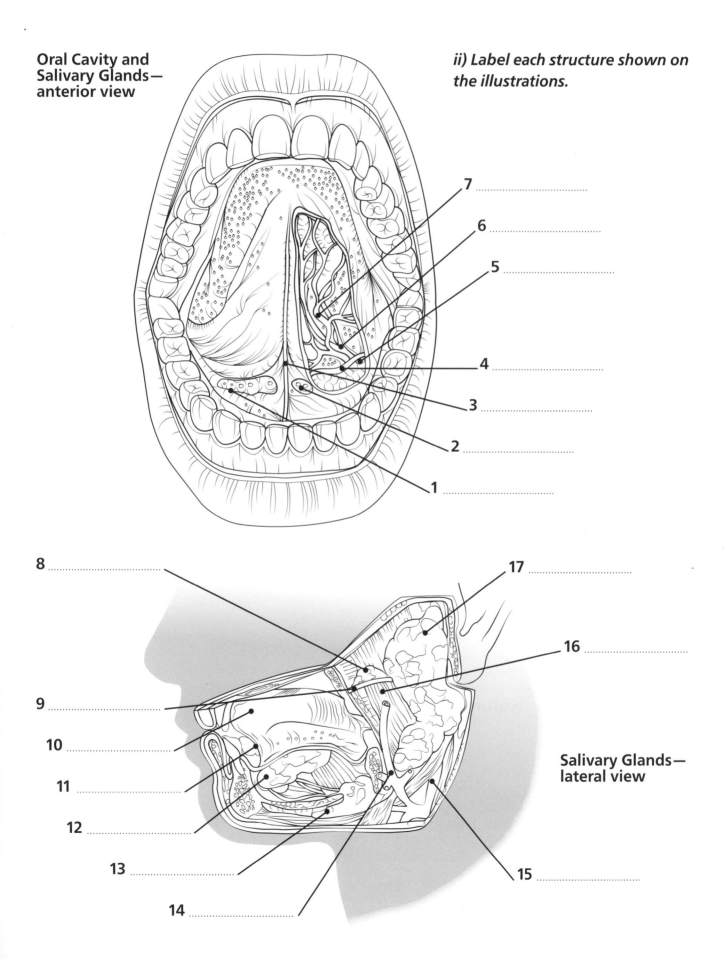

Oral Cavity and Salivary Glands— anterior view

ii) Label each structure shown on the illustrations.

7 ....

6 ....

5 ....

4 ....

3 ....

2 ....

1 ....

8 ....

17 ....

16 ....

9 ....

10 ....

11 ....

Salivary Glands— lateral view

12 ....

13 ....

15 ....

14 ....

Abdominal Organs

The bulk of the digestive system is located within the abdominal cavity. Components of the gut tube (stomach and small and large intestines) are mostly suspended from the abdominal wall by double-layered sheets of tissue, generically called mesenteries. Blood vessels, nerves, and lymphatic vessels run to the gut tube between the layers of the mesenteries. The liver and gallbladder lie beneath the right dome of the diaphragm, while a lymphoid organ (the spleen) lies beneath the left dome.

Key terms:

Abdominal organs Abdominal organs may either be intraperitoneal (i.e., supported by mesenteries, e.g., stomach or small intestine) or retroperitoneal (i.e., behind the peritoneal cavity, e.g., kidney, adrenal gland, or aorta).

Accessory pancreatic duct A small duct of the exocrine pancreas that drains juices from the head of the pancreas to the lesser duodenal papilla.

Anus The terminal part of the gastrointestinal tract. The anal canal is 1¼ inches (3 cm) long and runs from the upper pelvic diaphragm to the anal verge. It is surrounded by the internal and external anal sphincters, and its interior is marked by vertical folds of mucosa called anal columns.

Appendix See pp. 186–187.

Ascending colon See pp. 186–187.

Body of pancreas The part of the pancreas lying in front of the aorta.

Cecum See pp. 186–187.

Common bile duct The duct carrying bile from the liver and gallbladder to the second part of the duodenum. It is formed from the junction of the common hepatic duct and cystic duct, and it runs through the head of the pancreas to open with the main pancreatic duct at the greater duodenal papilla.

Common hepatic duct A duct carrying bile from the liver. It is formed from the junction of two hepatic ducts and joins with the cystic duct to form the bile duct.

Cystic duct The duct carrying bile between the neck of the gallbladder and the bile duct. It has spiral folds in the mucosa.

Diaphragm A sheet of muscle and tendon separating the thoracic and abdominal cavities. It is perforated by the aorta, inferior vena cava, and esophagus.

Duodenojejunal junction The junction of the terminal (fourth part of the) duodenum and the beginning of the jejunum. There may be a flexure produced by the attachment of the junction to the posterior abdominal wall by the suspensory ligament of Treitz.

Gallbladder See pp. 186–187.

Greater duodenal papilla The small protrusion of the mucosa of the medial wall of the second part of the duodenum, where the hepatopancreatic ampulla (and hence bile duct and main pancreatic duct) opens.

Head of pancreas The part of the pancreas encircled by the duodenum. It has the main and accessory pancreatic ducts passing through it, as well as the bile duct.

Hepatic (right colic) flexure The junction between the ascending and transverse colon. The flexure is in contact with the visceral surface of the liver.

Ileum See pp. 186–187.

Jejunum See pp. 186–187.

Left lobe of liver The left lobe of the liver is that part to the left of the falciform ligament and the fissures for the ligamentum venosum and teres. The left lobe is anterior to the stomach.

Main pancreatic duct The major duct draining the tail, body, neck, and head of the pancreas. It is often joined by the bile duct to form a hepatopancreatic ampulla that opens into the second part of the duodenum.

Portal vein The portal vein drains venous blood from the gut into the liver for processing. It is usually formed behind the neck of the pancreas from the junction of the splenic and superior mesenteric veins.

Pyloric sphincter A smooth muscle sphincter encircling the outlet from the stomach to the first part of the duodenum. It controls the movement of chyme to the small intestine.

Pylorus See pp. 186–187.

Rectum See pp. 186–187.

Right lobe of liver The right lobe of the liver is all parts to the right of the falciform ligament. It lies under the right dome of the diaphragm and has the right kidney, right adrenal gland, gallbladder, and right colic flexure in contact with it.

Sigmoid colon See pp. 186–187.

Spleen See pp. 186–187.

Stomach See pp. 186–187.

Transverse colon See pp. 186–187.

Abdominal Organs— anterior view

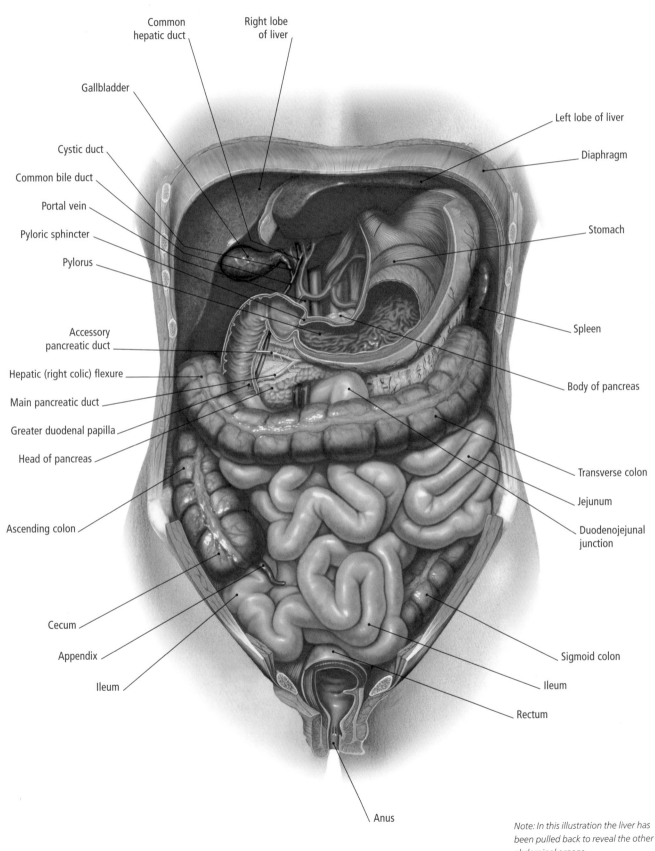

Common hepatic duct

Right lobe of liver

Gallbladder

Left lobe of liver

Cystic duct

Diaphragm

Common bile duct

Portal vein

Stomach

Pyloric sphincter

Pylorus

Accessory pancreatic duct

Spleen

Hepatic (right colic) flexure

Body of pancreas

Main pancreatic duct

Greater duodenal papilla

Head of pancreas

Transverse colon

Jejunum

Ascending colon

Duodenojejunal junction

Cecum

Appendix

Sigmoid colon

Ileum

Ileum

Rectum

Anus

Note: In this illustration the liver has been pulled back to reveal the other abdominal organs.

True or false?

1   The stomach fundus lies under the right dome of the diaphragm.

2   The arteries supplying the stomach lie along its omental borders.

3   The left kidney is in close contact with the body of the stomach.

4   Parietal or oxyntic cells produce stomach acid.

5   Zymogenic or chief cells make the enzyme chymotrypsin.

6   The pyloric antrum is separated from the body by the angular incisure.

7   Arterial supply to the stomach is mainly from the superior mesenteric artery.

8   The greater omentum attaches along the lesser curvature of the stomach.

9   The pyloric sphincter is composed of striated muscle in the muscularis externa layer.

10   The initial section of the first part of the duodenum has a smooth mucosal surface.

11   The bile duct opens into the third part of the duodenum.

12   The main pancreatic duct opens at the summit of the greater or major duodenal papilla.

13   The junction of the fourth part of the duodenum with the jejunum is anchored to the posterior abdominal wall by the suspensory ligament of Treitz.

14   Microvilli are features of enterocytes that increase the surface area for absorption.

15   Teniae coli are bands of muscle from the inner circular part of the muscularis externa.

16   *Appendices epiploicae are most commonly seen on the ileum and cecum.*

17   *The sigmoid colon is attached to the posterior abdominal wall by the sigmoid mesocolon.*

18   *The rectal ampulla is a capacious region that holds feces immediately before defecation.*

19   *Internal hemorrhoids are derived from dilated veins in the rectal ampulla.*

20   *The epithelium at the pecten of the anal canal is nonkeratinized stratified squamous epithelium.*

Hiatus hernia

A hernia is the abnormal protrusion of an organ through a body opening. In the case of hiatus hernia, the body opening is the esophageal hiatus in the diaphragm, and either the stomach cardia or fundus protrudes into the chest. Hiatus hernia is caused by weakness in the muscular fibers encircling the esophageal opening and excessive pressure in the abdominal cavity (often due to obesity). Symptoms of hiatus hernia include the taste of acid, heartburn (burning retrosternal pain), and trouble swallowing. It is usually treated by weight reduction, avoidance of spicy foods, eating small meals frequently, oral antacids, and, in some cases, surgery.

Stomach—internal structure

The stomach is a muscular bag for churning food and mixing it with stomach juices, which are acidic and also contain the proteolytic (protein-dissolving) enzyme pepsin.

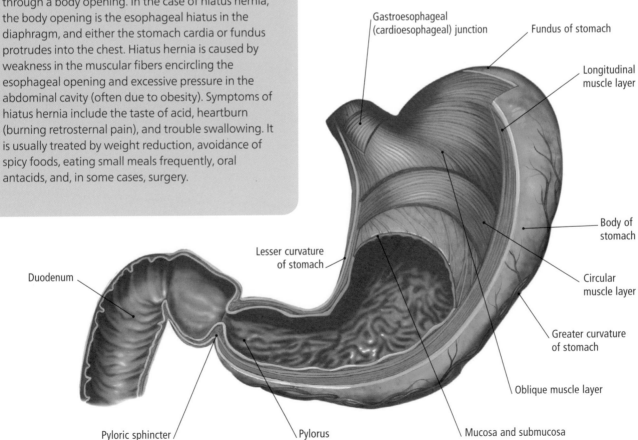

Gastroesophageal (cardioesophageal) junction

Fundus of stomach

Longitudinal muscle layer

Body of stomach

Circular muscle layer

Greater curvature of stomach

Oblique muscle layer

Mucosa and submucosa

Lesser curvature of stomach

Duodenum

Pyloric sphincter

Pylorus

Color and label

*i) Label each structure shown
on the illustrations.*

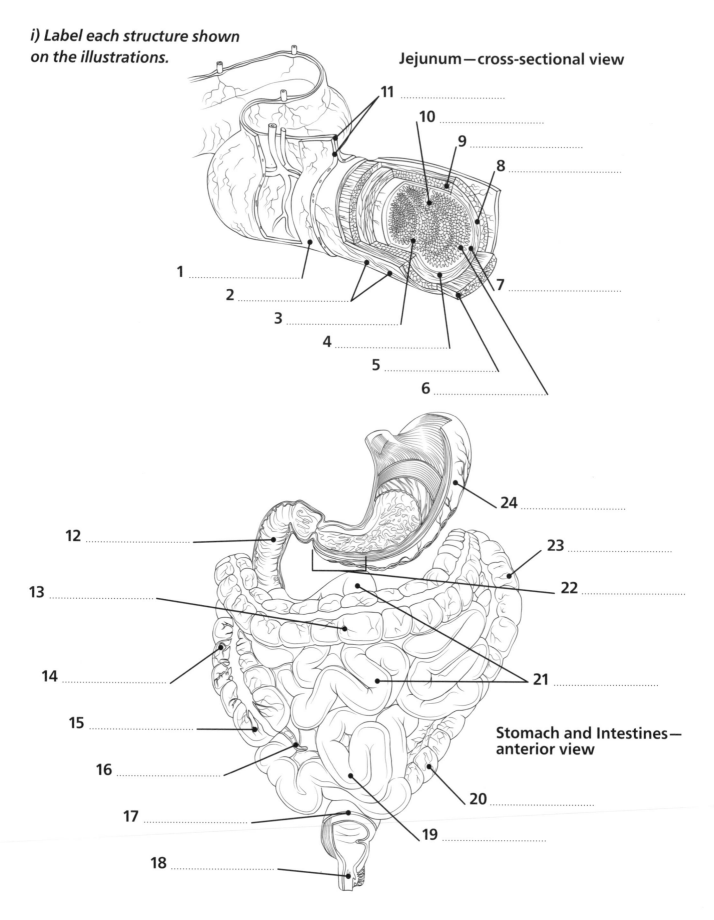

Jejunum—cross-sectional view

11 ....

10 ....

9 ....

8 ....

1 ....

2 ....

3 ....

4 ....

5 ....

6 ....

7 ....

24 ....

23 ....

22 ....

12 ....

13 ....

14 ....

15 ....

16 ....

17 ....

18 ....

21 ....

20 ....

19 ....

**Stomach and Intestines—
anterior view**

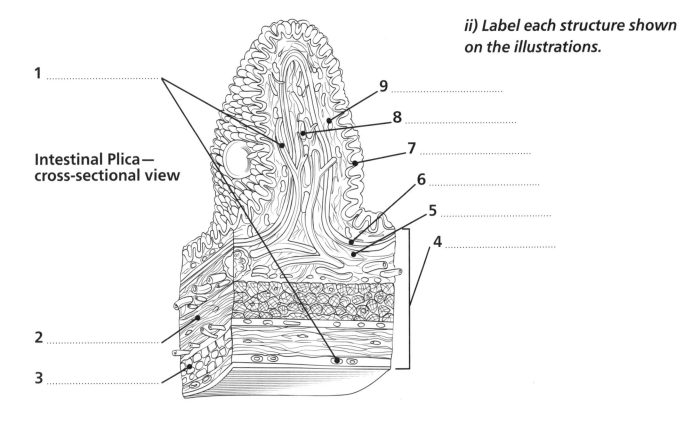

ii) Label each structure shown on the illustrations.

1 ....

**Intestinal Plica—
cross-sectional view**

2 ....

3 ....

9 ....

8 ....

7 ....

6 ....

5 ....

4 ....

Anus—coronal view

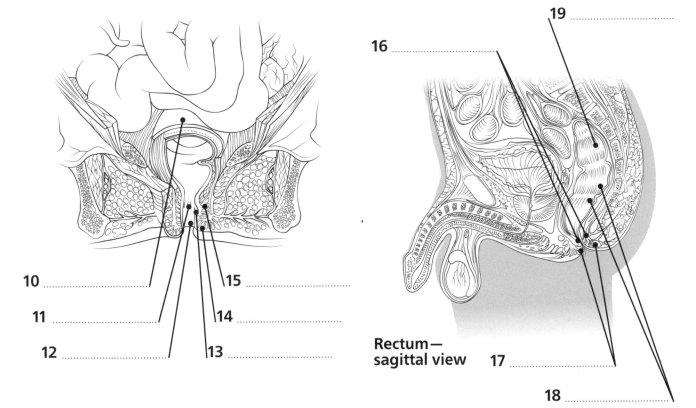

10 ....

11 ....

12 ....

15 ....

14 ....

13 ....

16 ....

19 ....

**Rectum—
sagittal view**

17 ....

18 ....

Multiple choice

1 Which structure attaches the stomach to the liver?
- (A) lesser omentum
- (B) greater omentum
- (C) the mesentery
- (D) falciform ligament
- (E) tenia mesocolica

2 Where does gas usually collect in the stomach?
- (A) cardia
- (B) pylorus
- (C) body
- (D) fundus
- (E) antrum

3 Which of the following arteries supplies the greater curvature of the stomach?
- (A) left gastric artery
- (B) right gastric artery
- (C) pyloric artery
- (D) left gastroepiploic artery
- (E) hepatic artery

4 Which of the following does not lie immediately posterior to the stomach?
- (A) left kidney
- (B) left suprarenal gland
- (C) splenic artery
- (D) pancreas
- (E) abdominal esophagus

5 A peptic ulcer eroding the posterior wall of the stomach could cause bleeding from the
- (A) splenic artery
- (B) portal vein
- (C) left gastric artery
- (D) gastroduodenal artery
- (E) superior mesenteric artery

6 Which part of the duodenum is attached to the liver by the hepatoduodenal ligament?
- (A) first
- (B) second
- (C) third
- (D) fourth
- (E) fifth

7 Which part of the duodenum is most directly anterior to the right kidney?
- (A) first
- (B) second
- (C) third
- (D) fourth
- (E) fifth

8 Which organ encircles the head of the pancreas?
- (A) jejunum
- (B) ileum
- (C) duodenum
- (D) stomach
- (E) gallbladder

9 Which of the following is a naked eye feature that increases the surface area of the small intestinal mucosa?
- (A) rugae
- (B) plicae circulares
- (C) microvilli
- (D) trabeculae
- (E) septa

10 Which of the following does not assist absorption from the jejunum?
- (A) villi
- (B) plicae circulares
- (C) microvilli
- (D) Peyer's patches
- (E) arterial vascular arcades

11 *What is the main function of the large intestine?*

Ⓐ absorb sugars

Ⓑ absorb water

Ⓒ excrete bile

Ⓓ absorb minerals

Ⓔ both B and D are correct

12 *Where does the junction of the sigmoid colon and rectum occur?*

Ⓐ at the third lumbar vertebra

Ⓑ at the iliac crest

Ⓒ at the arcuate line of the ilium

Ⓓ at the third sacral vertebra

Ⓔ at the tip of the coccyx

13 *Which of the following is continuous with the inner circular muscle layer of the anorectal canal?*

Ⓐ external anal sphincter

Ⓑ internal anal sphincter

Ⓒ conjoint longitudinal layer

Ⓓ internal hemorrhoidal plexus

Ⓔ external hemorrhoidal plexus

14 *What type of epithelium would be found at the pecten of the anal canal?*

Ⓐ simple columnar epithelium with goblet cells

Ⓑ pseudostratified columnar epithelium with cilia

Ⓒ stratified squamous epithelium, keratinized

Ⓓ stratified squamous epithelium, nonkeratinized

Ⓔ simple columnar epithelium, nonkeratinized

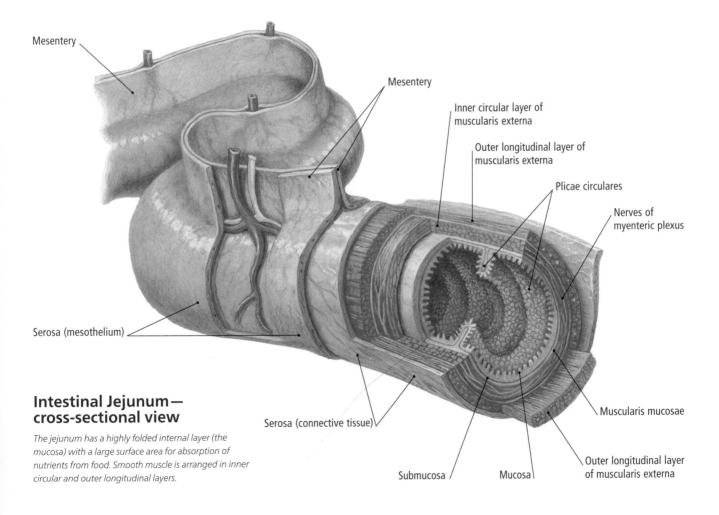

Mesentery

Mesentery

Inner circular layer of muscularis externa

Outer longitudinal layer of muscularis externa

Plicae circulares

Nerves of myenteric plexus

Serosa (mesothelium)

Muscularis mucosae

Outer longitudinal layer of muscularis externa

Serosa (connective tissue)

Submucosa

Mucosa

Intestinal Jejunum— cross-sectional view

The jejunum has a highly folded internal layer (the mucosa) with a large surface area for absorption of nutrients from food. Smooth muscle is arranged in inner circular and outer longitudinal layers.

Fill in the blanks

1   The _____ is the initial part of the stomach.

2   The _____ separates the cardia and fundus of the stomach.

3   When the stomach is contracted or small, the mucosa is thrown into folds called _____.

4   The _____ ligament of the _____ attaches the lesser curvature of the stomach to the liver.

5   Parietal cells of the stomach mucosa secrete _____ that binds with vitamin B12 to facilitate absorption of the vitamin in the small intestine.

6   The muscularis externa layer of the stomach has three layers of smooth muscle: _____, _____, and _____.

7   Endocrine G cells in the pyloric mucosa secrete _____, a peptide hormone that stimulates acid secretion by the rest of the stomach.

8   The epithelium of the stomach is _____ in type.

9   The initial part of the duodenum is called the _____ because of its smooth mucosal surface.

10   The _____ cells of the small intestine are responsible for the production of mucus.

11   The hormone _____ is produced by the mucosa of the duodenum and jejunum and controls the production of bicarbonate ions by the exocrine pancreas.

12   The hormone _____ is produced by the duodenum and jejunum and stimulates contraction of the gallbladder and enzyme secretion by the exocrine pancreas.

13   The lymphatic nodules found in the mucosa and submucosa of the small intestine are called _____.

14   There are abundant lymphatic channels (central lacteals) in the cores of villi of the small intestine, which are important in _____ .

15   GALT (short for _____) is the special lymphoid tissue that protects the intestinal wall against invasion by microorganisms.

16   The _____ is attached to the medial wall of the cecum.

17   The interior of the cecum has two openings on its medial wall: the _____ and _____ orifices.

18   The transverse colon is suspended from the posterior abdominal wall by the _____, which attaches along the anterior border of the _____ .

19   The sigmoid mesocolon attaches over the bifurcation of the left _____ .

20   The upper part of the rectum is traversed by three _____ .

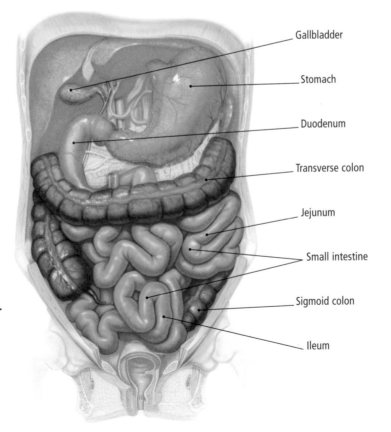

Stomach and Intestines— anterior view

The stomach begins the process of digestion of proteins. Digestion continues, and is accompanied by absorption, in the small intestine. The large intestine is the main site for absorption of water and minerals, and the formation of the stool.

Gallbladder

Stomach

Duodenum

Transverse colon

Jejunum

Small intestine

Sigmoid colon

Ileum

Match the statement to the reason

1   Removal of large parts of the stomach can cause pernicious anemia because…

a   it is a highly mobile apron of tissue suspended from the greater curvature of the stomach.

2   Peptic ulcers of the posterior stomach wall may involve the splenic artery because…

b   it runs along the superior border of the pancreas.

3   Peptic ulcers are most common in the first part of the duodenum because…

c   the pyloric smooth muscle controls the passage of stomach contents to the duodenum.

4   Hypertrophy of the pyloric sphincter may cause projectile vomiting in infants because…

d   this region is proximal to the point of release of bicarbonate from the pancreatic ducts.

5   The greater omentum may wall off infection in the lower two-thirds of the abdominal cavity because…

e   cells of the stomach mucosa produce a factor that aids absorption of vitamin B12.

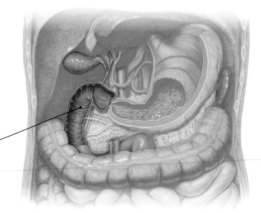

Duodenum

Duodenum, Ileum, and Jejunum

The duodenum, jejunum, and ileum are the three parts of the small intestine. The jejunum and ileum are suspended from the posterior abdominal wall by a peritoneal fold called the mesentery.

Duodenum

1 The interior of the jejunum looks very fluffy in a barium meal because…

a the mucosa is highly folded and contains many villi.

2 The vermiform appendix can be located by tracing the tenia coli of the cecum because…

b the organ is derived from the embryonic midgut, which refers sensation to that region.

3 Appendicitis usually starts with colicky periumbilical pain because…

c these muscle bands converge to the point of attachment of that organ.

4 Twisting of the sigmoid colon (sigmoid volvulus) can occur in some racial groups because…

d an anastomotic vessel (the marginal artery) connects the arteries.

5 Obstruction of one colic artery doesn't necessarily cause gangrene of the bowel because…

e they have an unusually long sigmoid mesocolon.

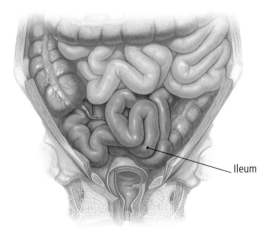

Ileum

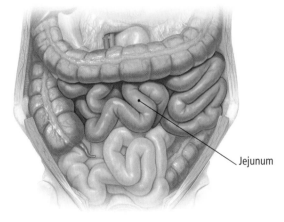

Jejunum

Liver, Gallbladder, and Pancreas

The liver is the largest gland in the body and processes all the nutrients from the gut. It also makes bile salts, plasma proteins, and clotting factors and stores carbohydrates (in glycogen). The gallbladder stores and concentrates the bile made by the liver, then releases it into the duodenum to emulsify fats when a fatty meal is consumed. The pancreas is a mixed exocrine (secreting to the gut) and endocrine (secreting to the bloodstream) gland.

Key terms:

Accessory pancreatic duct A small duct of the exocrine pancreas that drains juices from the head of the pancreas to the lesser duodenal papilla.

Bile ducts All the ducts that carry bile, including hepatic ducts.

Body of gallbladder The largest part of the gallbladder. It has smooth muscle in its wall to contract and release bile after a fatty meal.

Body of pancreas The part of the pancreas lying in front of the aorta.

Central vein The vein in the center of a liver lobule. It receives blood from the liver sinusoids and drains into one of the hepatic veins.

Common hepatic duct See pp. 196–197.

Coronary ligament A peritoneal reflection from the surface of the liver to the diaphragm. The line of the coronary ligament surrounds the bare area of the liver, a region that has no peritoneal covering.

Cystic artery A branch of the right hepatic artery that supplies the gallbladder and cystic duct.

Cystic duct The duct carrying bile between the neck of the gallbladder and the bile duct. It has spiral folds in the mucosa.

Falciform ligament A triangular membrane joining the anterior surface of the liver to the anterior abdominal wall. Its lower edge encircles the ligamentum teres (remnant of fetal left umbilical vein).

Fundus of gallbladder The lower part of the gallbladder. It protrudes below the inferior border of the liver.

Gallbladder See pp. 186–187.

Head of pancreas See pp. 196–197.

Hepatic artery (branch) A branch of the celiac trunk from the aorta. The hepatic artery proper supplies the liver and gallbladder.

Hepatic ducts Ducts carrying bile from each functional half of the liver.

Hepatocyte The basic cell type of the liver. Hepatocytes produce bile, store glycogen, make plasma proteins and clotting factors, and process nutrients and toxins from the gut.

Inferior border of liver The sharp inferior border may be felt on the anterior abdominal wall when the liver is enlarged.

Interlobular bile duct A bile duct running between liver lobules in the company of portal vein and hepatic artery branches (forming the portal triad).

Left hepatic duct A duct carrying bile from the left functional half of the liver.

Left lobe of liver See pp. 196–197.

Ligamentum teres A fibrous remnant of the left umbilical vein of prenatal life. It runs from the liver to the umbilicus.

Liver The largest gland of the body. It produces plasma proteins and clotting factors, processes nutrients and toxins from the gut, stores glycogen, and produces bile salts for the emulsification of fat in the gut.

Liver lobule A region of the liver. Each lobule has a central vein at its core, whereas the portal triads (bile duct, hepatic artery, and portal vein branches) are located at each corner.

Liver plate A sheet of hepatocytes within a liver lobule.

Main pancreatic duct The major duct draining the tail, body, neck, and head of the pancreas. It is often joined by the bile duct to form a hepatopancreatic ampulla that opens into the second part of the duodenum.

Neck of gallbladder The part of the gallbladder from which the cystic duct arises.

Neck of pancreas The narrow part of the pancreas between the head and body.

Pancreas See pp. 186–187.

Portal vein (branch) See pp. 196–197.

Right hepatic duct A duct carrying bile from the right functional half of the liver.

Right lobe of liver See pp. 196–197.

Sinusoid A fenestrated capillary between liver plates. The space around the sinusoid (space of Disse) enables the transfer of proteins between hepatocytes and the blood. Macrophages are found in the sinusoid wall.

Tail of pancreas The part of the pancreas that projects into the lienorenal ligament and is in contact with the hilum of the spleen.

Uncinate process of pancreas The projection of the head of the pancreas to the left, forming a pancreatic notch through which the superior mesenteric artery and vein pass.

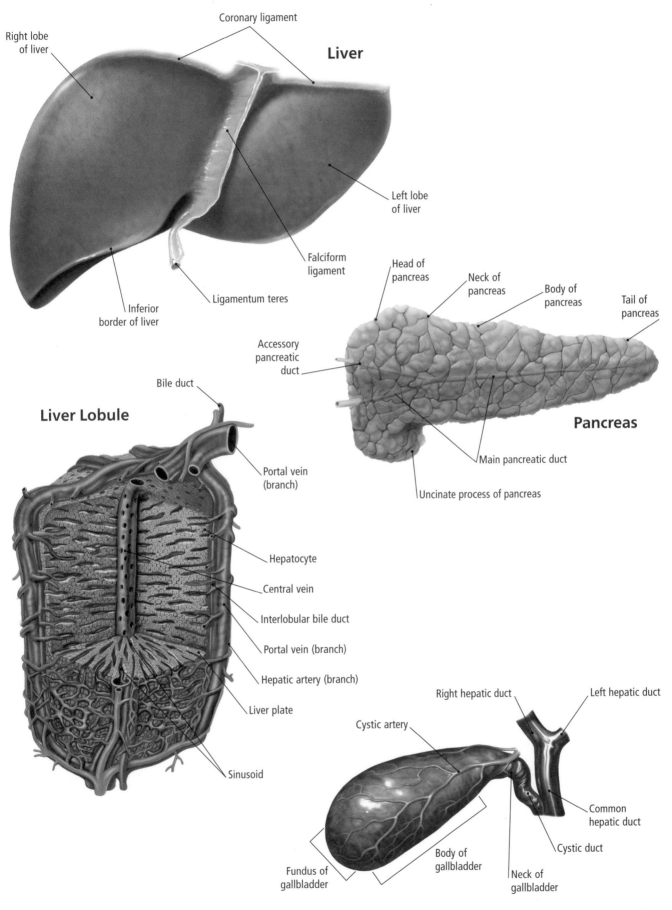

Liver

Right lobe of liver

Coronary ligament

Left lobe of liver

Falciform ligament

Ligamentum teres

Inferior border of liver

Pancreas

Head of pancreas

Neck of pancreas

Body of pancreas

Tail of pancreas

Accessory pancreatic duct

Main pancreatic duct

Uncinate process of pancreas

Liver Lobule

Bile duct

Portal vein (branch)

Hepatocyte

Central vein

Interlobular bile duct

Portal vein (branch)

Hepatic artery (branch)

Liver plate

Sinusoid

Gallbladder

Right hepatic duct

Left hepatic duct

Cystic artery

Common hepatic duct

Cystic duct

Fundus of gallbladder

Body of gallbladder

Neck of gallbladder

True or false?

1   One of the main roles of the liver is to store protein as glycogen.

2   The left and right hepatic ducts join to form the bile duct near the porta hepatis.

3   The portal vein is the most posterior structure in the porta hepatis.

4   The hepatic veins enter the inferior vena cava inferior to the liver.

5   The gallbladder is in contact with the visceral (inferior) surface of the liver.

6   The neck of the gallbladder may be dilated as Hartmann's pouch.

7   The spiral valve is formed by elevation of the cystic duct mucosa.

8   The bile duct is accompanied by the splenic artery.

9   The superior mesenteric vessels pass through the pancreatic notch.

10   The pancreas contains exocrine glandular tissue in the islets of Langerhans.

Carcinoma of the pancreas

Carcinoma (cancer) of the pancreas has one of the poorest life expectancies of any carcinoma. This is because the disease has often spread to distant sites (e.g., lymph nodes, liver) before the patient is aware that he or she has the disease. Cancers in the head of the pancreas may be detected early if they cause blockage of the bile duct, leading to obstructive jaundice, but tumors in the pancreas body or tail have usually spread to nerves and lymph vessels behind the pancreas before the patient is aware of the disease. Symptoms of pancreatic carcinoma include upper abdominal (epigastric) pain that bores through to the back and is often intense.

Multiple choice

1 *Which ligament attaches the liver to the anterior abdominal wall?*
(A) falciform ligament
(B) coronary ligament
(C) ligamentum venosum
(D) gastrohepatic ligament
(E) hepatoduodenal ligament

2 *Which of the following does not approach the liver at the porta hepatis?*
(A) bile duct
(B) hepatic artery proper
(C) portal vein
(D) lymphatics
(E) gastroduodenal artery

3 *Which of the following is not in direct contact with the liver?*
(A) spleen
(B) right kidney
(C) transverse colon
(D) first part of the duodenum
(E) right suprarenal (adrenal) gland

4 *Which of the following hepatic ligaments is a remnant of the left umbilical vein?*
(A) ligamentum venosum
(B) hepatoduodenal ligament
(C) falciform ligament
(D) coronary ligament
(E) ligamentum teres

5 *Which vessel directly gives rise to the cystic artery to the gallbladder?*
(A) splenic artery
(B) left hepatic artery
(C) right hepatic artery
(D) right gastric artery
(E) celiac artery

6 *Which of the following passes through the head of the pancreas?*
(A) superior mesenteric vein
(B) superior mesenteric artery
(C) portal vein
(D) bile duct
(E) both B and D are correct

7 *Where does the accessory pancreatic duct open into the duodenum?*
(A) greater (major) duodenal papilla
(B) lesser (minor) duodenal papilla
(C) pylorus
(D) duodenojejunal flexure
(E) plica fimbriata

Liver—microstructure

The liver is composed of multiple repeating structural units (hepatic lobules) in the shape of hexagonal prisms. The center of each lobule contains a tributary of the hepatic veins (central vein). Branches of the portal vein, hepatic arteries, and bile ducts lie at the corners.

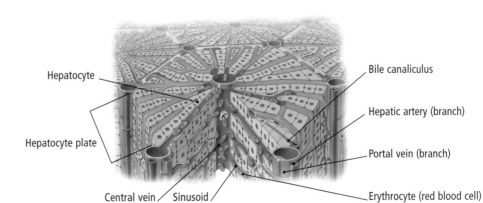

Hepatocyte
Hepatocyte plate
Central vein
Sinusoid
Bile canaliculus
Hepatic artery (branch)
Portal vein (branch)
Erythrocyte (red blood cell)

Color and label

*i) Label each structure shown on
the illustrations.*

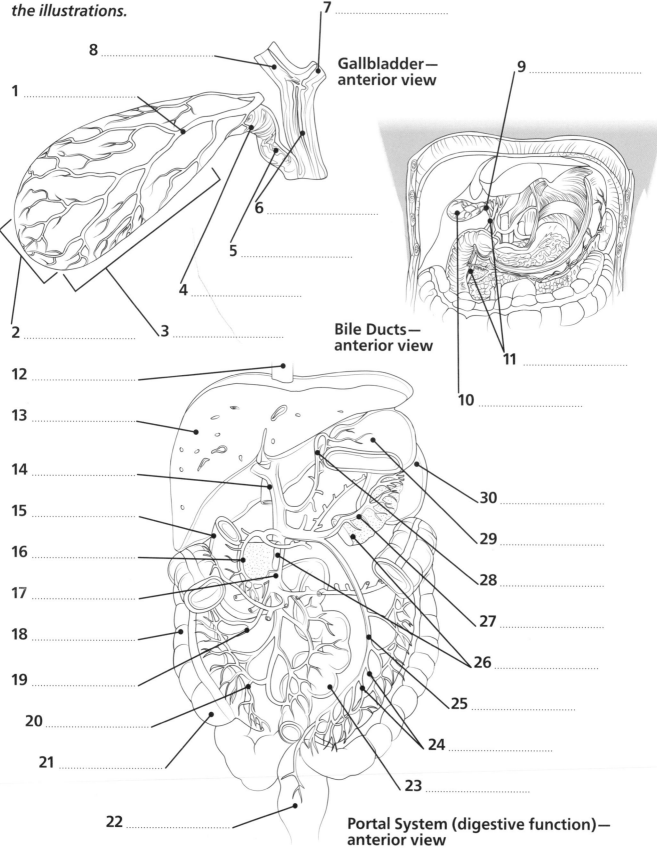

7 ....

8 ....

1 ....

**Gallbladder—
anterior view**

9 ....

6 ....

5 ....

4 ....

2 ....

3 ....

**Bile Ducts—
anterior view**

11 ....

10 ....

12 ....

13 ....

14 ....

15 ....

16 ....

17 ....

18 ....

19 ....

20 ....

21 ....

22 ....

30 ....

29 ....

28 ....

27 ....

26 ....

25 ....

24 ....

23 ....

**Portal System (digestive function)—
anterior view**

ii) Label each structure shown on the illustrations.

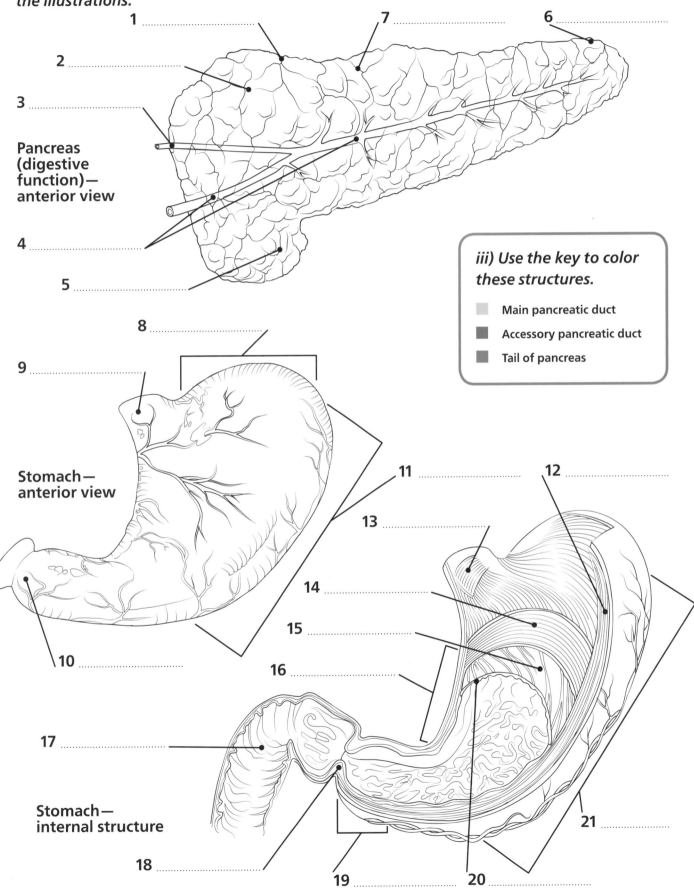

1 ....

2 ....

3 ....

Pancreas (digestive function)— anterior view

4 ....

5 ....

7 ....

6 ....

iii) Use the key to color these structures.

☐ Main pancreatic duct

■ Accessory pancreatic duct

■ Tail of pancreas

8 ....

9 ....

Stomach— anterior view

10 ....

11 ....

12 ....

13 ....

14 ....

15 ....

16 ....

17 ....

Stomach— internal structure

18 ....

19 ....

20 ....

21 ....

Fill in the blanks

1   The _____process of the liver encircles the inferior vena cava.

2   The _____ligament encircles the bare area of the liver.

3   The potential space between the liver and the diaphragm is called the _____space.

4   The space between the liver and the right kidney is called the _____.

5   The _____of the gallbladder projects below the inferior margin of the liver.

6   The _____lies posterior to the bile duct in the hepatoduodenal ligament.

7   The _____and the _____ open into the _____
beneath the summit of the greater (major) duodenal papilla.

8   The tail of the pancreas is in contact with the hilum of the _____.

9   Two vessels, the _____and _____, are related to the superior
border of the pancreas.

10   The _____process of the pancreas passes behind the superior mesenteric vessels.

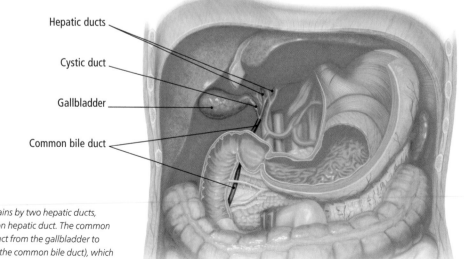

Hepatic ducts

Cystic duct

Gallbladder

Common bile duct

Bile Ducts

The bile fluid from the liver drains by two hepatic ducts,
which join to form the common hepatic duct. The common
hepatic duct joins the cystic duct from the gallbladder to
form the bile duct (also called the common bile duct), which
opens into the second part of the duodenum.

Match the statement to the reason

1 *The portal vein carries almost all blood from the gut to the liver because…*

a *the hepatic artery, portal vein, and bile duct have left and right branches to each side of that division.*

2 *The liver can be divided along the plane of the inferior vena cava and gallbladder fossa because…*

b *its venous drainage is usually by multiple channels directly into the liver.*

3 *The liver is a common site for secondary tumors to lodge because…*

c *the nutrients and toxins absorbed at the gut must be metabolically processed.*

4 *The gallbladder rarely becomes gangrenous because…*

d *obstruction of the main pancreatic duct may cause the release of digestive enzymes into the glandular tissue.*

5 *Gallstones in the bile duct may cause pancreatitis because…*

e *almost all venous blood from the gut passes through it.*

Pancreas—exocrine cells

The exocrine cells of the pancreas produce the enzymes required for the digestion of food. These include pancreatic amylase (for digestion of starch), lipase (for fats), and chymotrypsin, trypsin, and carboxypeptidase (for digestion of protein). Pancreatic juice is also alkaline, to neutralize stomach acid.

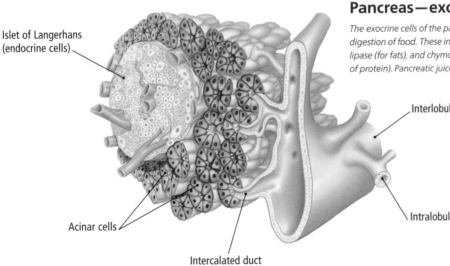

Islet of Langerhans (endocrine cells)

Interlobular duct

Intralobular duct

Acinar cells

Intercalated duct

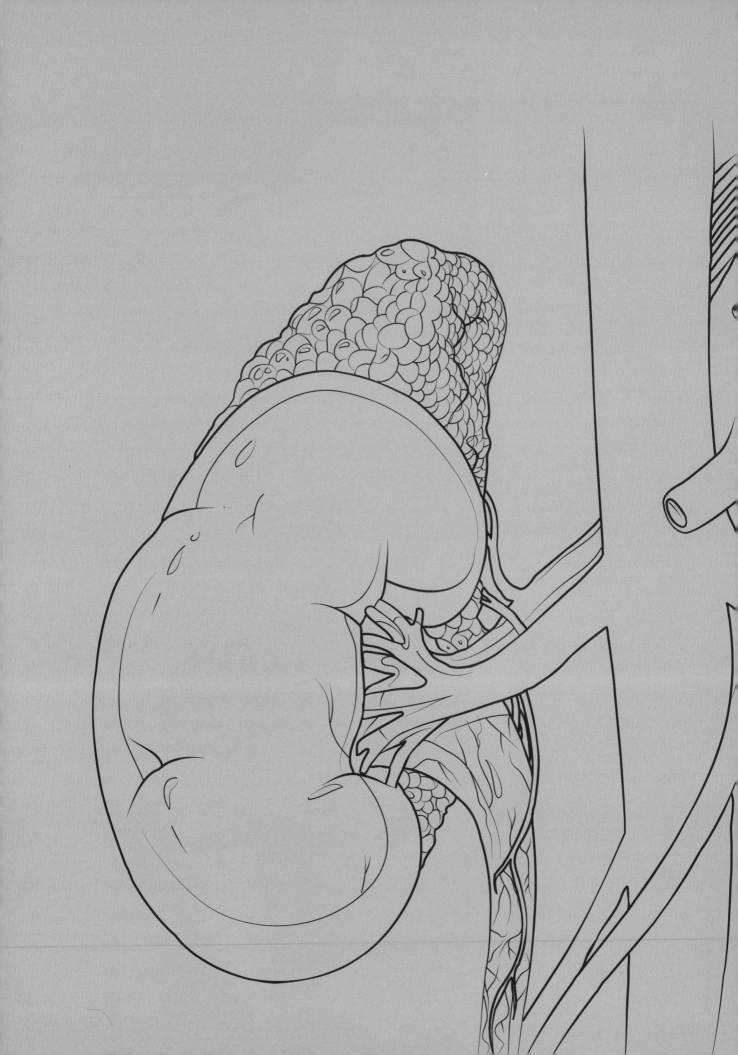

CHAPTER 7:
THE URINARY SYSTEM

Male and Female Urinary Systems

The urinary system is primarily concerned with the excretion of nitrogenous waste in the form of water-soluble urea. However, it also plays key roles in the regulation of the ionic balance of the blood (sodium, potassium, chloride, and calcium), the control of blood pH (through hydrogen and bicarbonate ion regulation), blood pressure maintenance (by the renin-angiotensin system), and red blood cell production (by erythropoietin). Key organs are the kidney, ureters, urinary bladder, and urethra.

Key terms:

Abdominal aorta The major artery of the body. It passes through the aortic opening (aortic hiatus) in the diaphragm.

Adrenal gland An endocrine gland on top of the kidney, hence its alternative name of suprarenal gland. It contains cortical and medullary tissue. The cortex produces cortisol and the mineralocorticoid hormone known as aldosterone; the medulla produces epinephrine and norepinephrine.

Bladder A muscular sac that stores urine. It consists of a lining of stretchable transitional epithelium surrounded by thick smooth muscle (the detrusor) and an adventitial layer.

Common iliac artery The two terminal branches of the abdominal aorta. They divide into external and internal iliac arteries.

Common iliac vein The vein formed from the junction of the external and internal iliac veins. The two common iliac veins join to form the inferior vena cava.

External iliac artery The branch of the common iliac artery that becomes the femoral artery at the inguinal ligament.

External iliac vein The continuation of the femoral vein after it passes under the inguinal ligament. The external iliac vein joins the internal iliac vein to form the common iliac vein.

Inferior vena cava The largest vein of the abdomen. It carries blood from the lower limb, pelvis, kidneys, and posterior abdominal wall. It passes through the diaphragm and enters the right atrium of the heart.

Internal iliac artery The branch of the common iliac artery that supplies the gluteal region, pelvic organs, medial pelvic wall, and perineum (space between the thighs).

Internal iliac vein The vein that drains the organs of the pelvis and the buttock region. It joins with the external iliac vein to form the common iliac vein.

Kidney An organ that excretes nitrogenous waste and regulates fluid balance, ionic concentration of body fluids, pH of the blood, and blood pressure. It also plays a key role in the control of erythrocyte (red blood cell) production.

Ovarian artery An artery arising from the aorta and descending the suspensory ligament of the ovary to supply the ovary and uterine tube.

Ovarian vein The vein draining the ovary. It ascends the suspensory ligament to enter the left renal vein (on the left) or inferior vena cava (on the right).

Testicular artery The artery supplying the testis. It arises from the aorta because the embryonic origin of the testis is the upper posterior abdominal wall.

Testicular vein The vein draining the testis. It passes through the inguinal canal and the posterior abdomen to drain into either the left renal vein (on the left) or the inferior vena cava (on the right).

Ureter The muscular tube about 10 inches (25 cm) long carrying urine from the renal pelvis to the urinary bladder. Its wall has an inner lining of stretchable transitional epithelium surrounded by smooth muscle and an adventitial connective tissue layer.

Male Urinary System— anterior view

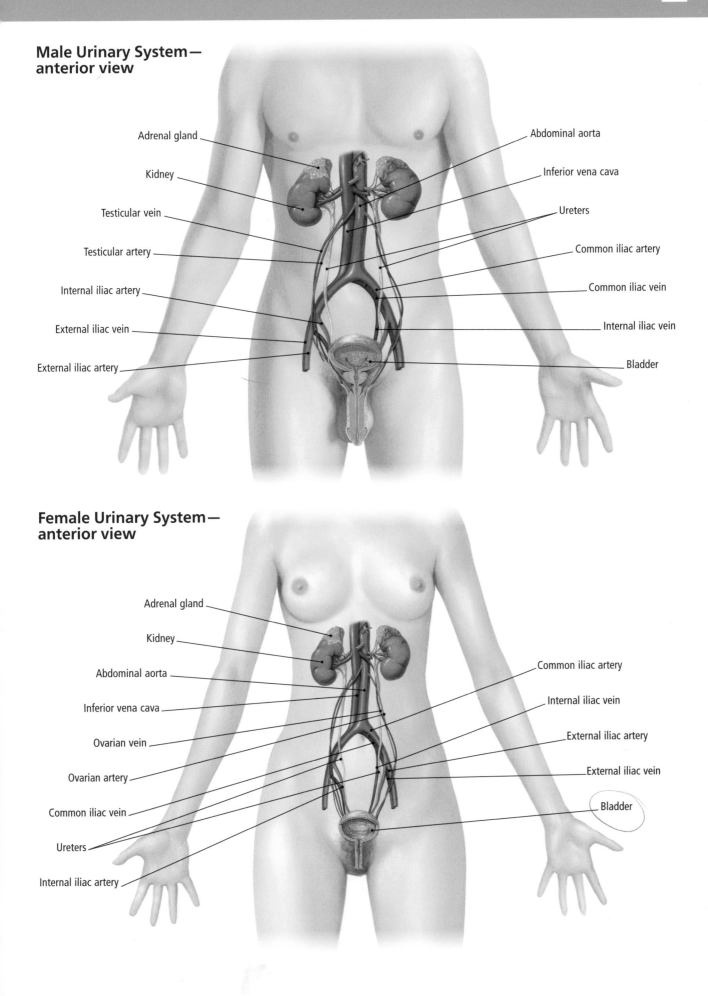

Adrenal gland

Kidney

Testicular vein

Testicular artery

Internal iliac artery

External iliac vein

External iliac artery

Abdominal aorta

Inferior vena cava

Ureters

Common iliac artery

Common iliac vein

Internal iliac vein

Bladder

Female Urinary System— anterior view

Adrenal gland

Kidney

Abdominal aorta

Inferior vena cava

Ovarian vein

Ovarian artery

Common iliac vein

Ureters

Internal iliac artery

Common iliac artery

Internal iliac vein

External iliac artery

External iliac vein

Bladder

True or false?

1   *The left kidney sits slightly lower than the right kidney on the posterior abdominal wall.*

2   *The kidneys are surrounded by perinephric (perirenal) fat and renal fascia.*

3   *The hilum of each kidney faces medially.*

4   *The renal artery is longer on the left than the right.*

5   *The renal veins run anterior to the renal arteries.*

6   *The kidneys are protected from behind by the 12th rib.*

7   *The ureter has abdominal and pelvic parts.*

8   *The ureter has a course through the peritoneal cavity.*

9   *The urinary bladder is always at the level of the sacral promontory.*

10   *The male urethra is shorter than the female urethra.*

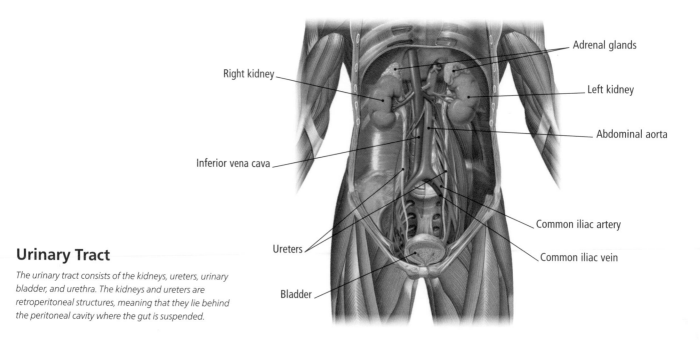

Urinary Tract

The urinary tract consists of the kidneys, ureters, urinary bladder, and urethra. The kidneys and ureters are retroperitoneal structures, meaning that they lie behind the peritoneal cavity where the gut is suspended.

Multiple choice

1 *Which of the following is not immediately anterior to the right kidney?*

(A) duodenum

(B) liver

(C) right suprarenal gland

(D) ascending colon

(E) head of pancreas

2 *Which of the following is not immediately anterior to the left kidney?*

(A) splenic artery

(B) body of the pancreas

(C) spleen

(D) liver

(E) left suprarenal gland

3 *Which of the following is immediately posterior to the kidneys?*

(A) dome of diaphragm

(B) stomach

(C) suprarenal gland

(D) pancreas

(E) liver

4 *A renal biopsy could be most easily taken by a needle passing through which of the following?*

(A) quadratus lumborum muscle

(B) umbilicus

(C) rectus abdominis muscle

(D) diaphragm

(E) external oblique muscle

5 *Which of the following vessels usually drains into the left renal vein?*

(A) common iliac vein

(B) portal vein

(C) common hepatic vein

(D) left testicular vein

(E) right ovarian vein

6 *Which of the following passes directly in front of the left ureter?*

(A) left common iliac artery

(B) left ovarian artery

(C) left internal iliac artery

(D) celiac artery

(E) superior mesenteric artery

7 *Which of the following organs is directly superior to the right kidney?*

(A) stomach

(B) spleen

(C) transverse colon

(D) duodenum

(E) suprarenal (adrenal) gland

Horseshoe kidney

Most people have two distinct kidneys, but in about one person in every 600, the kidneys are joined at their lower poles to form a horseshoe shape. This congenital disorder is more common in men and may have no adverse health effects at all because the amount of functional kidney tissue is unaffected. The kidneys usually ascend the posterior abdominal wall during development, so horseshoe kidneys may become stuck in an inferior position because the most ventral branch of the aorta, the inferior mesenteric artery, becomes caught inside the curve of the ascending horseshoe and stops any further rise.

Fill in the blanks

1   The _____ is a fat-filled compartment around the kidney.

2   The renal _____ is the fat-filled compartment within the kidney.

3   The _____ collects urine from the major calyces.

4   The _____ are the cortical tissue between pyramids.

5   The _____ is the apex of the renal pyramid.

6   The kidney tissue is divided into five _____ segments.

7   The _____ and _____ are related to the base of the bladder in males.

8   The medial fibers of the _____ muscle support the urinary bladder neck.

9   The _____ connects the bladder neck with the external environment.

10   The superior surface of the female urinary bladder is related to the body of the
     _____.

Inferior vena cava — Celiac trunk
Abdominal aorta
Superior mesenteric artery — Adrenal gland
Renal artery
Renal vein
Kidney
Ureter

Kidneys

The kidneys are bean-shaped organs situated behind the stomach, liver, and duodenum. They are surrounded by fat (the perinephric fat) and connective tissue membranes (the renal fascia). Kidneys have a rich vascular supply.

Match the statement to the reason

1 Infection around the kidney always spreads inferiorly because…

a this type of abnormal tissue tends to grow along venous channels.

2 The left renal vein is longer than the right because…

b the renal fascial compartment is open inferiorly but closed superiorly.

3 The kidneys are actually supplied by modified lumbar arteries because…

c the urachus connects the urinary bladder and the yolk sac during early fetal life.

4 Obstruction of the inferior vena cava may occur in renal cell carcinoma because…

d the organs ascend the posterior abdominal wall during prenatal life.

5 The apex of the bladder has a connection with the umbilicus because…

e the vessel must cross the aorta to reach the inferior vena cava.

Polycystic kidney

Polycystic kidney disease is a genetic disorder in which the renal tubules fail to connect properly. Fluid continues to pump into the tubules, dilating them to form non-functional cysts that enlarge progressively during childhood and adult life. The condition may be either genetically dominant (1:500 live births) or recessive (1:20,000 live births). Signs and symptoms include elevated blood pressure, hematuria (blood in urine), headaches, abdominal pain, and excessive urination (polyuria). Chronic renal failure and urinary tract infection are the most serious consequences, and treatment is by renal dialysis, ongoing antibiotics, and eventually renal transplantation.

The Kidneys (Macroscopic Features)

The kidneys are paired bean-shaped organs that lie behind the peritoneal cavity against the posterior abdominal wall. They are surrounded by fat in the perinephric space and connective tissue sheaths (renal fascia). The kidneys have fibrous capsules and are each divided into an outer cortex and an inner medulla composed of the renal pyramids. Urine is collected by the calyceal system (minor calyces, major calyces, and renal pelvis) and drains into the ureter.

Key terms:

Abdominal aorta See pp. 218–219.

Adrenal glands See pp. 218–219.

Arcuate artery Arterial branches running along the border between the cortex and medulla. They receive interlobar arteries and give off interlobular arteries.

Celiac trunk The first ventral branch of the abdominal aorta. It supplies the foregut derivatives (stomach to duodenum, liver, and gallbladder).

Cortex The outer layer of the kidney, immediately beneath the capsule. It contains the glomeruli, the proximal and distal convoluted tubules, and the initial parts of the collecting ducts.

Inferior vena cava See pp. 218–219.

Interlobular artery An artery ascending into the renal cortex from the arcuate artery. It gives off afferent glomerular arterioles.

Kidney An organ that excretes nitrogenous waste and regulates fluid balance, ionic concentration of body fluids, pH of the blood, and blood pressure. It also plays a key role in the control of erythrocyte (red blood cell) production.

Left adrenal gland An endocrine gland on top of the left kidney. It contains cortical and medullary tissue. The cortex produces cortisol and the mineralocorticoid hormone known as aldosterone; the medulla produces epinephrine and norepinephrine.

Left renal artery The left renal artery supplies the left kidney and gives branches to the left adrenal gland and left ureter.

Major calyx A primary branch of the renal pelvis. It receives urine from the minor calyces.

Minor calyx The terminal cuplike branch of the calyceal system. Each receives a renal papilla and joins with other minor calyces to form a major calyx.

Perirenal fat The kidneys are surrounded by a layer of fat between the renal capsule and the renal fascia.

Renal column Cortical tissue lying between two adjacent renal pyramids.

Renal papilla The tip of the medullary pyramid. Collecting ducts open at its surface to drain urine into the minor calyx.

Renal pelvis The dilated portion of the urinary tract between the major calyces and the ureter.

Renal pyramid (medulla) A conical structure in the medulla of each kidney. The bases of these structures are located at the corticomedullary junction, and their apices (renal papillae) open into the minor calyces.

Renal sinus The space surrounding the calyceal branches in the center of the kidney. It is filled with fat, blood vessels, and nerves.

Renal vein The vein draining the kidney. The left renal vein is much longer than the right and receives the gonadal (testicular or ovarian) and suprarenal veins.

Right adrenal gland An endocrine gland on top of the right kidney. It contains cortical and medullary tissue. The cortex produces cortisol and the mineralocorticoid hormone known as aldosterone; the medulla produces epinephrine and norepinephrine.

Segmental artery Branches of the renal artery that supply one of five renal vascular segments. These are discrete regions with minimal anastomotic connection to adjacent segments.

Superior mesenteric artery The branch of the abdominal aorta supplying derivatives of the embryonic midgut (distal duodenum, jejunum, ileum, cecum, appendix, ascending colon, and transverse colon).

Ureter See pp. 218–219.

blood is taken to the nephron which is part in the
Cortex and part medulla

nephrons key functions of kidneys

Plasma is dumped into the nephron
good stuff is pulled back in to blood vessels

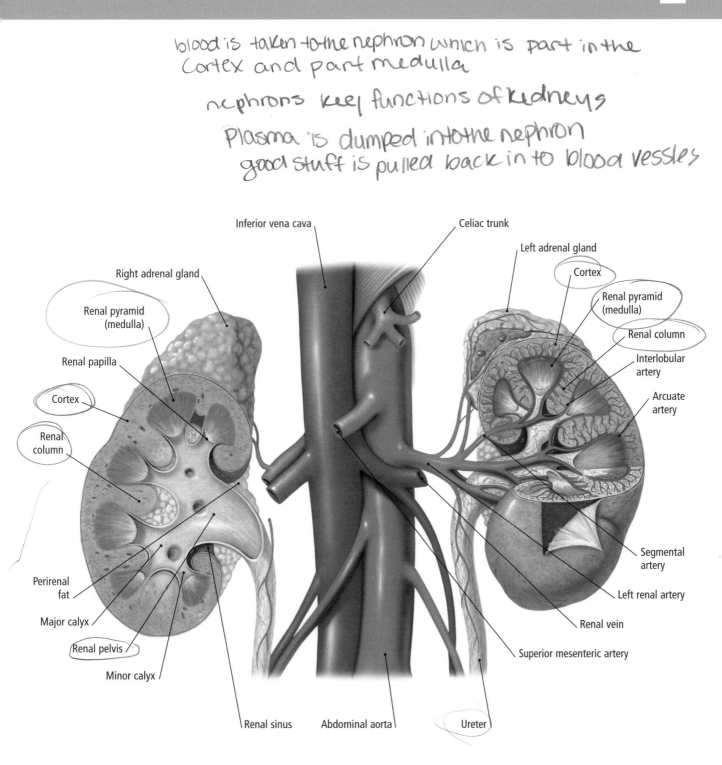

Inferior vena cava

Celiac trunk

Left adrenal gland

Cortex

Renal pyramid
(medulla)

Renal column

Interlobular
artery

Arcuate
artery

Right adrenal gland

Renal pyramid
(medulla)

Renal papilla

Cortex

Renal
column

Perirenal
fat

Major calyx

Renal pelvis

Minor calyx

Renal sinus

Abdominal aorta

Ureter

Segmental
artery

Left renal artery

Renal vein

Superior mesenteric artery

Kidneys and Adrenal Glands—anterior view

The Kidneys (Microscopic Features)

The fundamental functional unit of the kidney is the nephron, of which there are approximately 1 million in the two kidneys. Each nephron contains a renal corpuscle (with Bowman's capsule and glomerulus) and a series of tubules (proximal convoluted, loop of Henle with thick and thin segments, distal convoluted tubule, and connecting tubule). The corpuscle serves the function of ultrafiltration, whereas the tubules serve the functions of reabsorption, secretion, and excretion.

Key terms:

Afferent arteriole The arteriole branching from the interlobular artery and leading to the glomerulus.

Arcuate artery See pp. 224–225.

Arcuate vein Venous tributaries running along the border between the cortex and medulla.

Basal lamina A layer of substance on which epithelial cells sit. It is usually produced by the epithelial cells themselves.

Bowman's space The urinary space between the parietal and visceral layers of the capsule of Bowman, a cup-shaped structure surrounding the glomerulus. It receives the ultrafiltrate from the glomerulus.

Cell body of podocyte Podocytes have foot processes that encircle the glomerular capillaries. The processes interdigitate and are separated by filtration slits.

Collecting duct One of the ducts that collect urine from the connecting tubules and join together to open at the renal papilla into the calyceal system. It is also a significant site of sodium chloride reabsorption from the urine.

Connecting tubule A short duct segment joining the distal convoluted tubule with the collecting ducts.

Distal convoluted tubule The distal part of the distal convoluted tubule is permeable to water in the presence of antidiuretic hormone and reabsorbs sodium chloride.

Efferent arteriole The arteriole leaving the glomerular capillary bed.

Endothelial cell The flattened lining cell of blood and lymph vessels, including lymphatics, arteries, capillaries, and veins.

Foot process of podocytes These processes (also called pedicels) interdigitate to cover the basal lamina and are separated from each other by filtration slits. The slits have a diaphragm that impedes the movement of large molecules into the urinary space.

Glomerular tuft of capillaries Fenestrated capillaries in the glomerulus. They are surrounded by mesangial support cells and the foot processes of podocytes.

Glomerulus The glomerulus consists of glomerular capillaries, mesangial cells in their matrix, and podocytes forming the visceral layer of the surrounding capsule of Bowman.

Interlobular artery See pp. 224–225.

Interlobular vein A vein draining cortical tissue of the kidney.

Kidney See pp. 224–225.

Macula densa A specialized epithelial region at the junction of the thick ascending limb and distal convoluted tubule.

Medullary plexus of peritubular capillaries A fine network of capillaries in the renal medulla. It transports nutrients and oxygen to the loops of Henle and removes excess water and solutes.

Nephron This is the functional filtering unit of the kidney and consists of a capsule of Bowman surrounding a glomerulus.

Parietal layer of Bowman's capsule The parietal layer of Bowman's capsule consists of squamous (flattened) epithelial cells surrounding the urinary space.

Peritubular capillaries Capillaries supplying the proximal and distal convoluted tubules. They are downstream of the efferent glomerular arterioles.

Proximal convoluted tubule The tubule immediately after the glomerulus. It is the site of absorption of 70 percent of the water, glucose, sodium, chloride, and potassium that is filtered by the glomerulus.

Thick ascending limb of loop of Henle (distal straight tubule) This is continuous with the distal convoluted tubule and is lined by low cuboidal epithelium.

Thick descending limb of loop of Henle (proximal straight tubule) This is a continuation of the proximal convoluted tubule. It is lined by low cuboidal epithelium.

Thin ascending limb of loop of Henle Important site of urea reabsorption into the tubular fluid and sodium chloride reabsorption from the tubular fluid due to the presence of a sodium/potassium ATPase pump in the epithelium.

Thin descending limb of loop of Henle This part of the limb is permeable to water. As the tubular fluid passes deep into the hyperosmotic medulla, water is reabsorbed from the loop to make the tubular fluid more concentrated.

Urinary pole The end of the nephron (renal corpuscle) from which the proximal convoluted tubule arises.

Vasa recta A network of vessels running into and out of the medulla. They absorb excess water and solutes added to the interstitial fluid from the loops of Henle.

Vascular pole The end of the nephron where the afferent and efferent arterioles enter and leave, respectively.

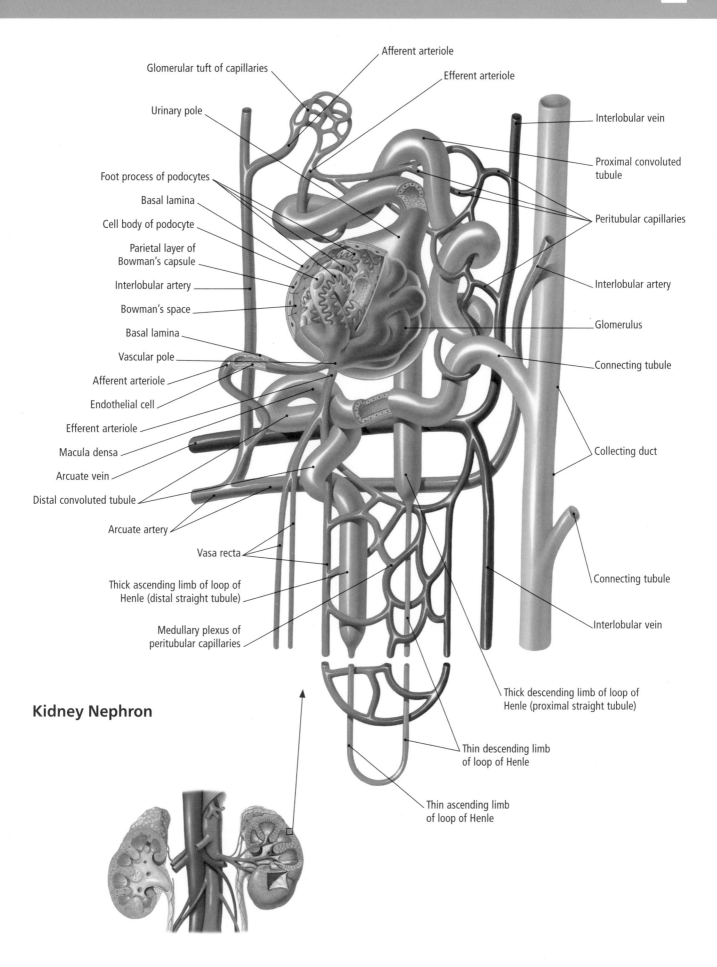

Afferent arteriole

Glomerular tuft of capillaries

Efferent arteriole

Urinary pole

Interlobular vein

Proximal convoluted tubule

Foot process of podocytes

Basal lamina

Cell body of podocyte

Peritubular capillaries

Parietal layer of Bowman's capsule

Interlobular artery

Bowman's space

Interlobular artery

Basal lamina

Glomerulus

Vascular pole

Connecting tubule

Afferent arteriole

Endothelial cell

Efferent arteriole

Macula densa

Collecting duct

Arcuate vein

Distal convoluted tubule

Arcuate artery

Vasa recta

Thick ascending limb of loop of Henle (distal straight tubule)

Medullary plexus of peritubular capillaries

Connecting tubule

Interlobular vein

Thick descending limb of loop of Henle (proximal straight tubule)

Thin descending limb of loop of Henle

Thin ascending limb of loop of Henle

Kidney Nephron

True or false?

1   Renal corpuscles are located in the renal pyramids.

2   Each renal corpuscle has a vascular and urinary pole.

3   There are about 3 million nephrons in each kidney.

4   The glomerulus is the primary site of renal absorption.

5   The three cell types in the glomerulus are the capillary endothelium, mesangial cells, and podocytes.

6   Mesangial cells have phagocytic capacity.

7   Each renal corpuscle has an efferent venule.

8   Proximal convoluted tubular epithelial cells have microvilli to assist absorption.

9   The proximal convoluted tubule attaches to the vascular pole of the renal corpuscle.

10   The driving force for water reabsorption in the proximal convoluted tubule is NaCl absorption.

11   Angiotensin II promotes NaCl and water reabsorption at the proximal convoluted tubule.

12   The ascending limb of the loop of Henle has only a thin segment.

13   The loop of Henle of all nephrons is confined to the renal cortex.

14   Urodilatin is secreted by the distal convoluted tubule and inhibits NaCl and water reabsorption at the medullary collecting tubules.

15 *Thiazide diuretics act to inhibit NaCl reabsorption at the proximal convoluted tubule.*

16 *Antidiuretic hormone (ADH) acts mainly at the proximal convoluted tubule.*

17 *Atrial natriuretic factor promotes urinary excretion of NaCl and water.*

18 *Arcuate arteries run parallel to the junction of the cortex and medulla.*

19 *The kidney consists of five renal vascular segments, each with its own segmental artery.*

20 *Necrosis of a renal papilla is likely to obstruct the minor calyx.*

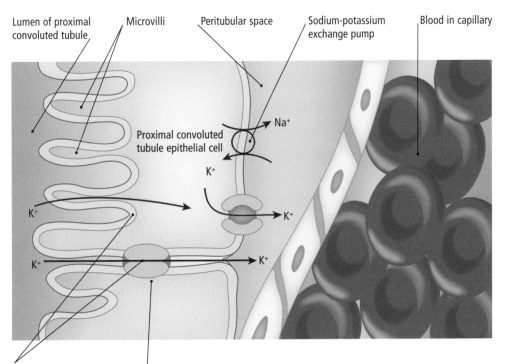

Mechanisms of Potassium Regulation

About 600 mmol of potassium passes into the glomerular filtrate each day, and most of that is reabsorbed at the level of the proximal convoluted tubules and loop of Henle. Some will be secreted in the distal convoluted tubule under the regulation of the steroid aldosterone. Potassium may be absorbed through the space between proximal convoluted tubule epithelial cells or through the cells themselves by an unknown mechanism. The normal concentration of potassium in the blood is 4.4 mM (range of 3.5 to 5.0 mM).

Color and label

Kidneys—anterior view

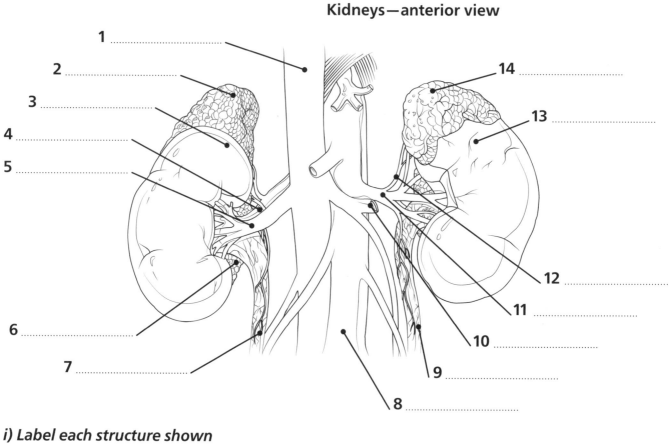

1 ....

2 ....

3 ....

4 ....

5 ....

6 ....

7 ....

8 ....

9 ....

10 ....

11 ....

12 ....

13 ....

14 ....

i) Label each structure shown on the illustrations.

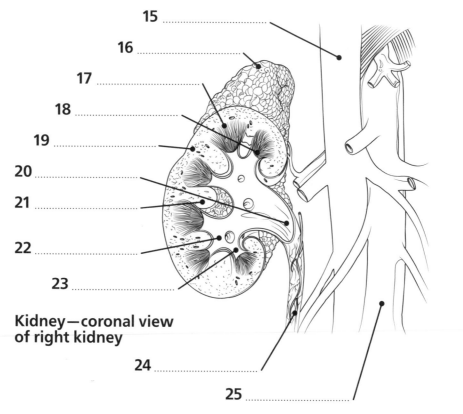

15 ....

16 ....

17 ....

18 ....

19 ....

20 ....

21 ....

22 ....

23 ....

24 ....

25 ....

Kidney—coronal view of right kidney

Urinary Tract—anterior view

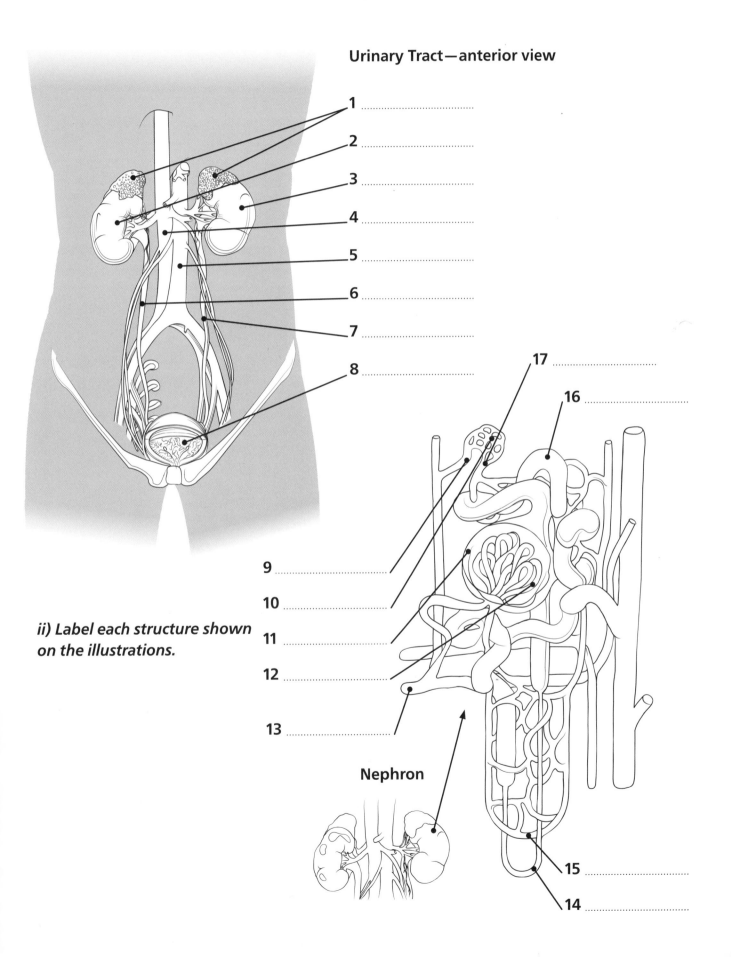

1 ....

2 ....

3 ....

4 ....

5 ....

6 ....

7 ....

8 ....

*ii) Label each structure shown
on the illustrations.*

9 ....

10 ....

11 ....

12 ....

13 ....

17 ....

16 ....

Nephron

15 ....

14 ....

Multiple choice

1   Glomerular capillaries are of which type?
(A)   fenestrated
(B)   continuous
(C)   discontinuous
(D)   basement
(E)   none of the above

2   Which of the following does not usually enter the ultrafiltrate?
(A)   albumin
(B)   glucose
(C)   potassium
(D)   cysteine
(E)   ethanol

3   Glomerular capillaries have fenestrations of which diameter?
(A)   5 to 10 nm
(B)   10 to 25 nm
(C)   50 to 100 nm
(D)   150 to 200 nm
(E)   250 to 300 nm

4   Which of the following cells secrete renin?
(A)   podocytes
(B)   mesangial cells
(C)   Bowman's capsular epithelium
(D)   juxtaglomerular cells
(E)   efferent arterioles

5   What is the approximate normal flow rate of the glomerular filtrate?
(A)   1 ml/min
(B)   5 ml/min
(C)   20 ml/min
(D)   60 ml/min
(E)   120 ml/min

6   Which specialization is found on the apex of brush border cells of the proximal convoluted tubule?
(A)   microvilli
(B)   cilia
(C)   stereocilia
(D)   villi
(E)   flagella

7   Which organelle makes the high metabolic activity of proximal convoluted tubule cells possible?
(A)   nucleolus
(B)   Golgi apparatus
(C)   centriole
(D)   rough endoplasmic reticulum
(E)   mitochondrion

8   Which of the following does not occur at the proximal convoluted tubule?
(A)   absorption of glucose
(B)   absorption of amino acids
(C)   ultrafiltration
(D)   absorption of potassium
(E)   absorption of water

9   Which of the following is found at the junction of the thick ascending loop of Henle and the distal convoluted tubule?
(A)   macula densa
(B)   podocyte
(C)   mesangial cells
(D)   papillary cell
(E)   none of the above

10 *Which of the following occurs at the loop of Henle?*

Ⓐ secretion of potassium into the lumen
Ⓑ reabsorption of 15 percent of the ultrafiltered water
Ⓒ secretion of bicarbonate ions
Ⓓ absorption of most drugs
Ⓔ excretion of calcium

11 *The distal convoluted tubule*

Ⓐ is a key site for water reabsorption
Ⓑ has squamous cells in its epithelium
Ⓒ reabsorbs most glucose
Ⓓ is resistant to antidiuretic hormone
Ⓔ has cuboidal cells with abundant microvilli

12 *What effect does aldosterone have on the ascending loop of Henle, distal convoluted tubule, and collecting tubules?*

Ⓐ stimulates potassium reabsorption
Ⓑ stimulates water loss
Ⓒ stimulates calcium reabsorption
Ⓓ stimulates NaCl reabsorption
Ⓔ none of the above

13 *Which type of epithelium lines the collecting tubules?*

Ⓐ simple squamous
Ⓑ simple cuboidal
Ⓒ stratified squamous
Ⓓ transitional urothelium
Ⓔ pseudostratified ciliated columnar

14 *Where do the collecting tubules eventually open?*

Ⓐ distal convoluted tubules
Ⓑ proximal convoluted tubules
Ⓒ apex of the renal papilla
Ⓓ Bowman's capsule
Ⓔ major calyx

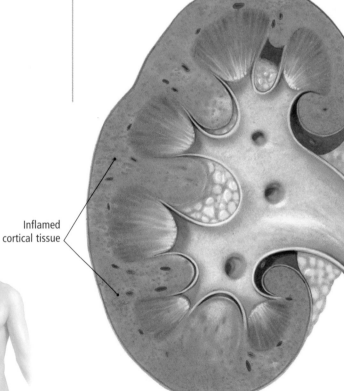

Inflamed cortical tissue

Nephritis

Nephritis is inflammation of the kidney; different types of nephritis affect the various structures of the kidney.

Fill in the blanks

1   The _____ have foot processes that interdigitate to make the filtration spaces.

2   The Bowman's space lies between the glomerulus and the _____.

3   The _____ clear trapped residues and protein from the basement membrane of the glomerulus.

4   The main site of reabsorption of water, nutrients, and electrolytes is the _____.

5   Adjacent epithelial cells of the proximal convoluted tubule are held together by _____.

6   Mitochondria are concentrated in the _____ between the _____ of proximal convoluted tubular cells.

7   The _____ of proximal convoluted tubular cells engulf large peptides.

8   The next segment of the nephron after the proximal convoluted tubule is the _____.

9   The thin segments of the loops of Henle are lined by _____ epithelium.

10   The thick segments of the loops of Henle are lined by _____ epithelium.

11   The _____ is a specialized region at the junction of the thick ascending limb of the loop of Henle and the distal convoluted tubule.

12   The macula densa is in contact with both _____ and _____ arterioles.

13   The _____ cells of the juxtaglomerular apparatus produce renin.

14   _____ nephrons have loops of Henle entirely confined to the renal cortex.

15 _____ are blood vessels that accompany the loops of Henle in the medulla.

16 The hormone _____ is secreted by cardiac cells of the right atrium and increases urinary excretion of NaCl and water.

17 The fluid produced at the Bowman's capsule is called _____.

18 Urine exits the nephron at the openings of the _____.

19 _____ arteries run in the space that separates the renal lobes.

20 Interlobular arteries ascend from the _____ arteries into the renal cortex.

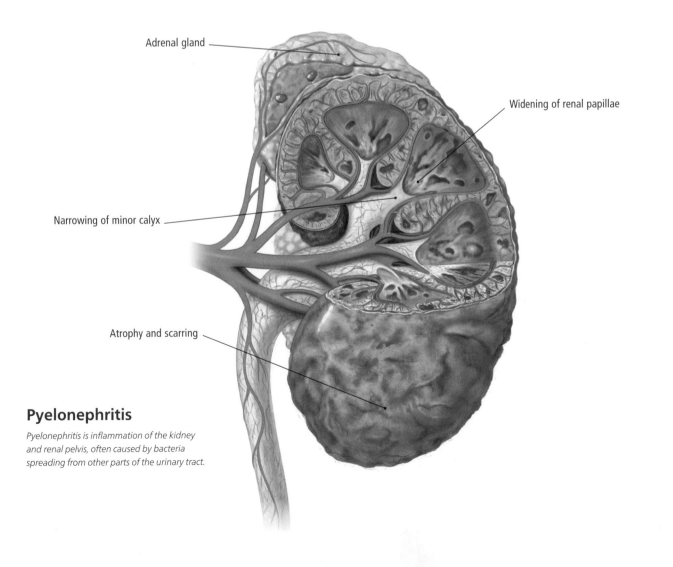

Adrenal gland

Widening of renal papillae

Narrowing of minor calyx

Atrophy and scarring

Pyelonephritis

Pyelonephritis is inflammation of the kidney and renal pelvis, often caused by bacteria spreading from other parts of the urinary tract.

Match the statement to the reason

1 The kidneys receive about 20 percent of the cardiac output because…

a these organs play a key role in filtering the blood of the body.

2 Desert mammals have particularly large medullary volumes because…

b low flow in the afferent glomerular arteriole triggers renin release.

3 Failure of water reabsorption by the kidney would lead to catastrophic water loss because…

c this is an important site of water reabsorption from the glomerular filtrate.

4 Obstruction of the renal artery leads to elevated blood pressure because…

d the kidney produces erythropoietin.

5 Kidney disease is often accompanied by reduced red blood cell mass because…

e 180 liters of glomerular ultrafiltrate is produced per day.

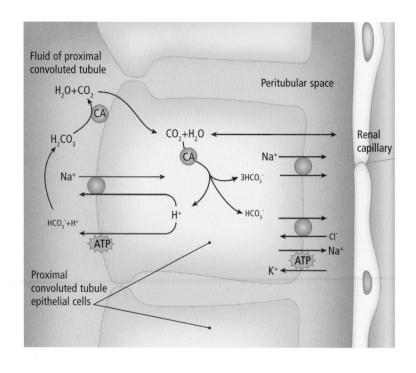

The Process of Renal Regulation

The kidneys control the acid–base balance of the body by a process of regulated bicarbonate ion reabsorption in the proximal convoluted tubule. This requires the ejection of hydrogen ions into the tubule to capture carbon dioxide, which is then converted to bicarbonate ions, which in turn are shunted to the interstitial fluid outside the cell and then eventually absorbed into the bloodstream.

1 *Abnormal calcium metabolism can accompany renal disease because…*

a the ability to maintain a concentration gradient in the medulla is lost.

2 *Glomerulonephritis leads to hematuria because…*

b the kidney activates 1,25-hydroxycholecalciferol, a vitamin D derivative.

3 *Sympathetic stimulation of the kidney leads to increased renin release because…*

c there has been a failure of the integrity of the glomerular filtration system.

4 *Necrosis of the renal papilla causes major problems with the ability to concentrate urine because…*

d there are abundant adrenergic nerve fibers innervating the juxtaglomerular cells.

5 *Kidney disease may involve loss of plasma proteins in the urine (nephrotic syndrome) because…*

e the inflammation leads to failure of the glomerular filtration membrane and passage of red blood cells.

Chronic renal failure

Chronic renal failure occurs when kidney function is lost over a period of months to years. Common causes include diabetes mellitus, hypertension, vascular disease, glomerulonephritis, polycystic kidney disease, and urinary tract obstruction. Symptoms and signs include vague malaise, loss of appetite, cardiac arrhythmias, headaches (from elevated blood pressure), tiredness (from anemia), and sexual dysfunction. Patients will have elevated levels of serum urea and creatinine, elevated serum potassium and phosphate, reduced serum calcium, and metabolic acidosis. Treatment is by dialysis, either peritoneal or by the vascular system using an arteriovenous shunt in the forearm, and transplantation of a kidney.

Ureters, Bladder, and Urethra

The ureters are muscular tubes that carry urine from the kidneys to the urinary bladder for storage. The urinary bladder is tetrahedral in shape when empty but can expand to a globular shape when full (300 to 500 ml). The urine reaches the external environment through the urethra, which is only 1.5 inches (4 cm) long in women, but up to about 8 inches (20 cm) long in men. The male urethra passes through the center of the prostate, where it can be obstructed in benign prostatic hypertrophy.

Key terms:

Abdominal aorta See pp. 218–219.

Bladder See pp. 218–219.

Bladder lining A type of epithelium called transitional epithelium or urothelium. It consists of columnar epithelium that can stretch to become squamous as the bladder fills.

Bulb of penis Part of the erectile tissue of the penis. It is continuous with the corpus spongiosum of the penile body. The spongy urethra passes through the center of the bulb.

Bulbourethral (Cowper's) gland This mucus-secreting gland drains into the intrabulbar part of the spongy (penile) urethra. It forms the initial part of the ejaculate.

Colliculus seminalis A tiny elevation on the posterior wall of the prostatic urethra. It receives the openings of the ejaculatory ducts and the prostatic utricle.

Common iliac artery See pp. 218–219.

Common iliac vein See pp. 218–219.

Corona glandis The rim of the glans of the penis.

Corpus cavernosum One of a pair of erectile tissue structures in the pendulous body of the penis. Each is surrounded by a tough tunica albuginea so that they develop great rigidity when filled with blood during erection.

Corpus spongiosum The continuation of the erectile tissue of the bulb into the body of the penis. It expands distally as the glans of the penis. The spongy urethra passes down its center.

External iliac artery See pp. 218–219.

External iliac vein See pp. 218–219.

External urethral orifice The opening of the spongy (penile) urethra at the tip of the penis.

Glans penis See pp. 282–283.

Inferior vena cava See pp. 218–219.

Internal iliac artery See pp. 218–219.

Internal iliac vein See pp. 218–219.

Neck of bladder Region of the bladder around the internal urethral orifice. Smooth muscle fibers inserted radially here may assist opening of the neck during micturition (urination).

Opening of ejaculatory duct The ductus deferens and duct of the seminal vesicle join to form an ejaculatory duct that opens on the colliculus seminalis.

Opening of ureters The openings of the two ureters are at two of the apices of the trigone, with the internal urethral orifice as the third apex.

Opening (meatus) of left ureter Each ureter passes obliquely through the wall of the bladder before opening. This provides a valve mechanism to limit retrograde flow into the ureter.

Ovarian artery An artery arising from the aorta and descending the suspensory ligament of the ovary to supply the ovary and uterine tube.

Ovarian vein The vein draining the ovary. It ascends the suspensory ligament to enter the left renal vein (on the left) or inferior vena cava (on the right).

Prepuce (foreskin) A fold of skin that covers the glans penis. Beneath it lies the preputial sac. The prepuce attaches to the ventrum of the penis by a frenulum near the external urethral orifice.

Prostate gland See pp. 282–283.

Prostatic utricle An embryonic remnant of the ducts that form the uterus and vagina in females.

Scrotum A double pouch that holds the testes, epididymides, and the lower parts of each spermatic cord. It consists of skin and dartos muscle.

Testicular artery The artery supplying the testis. It arises from the aorta because the embryonic origin of the testis is the upper posterior abdominal wall.

Testicular vein The vein draining the testis. It passes through the inguinal canal and the posterior abdomen to drain into either the left renal vein (on the left) or the inferior vena cava (on the right).

Trigone The smooth, triangular region on the inside of the bladder base. It is bounded by the entrances of the two ureters and the internal urethral orifice.

Ureter See pp. 218–219.

Urethra A tube carrying urine from the bladder to the external environment. In males it has prostatic, membranous, and spongy (penile) parts.

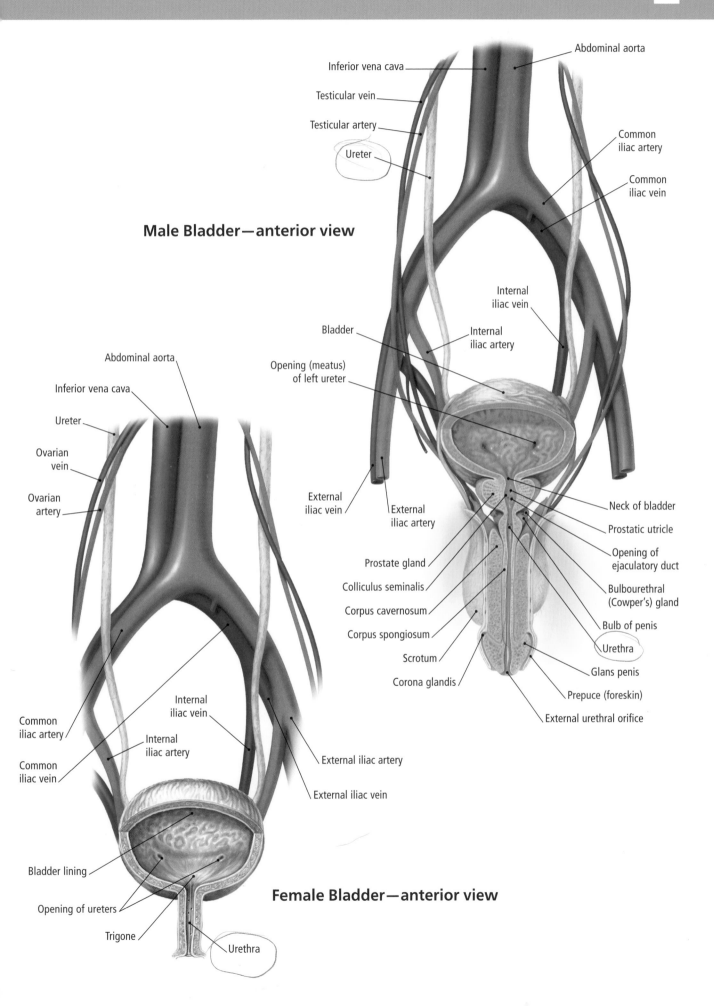

Male Bladder—anterior view

Abdominal aorta

Inferior vena cava

Testicular vein

Testicular artery

Ureter

Common
iliac artery

Common
iliac vein

Internal
iliac vein

Bladder

Internal
iliac artery

Opening (meatus)
of left ureter

External
iliac vein

External
iliac artery

Neck of bladder

Prostatic utricle

Opening of
ejaculatory duct

Prostate gland

Colliculus seminalis

Corpus cavernosum

Corpus spongiosum

Scrotum

Corona glandis

Bulbourethral
(Cowper's) gland

Bulb of penis

Urethra

Glans penis

Prepuce (foreskin)

External urethral orifice

Female Bladder—anterior view

Abdominal aorta

Inferior vena cava

Ureter

Ovarian
vein

Ovarian
artery

Common
iliac artery

Common
iliac vein

Internal
iliac vein

Internal
iliac artery

External iliac artery

External iliac vein

Bladder lining

Opening of ureters

Trigone

Urethra

True or false?

1   *The ureters run in front of the lumbar vertebral bodies.*

2   *The internal diameter of the ureter is about 5 mm.*

3   *The bulk of the wall of the ureter is made up of smooth muscle.*

4   *The ureter crosses the sacroiliac joint.*

5   *The urinary bladder is always a pelvic organ.*

6   *The base of the male urinary bladder is related to the rectum.*

7   *The bulb of the vestibule surrounds the neck of the female bladder.*

8   *The urethra passes through the external urethral sphincter in both sexes.*

9   *The male urethra passes down the length of the crus and corpus cavernosum of the penis.*

10   *The male urethra is dilated at its end as the navicular fossa.*

Renal calculi (stones)

Kidney stone disease (urolithiasis) is a condition where salts (e.g., calcium, phosphate) or organic compounds (e.g., oxalate, cysteine, uric acid) in the urine crystallize out of solution to form solid stones. Preventable risk factors include dehydration, obesity, and excessive consumption of sugars or calcium supplements. Stones usually form in the kidney and (if they are small enough) pass down the ureter, through the bladder, and out the urethra. If they do not pass, they may grow in the calyceal system or urinary bladder. The passage of a kidney stone can be intensely painful. Treatment is by hydration, pain relief, ultrasonic shock wave fragmentation (lithotripsy), or surgical removal if necessary.

Multiple choice

1 *Which of the following is a site where the ureter can be compressed or obstructed?*

Ⓐ the vesicoureteric junction
Ⓑ the iliac crest
Ⓒ the pelvic diaphragm
Ⓓ the ischial spine
Ⓔ the urogenital diaphragm

2 *Which vessel crosses the ureter in the female pelvis?*

Ⓐ middle rectal artery
Ⓑ uterine artery
Ⓒ internal pudendal artery
Ⓓ internal iliac artery
Ⓔ obturator artery

3 *Which structure opens at the superior corner(s) of the trigone?*

Ⓐ urethra
Ⓑ urachus
Ⓒ allantois
Ⓓ ureter
Ⓔ ejaculatory duct

4 *Which of the following could be mistaken for a ureteric calculus in an abdominal X-ray?*

Ⓐ seminal vesicles
Ⓑ calcified pelvic veins
Ⓒ ovarian cysts
Ⓓ feces in the rectum
Ⓔ enlarged prostate

5 *Which structure is found on the posterior wall of the prostatic urethra?*

Ⓐ urethral crest
Ⓑ seminal vesicle
Ⓒ interureteric crest
Ⓓ ureteric orifice
Ⓔ navicular fossa

6 *Which duct enters the prostatic urethra?*

Ⓐ bulbourethral gland duct
Ⓑ allantoic duct
Ⓒ navicular duct
Ⓓ ureteric duct
Ⓔ ejaculatory duct

7 *Which voluntary muscle directly controls the passage of urine?*

Ⓐ puborectalis
Ⓑ pubococcygeus
Ⓒ external urethral sphincter
Ⓓ detrusor
Ⓔ sphincter vesicae

Bladder

Internal urethral orifice

Urethra

External urethral orifice

Male Urethra

The male urethra is approximately 8 inches (20 cm) long and extends from the neck of the urinary bladder to the external urethral orifice. It consists of preprostatic, prostatic, membranous, and spongiose parts.

Color and label

**Urinary Tract (male)—
sagittal view**

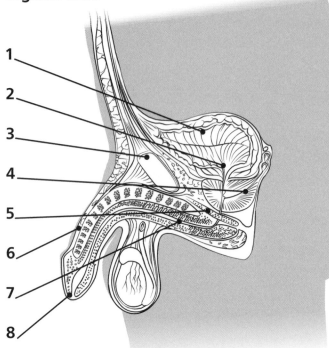

1
2
3
4
5
6
7
8

**Urinary Tract (female)—
sagittal view**

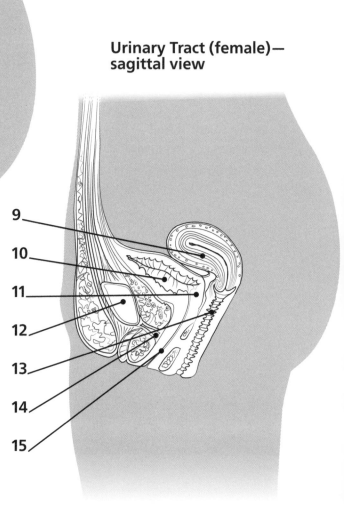

9
10
11
12
13
14
15

*i) Add numbers to the boxes below to
match each label to the correct part of
the illustrations.*

Penis	☐
Urethra	☐
Symphysis pubis	☐
Prostate gland	☐
Vagina	☐
Bladder	☐
Region of internal urethral sphincter	☐

Urethral meatus (opening)	☐
External urethral sphincter	☐
Bladder	☐
Internal urethral sphincter	☐
Symphysis pubis	☐
External urethral sphincter	☐
Uterus	☐
Urethra	☐

*ii) Label each structure shown
on the illustrations.*

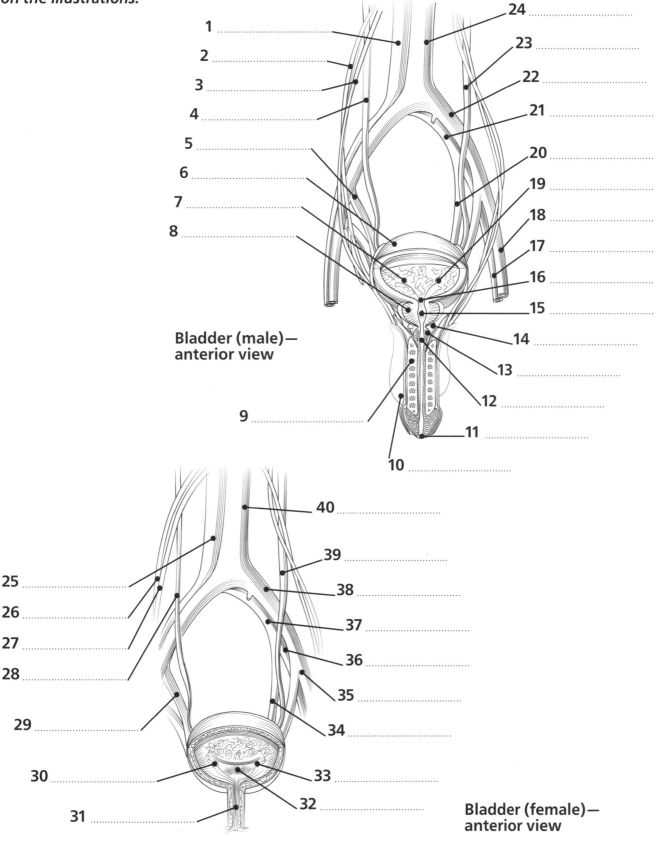

1 ....

2 ....

3 ....

4 ....

5 ....

6 ....

7 ....

8 ....

**Bladder (male)—
anterior view**

9 ....

10 ....

11 ....

12 ....

13 ....

14 ....

15 ....

16 ....

17 ....

18 ....

19 ....

20 ....

21 ....

22 ....

23 ....

24 ....

25 ....

26 ....

27 ....

28 ....

29 ....

30 ....

31 ....

32 ....

33 ....

34 ....

35 ....

36 ....

37 ....

38 ....

39 ....

40 ....

**Bladder (female)—
anterior view**

Fill in the blanks

1   The point of origin of the ureter is called the _____.

2   The first site of potential obstruction of the ureter is at the _____.

3   The process that helps move urine or calculi down the ureter is called _____.

4   When the ureters enter the pelvis, they turn medially at the level of the _____.

5   The arterial supply of the proximal ureter is by branches from the _____.

6   The ureter may kink when it crosses the pelvic brim because of the _____.

7   In women, the ureter passes lateral to the uterine _____.

8   The proximity of the ureter to the _____ makes the ureter vulnerable to accidental ligation in hysterectomy.

9   The point where the ureter enters the bladder is called the _____.

10   The smooth-walled region inside the bladder base is called the _____.

11   The urinary bladder is lined with a type of _____ epithelium.

12   The apex of the bladder is attached to the umbilicus by the _____.

13   The blood supply of the urinary bladder is mainly delivered by the superior and inferior _____ arteries.

14   The control of the detrusor is by parasympathetic axons in the pelvic _____ nerves.

15   The _____ are smooth muscle fibers at the bladder neck that prevent urine flow.

16 *The external urethral sphincter is part of the _____ diaphragm.*

17 *In males, ducts of the bulbourethral glands enter the urethra at the _____.*

18 *The dilated segment of the urethra in the bulb of the penis is called the _____.*

19 *The male urethra has small depressions or recesses in its mucosa called _____.*

20 *The urethra meets the external environment at the _____.*

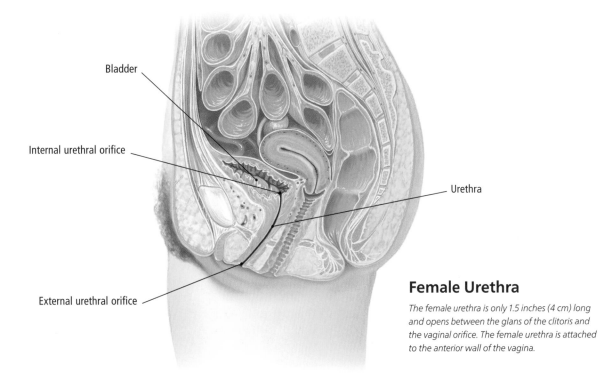

Bladder

Internal urethral orifice

Urethra

External urethral orifice

Female Urethra

The female urethra is only 1.5 inches (4 cm) long and opens between the glans of the clitoris and the vaginal orifice. The female urethra is attached to the anterior wall of the vagina.

Carcinoma of the bladder

Cancer in the urinary bladder usually arises from the lining epithelium. Risk factors include smoking, family history, prior radiation treatment, recurrent urinary tract infections, and exposure to industrial chemicals (dyes like benzidine, or 2-naphthylamine). Symptoms include blood in the urine (hematuria), pain when urinating, and lower back pain. Shaving at cystoscopy (insertion of instruments into the bladder) can treat small superficial tumors, but large advanced tumors require removal of the entire bladder (cystectomy).

Match the statement to the reason

1	The epithelium of the ureter is called transitional because…	**a**	this region can be narrower than other parts of the ureter.
2	The pelviureteric junction may be a site of urinary tract obstruction because…	**b**	buildup of fluid upstream dilates the calyceal system.
3	The ureter is capable of peristalsis because…	**c**	the ureter takes an oblique course through the wall of the urinary bladder.
4	Reflux of urine into the terminal ureter is prevented because…	**d**	it can change from columnar to squamous depending on the amount of dilation of the lumen.
5	Congenital defects in the ureter can cause hydronephrosis because…	**e**	its wall contains a thick layer of smooth muscle.

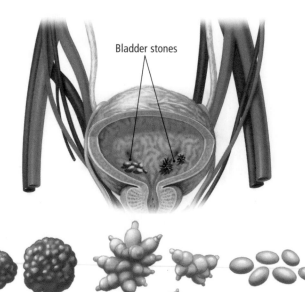

Bladder stones

Bladder Stones

A buildup of salts, cholesterol, and some proteins are the main cause of the formation of bladder stones. These bladder stones can develop without symptoms, until they cause blockage to the outlet of the bladder. Surgery or ultrasound treatment is usually required.

1 Hypertrophy of the central glands of the prostate can cause urinary tract obstruction because…

a a highly dilated bladder rises above the level of the pubic symphysis.

2 Catheterization of the male urethra must be done with care because…

b the urethra passes through the center of the prostate.

3 Urinary tract infection is more common in women because…

c the median lobe of the prostate lies immediately inferior to the bladder wall.

4 Prostatic hypertrophy can cause the formation of a lump in the bladder called the uvula because…

d the female urethra is shorter than the male.

5 Evacuation of a distended bladder can be done with a suprapubic catheter because…

e the urethra takes a right-angle turn at the intrabulbar fossa.

Bladder calculi

Bladder stones, or calculi, are crystals that have precipitated out of the urine to form hard objects. They may be composed of calcium phosphate, struvite, calcium oxalate, cystine, and urate. Bladder stones may abrade the bladder mucosa, leading to blood in the urine (hematuria), and may also cause painful urination, particularly at the end of the flow. If the stones have reached a size that exceeds the internal diameter of the urethra, they cannot be passed naturally and must be removed either by surgery or by fragmentation with ultrasound (lithotripsy). The crystalline nature of bladder calculi means that they can be shattered by ultrasound frequencies that are harmless to living tissue.

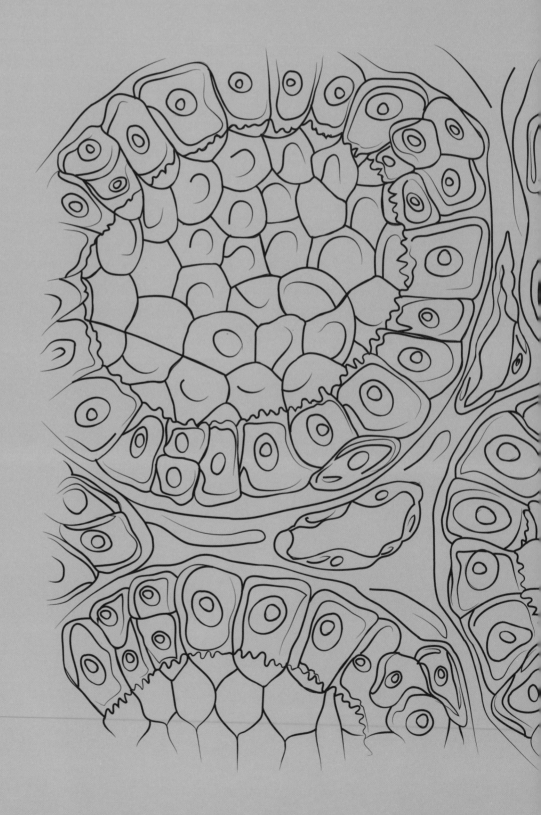

CHAPTER 8:
THE ENDOCRINE SYSTEM

Endocrine System Overview

The endocrine system is a group of ductless glands that are concerned with regulation of a constant internal state of the body (homeostasis), with the body clock, or with growth and reproductive function. They secrete their products (hormones) either directly into the bloodstream or into body cavities. Hormones may be small protein molecules (peptides) or may be derived from the cholesterol molecule (steroids) and exert their actions by changing the metabolic function of cells throughout the body.

Key terms:

Adrenal glands These are endocrine glands on top of the kidneys, hence their alternative name of suprarenal glands. They contain cortical and medullary tissue. The cortex produces cortisol and the mineralocorticoid hormone known as aldosterone; the medulla produces epinephrine and norepinephrine.

Ovaries Each ovary produces female sex cells, secretes estrogen and progesterone, and regulates the postnatal growth of reproductive organs and the development of secondary sexual characteristics (e.g., pubic hair and female fat distribution).

Pancreas A mixed endocrine and exocrine organ. The islets of Langerhans, which make insulin, glucagon, and somatostatin, are the endocrine component. The exocrine component produces proteases, amylase, and lipase for digestion.

Parathyroid glands Four small glands posterior to the thyroid gland lobes. They produce parathyroid hormone for regulation of calcium metabolism.

Pituitary gland The master gland of the endocrine system. It receives input from the hypothalamus and secretes regulatory hormones to control other endocrine glands. It is divided into an anterior part (adenohypophysis), which is controlled by secreted chemical factors from the hypothalamus, and a posterior part (neurohypophysis), which receives axons from nerve cells in the hypothalamus.

Testes Testes are endocrine organs (producing testosterone) and the site of production of sperm cells (spermatogenesis).

Thymus A lymphoid organ in the anterior mediastinum of the chest. It is active before puberty to produce T lymphocytes but becomes a fatty-fibrous remnant in adult life.

Thyroid gland A gland that concentrates iodine from the blood to produce the hormones thyroxine and triiodothyronine for regulation of the body's metabolic rate.

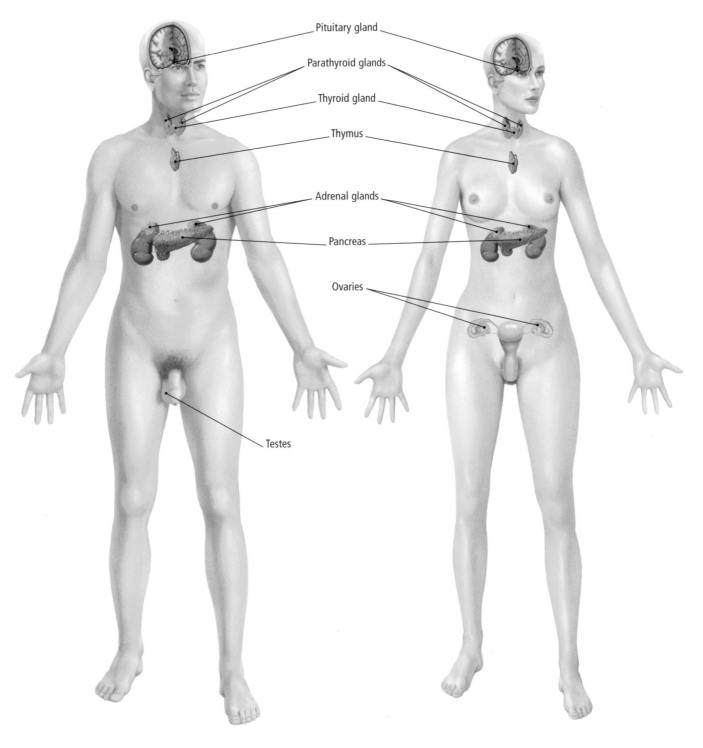

Pituitary gland

Parathyroid glands

Thyroid gland

Thymus

Adrenal glands

Pancreas

Ovaries

Testes

**Male Endocrine System—
anterior view**

**Female Endocrine System—
anterior view**

True or false?

1   *All hormones are steroid molecules.*

2   *Steroid hormones exert their effect by binding to receptors on the cell surface.*

3   *The pituitary develops exclusively from the roof of the mouth.*

4   *Steroid hormones are derived from the cholesterol molecule.*

5   *Peptide hormones are resistant to degradation by stomach juices.*

6   *The adrenal cortex and gonads are the main producers of steroid hormones.*

7   *Hormones exert their main effects over seconds to minutes.*

8   *Steroid hormones are lipid soluble.*

9   *Steroid hormones are usually transported in the blood on carrier proteins.*

10   *All endocrine glands are derived from the surface layers of the embryo (endoderm and ectoderm).*

11   *All hormones are carried by the bloodstream.*

12   *Peptide hormones are less than four amino acids long.*

13   *A key component of endocrine function is the positive feedback loop.*

14   *Nonsteroidal hormones exert their effects by a second messenger system embedded in the target cell membrane.*

15   *Steroidal hormones bind to a receptor in the cytoplasm that then moves to the nucleus.*

Multiple choice

1 Which of the following is the master gland of the endocrine system?

- (A) thyroid
- (B) parathyroid
- (C) pituitary
- (D) pancreas
- (E) adrenal

2 Which endocrine gland has the most direct connection with the hypothalamus?

- (A) thyroid
- (B) parathyroid
- (C) adrenal
- (D) pancreas
- (E) pituitary

3 Which endocrine gland lies in the center of the cranial cavity?

- (A) thyroid
- (B) parathyroid
- (C) pineal
- (D) pituitary
- (E) adrenal

4 Which endocrine gland would produce a lump in the throat when enlarged?

- (A) thyroid
- (B) parathyroid
- (C) pineal
- (D) pituitary
- (E) adrenal

5 Which endocrine gland is combined with an exocrine gland?

- (A) pituitary
- (B) pineal
- (C) testis
- (D) pancreas
- (E) adrenal

6 Which endocrine gland lies against the posterior abdominal wall?

- (A) ovary
- (B) parathyroid
- (C) adrenal
- (D) testis
- (E) pineal

7 Which endocrine gland has discrete cortex and medullary regions?

- (A) pituitary
- (B) testis
- (C) pancreas
- (D) thyroid
- (E) adrenal

Mechanisms of Action of Hormones

Steroidal hormones are able to enter the cell and bind to receptors, which translocate to the nucleus to affect protein synthesis. Nonsteroidal hormones act via receptors at the cell surface. Binding of the peptide hormone to the receptor activates an enzyme cascade that generates secondary messengers like cAMP to affect cellular function.

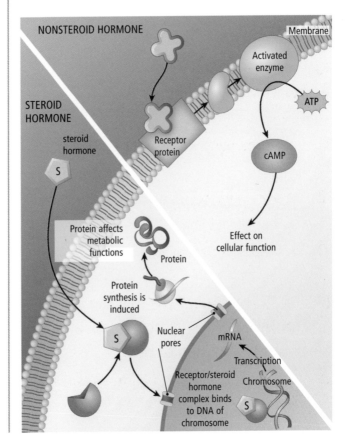

NONSTEROID HORMONE — Membrane — Activated enzyme — ATP — cAMP — Effect on cellular function — STEROID HORMONE — steroid hormone — S — Receptor protein — Protein affects metabolic functions — Protein — Protein synthesis is induced — S — Nuclear pores — mRNA — Transcription — Receptor/steroid hormone complex binds to DNA of chromosome — Chromosome — S

Fill in the blanks

1 Surgery on the pituitary could be done by accessing the gland through the _____ cavity.

2 The _____ are found on the posterior surface of the thyroid gland.

3 The neuroendocrine system includes the _____ and _____ glands.

4 The thyroid gland consists of two _____ joined by an isthmus.

5 The thyroid gland is located around the _____ and _____.

6 The _____ is a mixed endocrine and exocrine gland.

7 The _____ could be accessed surgically through the posterior abdominal wall.

8 The right _____ gland is in direct contact with the liver.

9 The adrenal _____ is the source of much of the adrenaline (epinephrine) released in the body.

10 The adrenal _____ is the source of cortisol and mineralocorticoids.

Pituitary gland

Anterior pituitary

BONE AND
MUSCLE GROWTH
Growth hormone (GH)

ADRENAL CORTEX
Adrenocorticotropic
hormone (ACTH)

THYROID GLAND
Thyroid-stimulating
hormone (TSH)

TESTIS AND OVARY
Follicle-stimulating hormone (FSH)
and luteinizing hormone (LH)

SKIN
Melanocyte-stimulating
hormone (MSH)

MAMMARY GLANDS
Prolactin (PRL)

Mechanism of Action of the Anterior Pituitary

The cells of the anterior pituitary produce a variety of hormones with diverse effects throughout the body. The anterior pituitary cells that produce these hormones are themselves regulated by releasing inhibitory factors secreted by the cells of the hypothalamus.

Match the statement to the reason

1 A tumor of the pituitary gland can cause loss of temporal field vision because…

2 Careless removal of the thyroid gland can cause problems with calcium metabolism because…

3 Failure of migration of the thyroid gland can cause a lump on the tongue because…

4 The pineal gland can be seen in a skull X-ray because…

5 Repeated bouts of pancreatitis may lead to diabetes mellitus because…

6 Peptide hormones must bind to the target cell surface because…

7 Tumors of the hypothalamus may greatly affect endocrine function because…

a calcification of the gland occurs during middle age.

b the parathyroid glands may also be inadvertently removed.

c autodigestion and inflammation of the pancreas may damage the islets of Langerhans.

d this gland arises at the lingual foramen cecum.

e the optic chiasm lies directly superior to the gland.

f that structure directs the activity of the different cell types in the pituitary.

g they are not lipid soluble and cannot diffuse across the lipid bilayer cell membrane.

The Pituitary, Pineal, Thyroid, and Parathyroid Glands

Two endocrine glands lie within the cranial cavity. The pituitary gland is situated in a fossa beneath the hypothalamus, and the pineal gland is situated in the center of the cranial cavity. The pituitary gland is the master endocrine gland because it directs the activities of the other glands, whereas the pineal gland is concerned with circadian rhythms. The thyroid and parathyroid glands are located in the lower neck. The parathyroids are small pea-sized glands attached to the back of the thyroid.

Key terms:

Anterior pituitary (adenohypophysis) The anterior lobe of the pituitary gland. It is highly vascularized and produces hormones in epithelial cells, where they are stored in granules and released as required.

Axon The long process of a nerve cell that transmits the impulse to another part of the brain or body.

Corpus callosum A large fiber bundle joining the two cerebral hemispheres. It carries over 300 million axons to permit cross-talk between the two sides of the forebrain.

Endocrine glands Endocrine glands secrete chemical messengers called hormones into the bloodstream or body cavities. The major endocrine glands are the pituitary, pineal, adrenal, thyroid, parathyroid, and endocrine pancreas glands.

Hypophyseal artery Branches of the internal carotid artery that supply the pituitary gland.

Hypophyseal portal system A portal venous system that carries blood with releasing and regulatory factors from some hypothalamic nuclei to the anterior lobe of the pituitary.

Hypothalamus The inferior part of the diencephalon. It is involved in homeostasis (maintaining a constant internal environment), motivation, and control of the autonomic nervous system.

Mammillary body Part of the hypothalamus. The two mammillary bodies are part of a circuit involved in laying down memories. They are named because of their similarity to small breasts.

Neurosecretory cells Cells of the hypothalamus that produce oxytocin and vasopressin and send these down their axons to the posterior pituitary.

Optic chiasm The region where the axons from the medial (nasal) retina cross to the other side of the brain.

Parathyroid glands Four small glands posterior to the thyroid gland lobes. They produce parathyroid hormone for regulation of calcium metabolism.

Pineal gland A gland attached to the superior surface of the diencephalon. It secretes the hormone melatonin and is involved in biological rhythms. It is also known as the pineal body.

Pituitary gland See pp. 250–251.

Pituitary stalk A stalk connecting the pituitary gland to the underside of the hypothalamus. It carries the portal system to the anterior pituitary and the axons of the hypothalamo-hypophyseal tract to the posterior pituitary.

Posterior pituitary (neurohypophysis) The posterior lobe of the pituitary gland. It contains the axon terminals of neurosecretory cells in the hypothalamus.

Superior and inferior colliculi Paired elevations on the dorsal (posterosuperior) surface of the midbrain. The superior colliculus is involved in vision. The inferior colliculus is involved in hearing.

Thalamus An egg-shaped region on each side of the third ventricle. It acts as a relay station between the brain stem and the cortex. The thalamus contains nuclei involved in vision, hearing, and touch, as well as nuclei engaged in feedback motor circuits.

Thyroid gland See pp. 250–251.

Vein A low-pressure vessel returning blood to the heart. Veins have a relatively thin tunica media.

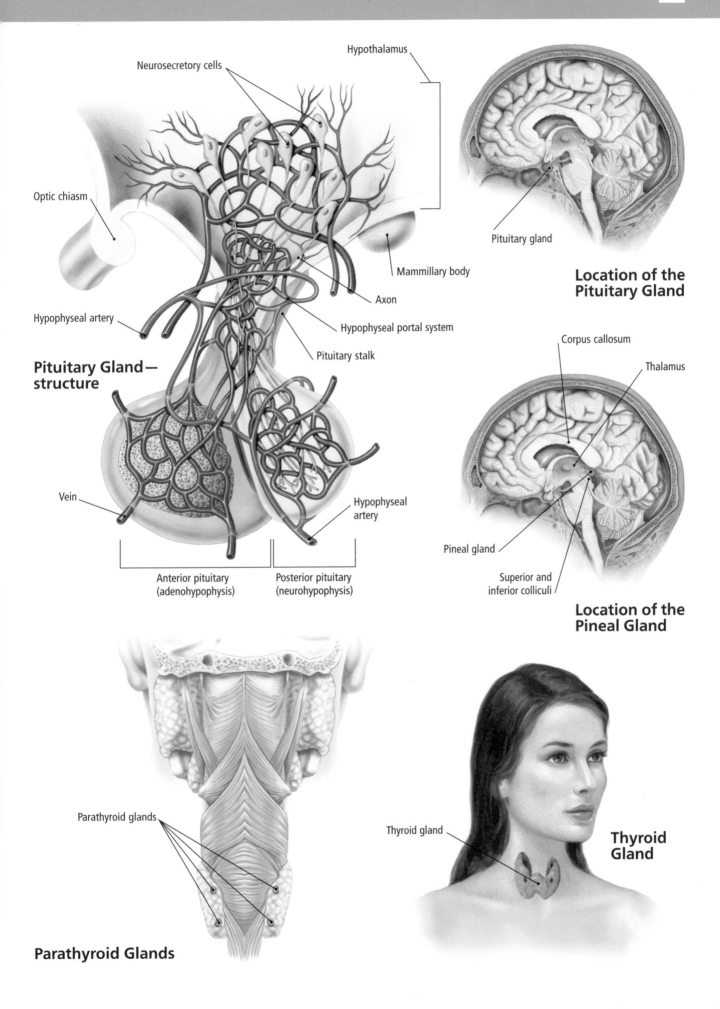

Pituitary Gland—structure

Neurosecretory cells

Hypothalamus

Optic chiasm

Mammillary body

Axon

Hypophyseal artery

Hypophyseal portal system

Pituitary stalk

Vein

Hypophyseal artery

Anterior pituitary (adenohypophysis)

Posterior pituitary (neurohypophysis)

Location of the Pituitary Gland

Pituitary gland

Corpus callosum

Thalamus

Pineal gland

Superior and inferior colliculi

Location of the Pineal Gland

Parathyroid glands

Parathyroid Glands

Thyroid gland

Thyroid Gland

True or false?

1   ACTH is produced in the posterior pituitary gland.

2   Acidophil cells of the anterior pituitary produce luteinizing hormone.

3   Prolactin is made by acidophil cells of the anterior pituitary.

4   Basophil cells of the anterior pituitary produce follicle-stimulating hormone.

5   The hormone oxytocin is released from the axons of the hypothalamo-neurohypophyseal tract.

6   The interstitial cells of the pineal gland are secretory in function.

7   Thyroid hormones are synthesized from iodine and the amino acid cysteine.

8   Parafollicular C cells are the source of calcitonin.

9   The parathyroid glands contain follicles filled with colloid.

10   The oxyphil cells of the parathyroid gland secrete parathyroid hormone.

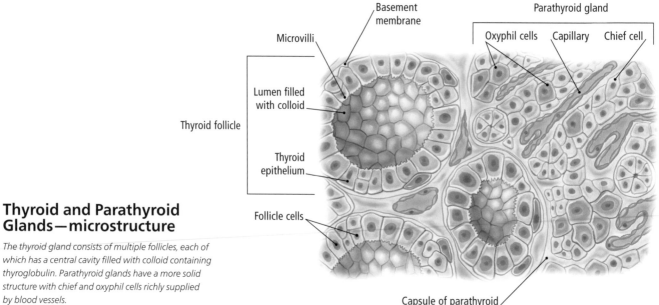

Thyroid and Parathyroid Glands—microstructure

The thyroid gland consists of multiple follicles, each of which has a central cavity filled with colloid containing thyroglobulin. Parathyroid glands have a more solid structure with chief and oxyphil cells richly supplied by blood vessels.

Multiple choice

1 *Which tract connects the hypothalamus to the posterior pituitary?*

(A) fornix
(B) hypothalamo-neurohypophyseal tract
(C) hypothalamo-induseum tract
(D) corpus callosum
(E) pyramidal tract

2 *Which anterior pituitary hormone is directly involved with mammary gland growth?*

(A) thyrotropin
(B) luteinizing hormone
(C) follicle-stimulating hormone
(D) prolactin
(E) oxytocin

3 *Which anterior pituitary hormone stimulates production of insulin-like growth factor 1 by the liver?*

(A) prolactin
(B) thyrotropin
(C) follicle-stimulating hormone
(D) luteinizing hormone
(E) growth hormone

4 *Which is the most common type of hormone-secreting cell in the anterior pituitary?*

(A) growth hormone
(B) luteinizing hormone
(C) prolactin
(D) follicle-stimulating hormone
(E) thyrotropin

5 *Which hormone causes uterine contraction during labor?*

(A) antidiuretic hormone
(B) oxytocin
(C) follicle-stimulating hormone
(D) prolactin
(E) growth hormone

6 *Which of the following is a direct effect of thyroid hormones?*

(A) decreased serum calcium
(B) increased muscle blood flow
(C) increased heart rate
(D) increased metabolic rate
(E) increased blood glucose

7 *Which hormone acts to elevate calcium levels in the blood?*

(A) parathyroid hormone
(B) calcitonin
(C) thyrotropin
(D) luteinizing hormone
(E) growth hormone

Pituitary adenomas

The pituitary gland is situated below the hypothalamus in the pituitary fossa of the sphenoid bone. Benign tumors may arise in the gland and compress neighboring structures such as the optic chiasm, causing blindness in the temporal fields of both eyes (bitemporal hemianopia). These tumors may also secrete hormones such as growth hormone and adrenocorticotropic hormone (ACTH), which induce physiological changes in the patient. Treatment is often by surgical removal of the tumor through the nasal cavity and sphenoid sinus.

Color and label

i) Label each structure shown on the illustrations.

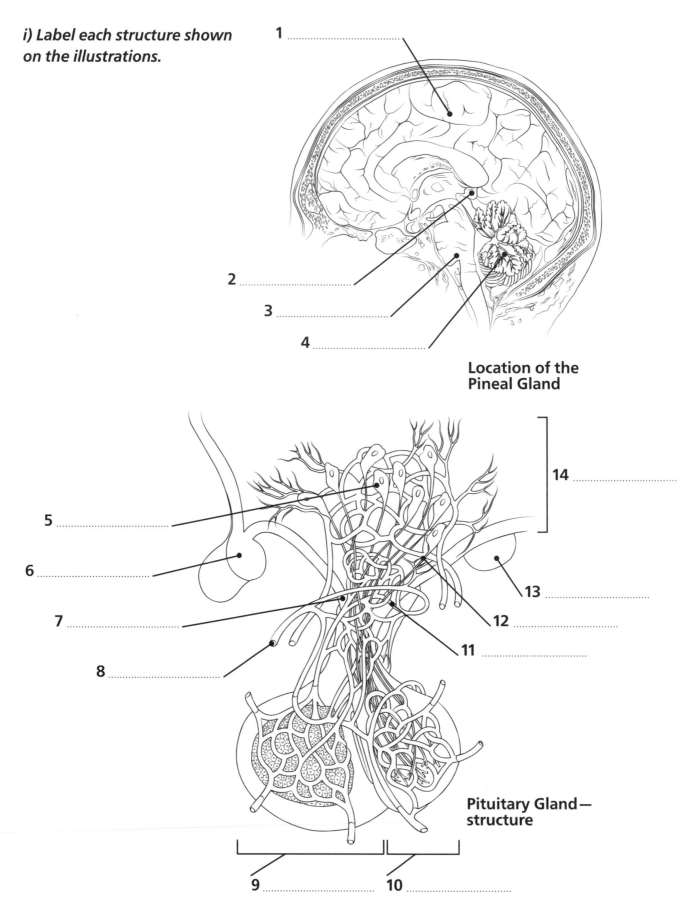

1 ....

2 ....

3 ....

4 ....

Location of the Pineal Gland

5 ....

6 ....

7 ....

8 ....

14 ....

13 ....

12 ....

11 ....

Pituitary Gland— structure

9 ....

10 ....

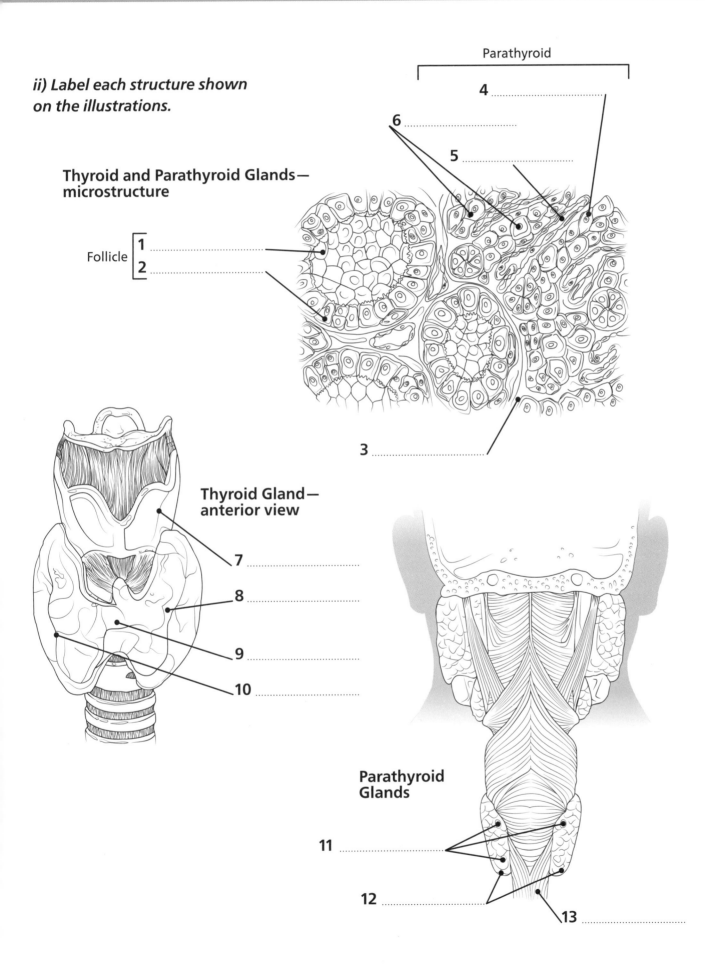

ii) Label each structure shown on the illustrations.

Parathyroid

4 ....

6 ....

5 ....

Thyroid and Parathyroid Glands— microstructure

Follicle
1 ....
2 ....

3 ....

Thyroid Gland— anterior view

7 ....

8 ....

9 ....

10 ....

Parathyroid Glands

11 ....

12 ....

13 ....

Fill in the blanks

1   Growth hormone secretion from the anterior pituitary is inhibited by _____ from the hypothalamus.

2   The _____ system carries blood from the hypothalamus to the anterior pituitary.

3   The _____ and _____ nuclei of the hypothalamus give rise to the hypothalamo-neurohypophyseal tract.

4   With routine hematoxylin-eosin staining, three types of cells can be seen in the anterior pituitary: _____, _____, and _____.

5   Peak production of growth hormone occurs in the _____ before _____.

6   Excess growth hormone in adults causes _____.

7   Decreased secretion of growth hormone during development causes _____.

8   In males, luteinizing hormone stimulates the production of _____ by _____ cells.

9   In women, _____ stimulates the development of the ovarian follicle.

10   ACTH or corticotropin acts on the adrenal cortex to stimulate _____ release.

11   The terminals of axons of the hypothalamo-neurohypophyseal tract are called _____.

12   Neurogenic diabetes insipidus is due to inadequate secretion of _____ by the posterior pituitary.

13   The nerve supply to the pineal gland is by sympathetic fibers from the _____.

14   The embryonic origin of parafollicular (C cells) is the _____.

15 Dietary iodide deficiency leads to enlargement of the thyroid gland called _____.

16 The most potent thyroid hormone is _____.

17 Pituitary control of the thyroid gland is by _____ (_____).

18 The factor that promotes absorption of calcium from the gut is _____.

19 The _____ cells of the parathyroid gland may be transitional chief cells.

20 The hormone _____ suppresses the mobilization of calcium from bone by osteoclasts.

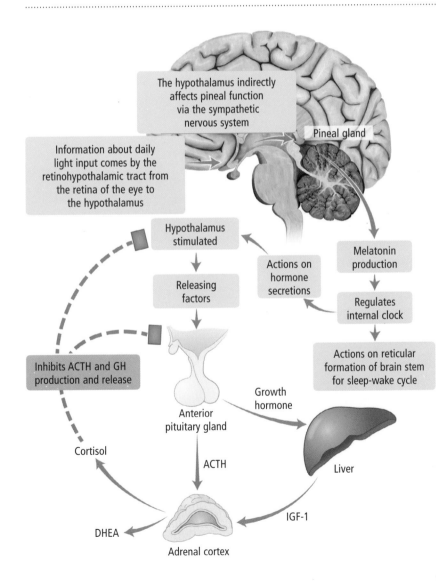

Regulation of Daily Cycles by Hormones

Many functions of the brain and endocrine system are regulated to follow a diurnal or daily cycle by the secretion of melatonin from the pineal gland above the thalamus of the brain. Both growth hormone and corticosteroids follow regular daily cycles of peaks and troughs of secretion. Melatonin secretion is itself regulated by light input from the retina to the hypothalamus. Note: DHEA = dehydroepiandrosterone; IGF-1 = insulin-like growth factor 1.

Match the statement to the reason

1 Hypersecretion of prolactin in women can cause infertility because…

a its secretion is highest during night hours.

2 An adenoma of growth hormone–secreting acidophil cells will cause gigantism because…

b this hormone inhibits ovulation and causes few or irregular menstrual periods.

3 The anterior pituitary does not have a blood-brain barrier because…

c the retinohypothalamic tract activates a pathway that modulates pineal secretion of melatonin.

4 Melatonin is often called the hormone of darkness because…

d that would obstruct the free movement of hormones into the bloodstream.

5 The circadian rhythm is set by light exposure because…

e exposure to the hormone occurs before the closure of the skeletal epiphyseal plates.

Graves' disease and thyrotoxicosis

In Graves' disease, the thyroid gland is hyperfunctional due to stimulation by abnormal autoantibodies. The body's T cells produce these against the thyrotropin receptor, so the autoantibodies mimic the effect of thyrotropin and increase thyroid activity. The thyroid gland enlarges (producing a goiter), and the cells of the follicles become columnar in shape. This hyperactive gland produces more thyroid hormone, the eyes bulge (exophthalmos) due to accumulation of material behind the eyeball, and the patient experiences warm skin, increased heart rate (tachycardia), difficulty sleeping, and fine finger tremors.

1   Dietary iodine deficiency causes enlargement of the thyroid gland because…

a   the hormone controls the differentiation and activity of osteoclasts.

2   A tumor of the parathyroid gland can cause hypercalcemia because…

b   excess parathyroid hormone causes bone demineralization.

3   Autoimmune damage to the thyroid gland in Hashimoto's disease can cause hypothyroidism because…

c   thyroid hormone is essential for proper development of brain tissue.

4   Lack of thyroid hormone during fetal development causes intellectual disability because…

d   the gland accumulates large amounts of colloid.

5   Parathyroid hormone directly regulates bone reabsorption because…

e   antibodies in this disease damage thyroid peroxidase, the enzyme responsible for linking iodine with thyroglobulin.

6   Patients with Graves' disease will have reduced secretion of thyroid-stimulating hormone from the anterior pituitary gland because…

f   vasopressin/antidiuretic hormone release is impaired.

7   Neurogenic diabetes insipidus can follow damage to the posterior pituitary gland because…

g   the excess release of thyroid hormone causes feedback inhibition.

The Adrenal Glands and Pancreas

The adrenal or suprarenal glands are located on the superior poles of the kidneys. They are each separated from the kidney by only a thin layer of fascia and loose connective tissue. Each adrenal gland has a cortex and a medulla with distinct functions and secretions. The pancreas is a mixed endocrine/exocrine gland in the upper abdomen. The endocrine component of the pancreas is composed of the islets of Langerhans, which are dispersed throughout the gland.

Key terms:

Accessory pancreatic duct A small duct of the exocrine pancreas that drains juices from the head of the pancreas to the lesser duodenal papilla.

Adrenal cortex The adrenal cortex is the outer zone of the gland. It has three regions: zona glomerulosa (making aldosterone), zona fasciculata (making cortisol), and zona reticularis (making steroid sex hormones).

Adrenal gland See pp. 250–251.

Adrenal medulla The adrenal medulla contains modified postganglionic sympathetic nerve cells that produce epinephrine and norepinephrine.

Body of pancreas The part of the pancreas lying in front of the aorta.

Capsular artery One of the arteries running through the capsule of the adrenal gland.

Capsule The connective tissue layer that surrounds the adrenal gland.

Deep plexus A network of capillaries within the zona reticularis before the blood enters the medullary plexus.

Endocrine glands Endocrine glands secrete chemical messengers called hormones into the bloodstream or body cavities. The major endocrine glands are the the pituitary, pineal, adrenal, thyroid, parathyroid, and endocrine pancreas glands.

Head of pancreas The part of the pancreas encircled by the duodenum. It has the main and accessory pancreatic ducts passing through it, as well as the bile duct.

Kidney See pp. 218–219.

Main pancreatic duct The major duct draining the tail, body, neck, and head of the pancreas. It is often joined by the bile duct to form a hepatopancreatic ampulla that opens into the second part of the duodenum.

Medulla The central part of the adrenal gland that contains modified sympathetic nerve cells. These produce the hormones epinephrine and norepinephrine.

Medullary plexus A plexus of veins in the medulla to carry away the hormones from both the adrenal cortex and medulla.

Medullary vein Large veins of the adrenal medulla drain blood and hormones to the suprarenal (adrenal) veins.

Neck of pancreas The narrow part of the pancreas between the head and body.

Pancreas See pp. 250–251.

Sinusoidal vessel Fenestrated capillaries passing through the zona glomerulosa and zona fasciculata collecting secreted hormones.

Subcapsular plexus Blood vessels derived from the superior and middle suprarenal arteries. They give rise to fenestrated sinusoidal vessels.

Suprarenal artery The adrenal gland has a rich arterial supply from the aorta and inferior phrenic and renal arteries.

Tail of pancreas The part of the pancreas that projects into the lienorenal ligament and is in contact with the hilum of the spleen.

Uncinate process of pancreas The projection of the head of the pancreas to the left, thereby forming a pancreatic notch through which the superior mesenteric artery and vein pass.

Zona fasciculata This region makes up 75 percent of the cortex and is formed of cuboid cells arranged in longitudinal cords. It makes cortisol.

Zona glomerulosa This region makes up only 10 to 15 percent of the cortex. Its cells are aggregated into glomerulus-like arrangements and have lipid droplets in their cytoplasm. It makes aldosterone.

Zona reticularis This region makes up 5 to 10 percent of the cortex. Its cells are arranged in a network of short cords penetrated by fenestrated capillaries. It makes sex steroid hormones.

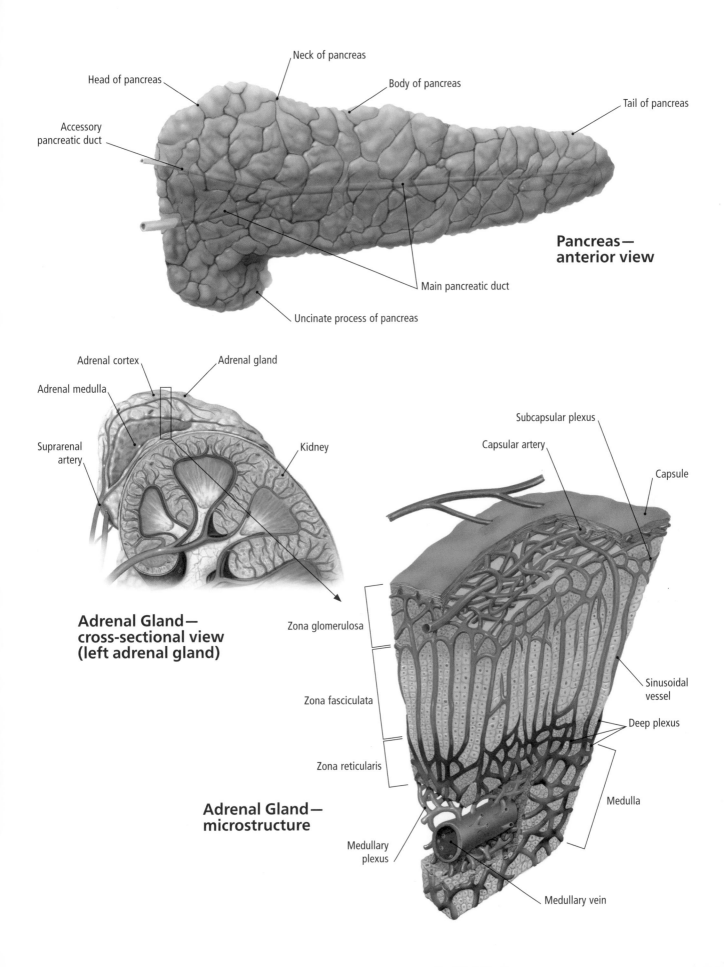

Neck of pancreas

Head of pancreas

Body of pancreas

Tail of pancreas

Accessory
pancreatic duct

**Pancreas—
anterior view**

Main pancreatic duct

Uncinate process of pancreas

Adrenal cortex

Adrenal gland

Adrenal medulla

Suprarenal
artery

Kidney

**Adrenal Gland—
cross-sectional view
(left adrenal gland)**

Subcapsular plexus

Capsular artery

Capsule

Zona glomerulosa

Zona fasciculata

Sinusoidal
vessel

Deep plexus

Zona reticularis

**Adrenal Gland—
microstructure**

Medulla

Medullary
plexus

Medullary vein

True or false?

1   *The adrenal medulla is derived from modified autonomic neurons.*

2   *The outermost layer of the adrenal cortex is the zona reticularis.*

3   *Secretion of aldosterone by the adrenal cortex is under the control of angiotensin II.*

4   *Adrenal sex steroids are made by the adrenal medulla.*

5   *Adrenal cortex cells contain large droplets of lipid (fat).*

6   *Adrenal medullary hormones are synthesized from the amino acid tyrosine.*

7   *Most of the chromaffin cells of the adrenal medulla secrete noradrenaline (norepinephrine).*

8   *Beta cells of the pancreatic islets produce glucagon.*

9   *Delta cells of the pancreatic islets make gastrin and somatostatin.*

10   *The endocrine pancreas makes up about 2 percent of the mass of the pancreas.*

Cushing's disease and Cushing's syndrome

Cushing's syndrome is a condition in which the body is exposed to excessive levels of corticosteroids. This may be due to the use of corticosteroids as medication for inflammatory conditions and is characterized by a moon face, "buffalo hump" between the shoulders, stretch marks on the abdomen, elevated blood pressure, acne, obesity, bone weakness (osteoporosis), and muscle wasting. One particular type is called Cushing's disease, in which a tumor in the anterior pituitary secretes excessive amounts of adrenocorticotropic hormone (ACTH). In Cushing's disease, there is increased production of cortisol by the zona fasciculata of the adrenal cortex. The condition can be treated by removing the tumor.

Multiple choice

1 *Where are mineralocorticoids normally synthesized?*
(A) adrenal medulla
(B) zona glomerulosa of adrenal cortex
(C) zona fasciculata of adrenal cortex
(D) zona reticularis of adrenal cortex
(E) oxyphil cells of adrenal capsule

2 *Where are steroid sex hormones normally synthesized in the adrenal gland?*
(A) adrenal medulla
(B) zona glomerulosa of adrenal cortex
(C) zona fasciculata of adrenal cortex
(D) zona reticularis of adrenal cortex
(E) none of the above

3 *Which steroid sex hormones are most commonly produced by the adrenal cortex?*
(A) estrogen and progesterone
(B) dehydroepiandrosterone and androstenedione
(C) testosterone and inhibin
(D) estrogen and testosterone
(E) none of the above are correct

4 *Which zone makes up the largest proportion of the adrenal cortex?*
(A) zona glomerulosa
(B) zona fasciculata
(C) zona reticularis
(D) zona chromaffinoma
(E) zona orbicularis

5 *The adrenal gland is supplied by which of the following arteries?*
(A) inferior phrenic artery
(B) superior mesenteric artery
(C) renal artery
(D) splenic artery
(E) both A and C are correct

6 *Which pancreatic islet hormone directly increases blood glucose levels?*
(A) insulin
(B) somatostatin
(C) gastrin
(D) glucagon
(E) pancreatic polypeptide

7 *Which of the following is made by the alpha cells of the pancreatic islets?*
(A) glucagon
(B) insulin
(C) somatostatin
(D) gastrin
(E) pancreatic polypeptide

Islets of Langerhans

The islets of Langerhans are clusters of cells within the pancreas. These islets are composed of three types of cells: alpha cells, beta cells, and delta cells. Hormones produced by the alpha and beta cells raise and lower blood sugar levels, respectively, while the delta cells produce somatostatin, which can inhibit the release of both glucagon and insulin.

Insuloacinar portal vessels

Beta cells (producing insulin)

Islet of Langerhans

Alpha cells (producing glucagon)

Delta cells (producing somatostatin)

Color and label

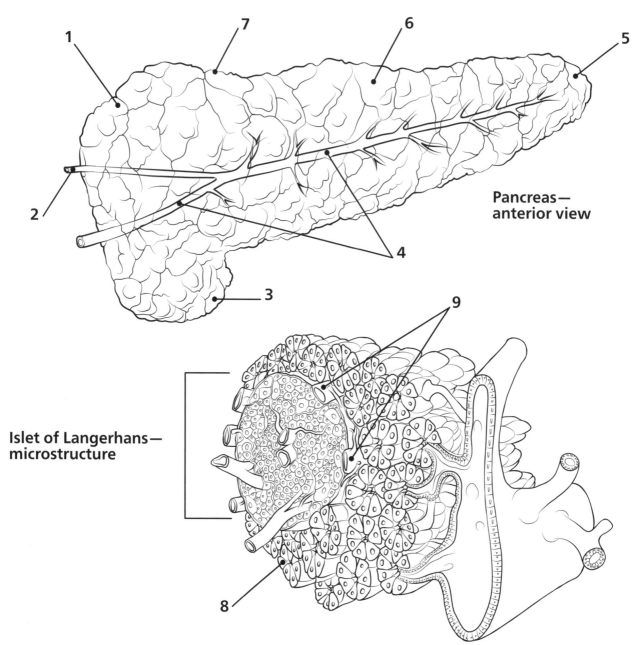

Pancreas— anterior view

Islet of Langerhans— microstructure

i) Add numbers to the boxes below to match each label to the correct part of the illustrations.

Main pancreatic duct	☐	
Tail (of pancreas)	☐	
Neck (of pancreas)	☐	
Head (of pancreas)	☐	
Accessory pancreatic duct	☐	

Insuloacinar portal vessels	☐
Uncinate process	☐
Exocrine pancreas	☐
Body (of pancreas)	☐

ii) Label each structure shown on the illustrations.

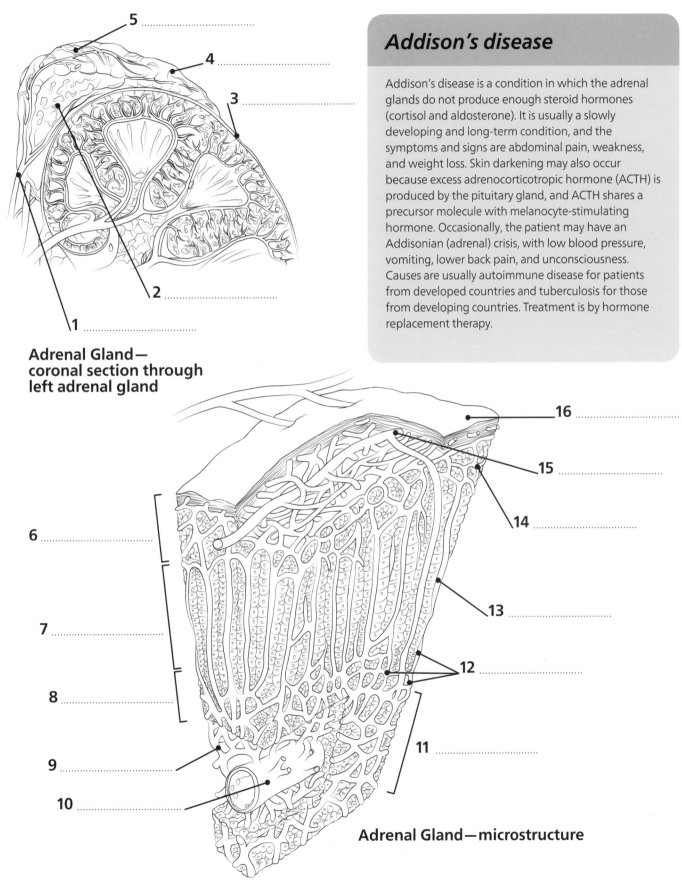

Adrenal Gland— coronal section through left adrenal gland

Adrenal Gland—microstructure

Fill in the blanks

1 The pituitary hormone that stimulates cortisol production by the adrenal cortex is
_____.

2 In adrenogenital secretion, women may develop masculinization signs such as excess sexual hair
(_____) and enlargement of the _____.

3 Cortisol may increase blood glucose by stimulating _____ in the liver.

4 The actions of the hormones from the adrenal medulla are mediated by their action on adrenergic
_____ and _____ receptors.

5 Adrenal medullary hormones have a _____ effect on arterioles in cardiac muscle.

6 All venous drainage of the adrenal gland is by the _____ of the adrenal medulla.

7 The pancreatic hormone _____ inhibits release of both glucagon and insulin.

8 Venules drain blood from the _____ to the exocrine pancreas to allow endocrine
control of exocrine function.

9 _____ cells are the main cell type in the core of pancreatic islets.

10 Glucagon increases blood glucose levels by stimulating _____ in the liver.

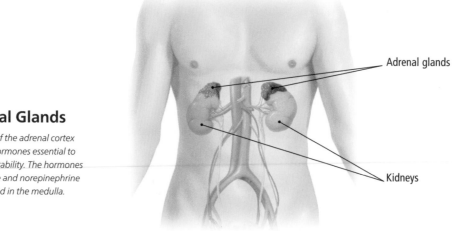

Adrenal Glands

*Each zone of the adrenal cortex
produces hormones essential to
metabolic stability. The hormones
epinephrine and norepinephrine
are produced in the medulla.*

Match the statement to the reason

1 Adrenogenital syndrome causes masculinization of females because…

a this hormone promotes urinary retention of sodium and excretion of potassium and hydrogen ions.

2 Adrenal cortex cells contain large droplets of lipid because…

b excess catecholamines are released and cause arterial vasoconstriction.

3 Failure of aldosterone secretion leads to excess sodium loss from the kidney because…

c steroid hormones are made from cholesterol molecules.

4 Tumors of the adrenal medulla (pheochromocytoma) cause elevated blood pressure because…

d excess androgens are made by the adrenal gland.

5 A tumor of the zona glomerulosa can cause excess aldosterone secretion because…

e that is the region where mineralocorticoids are usually produced.

Diabetes mellitus

Diabetes mellitus (DM) is a condition of inadequate control of blood glucose, which may be due to inadequate secretion of insulin from the pancreatic islets. This may be caused by autoimmune, toxic, or viral damage to the beta cells (type 1 DM), mostly but not always arising before age 25. The more common type 2 DM (80 percent of cases) arises when the body tissues develop resistance to insulin and do not take up glucose from the blood. Type 2 DM is usually seen in obese or genetically predisposed adults.

Another type of DM is gestational, seen in pregnant women. The symptoms and signs of DM include increased frequency of urination (polyuria) and thirst and increased fluid intake (polydipsia), which are both due to the osmotic effects of the elevated blood glucose on water loss at the kidney. Treatment is by weight loss and exercise (type 2 DM), oral hypoglycemic drugs (type 2 DM), or regular insulin injection (type 1 DM).

Male and Female Endocrine Glands

Apart from making the gametes (spermatozoon and oocytes), the male and female gonads also have important endocrine functions. The testes produce testosterone, activin, and inhibin; the ovaries produce estrogen, progesterone, activin, inhibin, and small amounts of androgens. Hormonal activities of the testes and ovaries are largely under the control of the anterior pituitary gland. Ovarian function can also be controlled by human chorionic gonadotropin (hCG) from the placenta.

Key terms:

Broad ligament A double-layered fold of peritoneum that encloses the uterine body, Fallopian (uterine) tube, and ovary. Blood vessels lie between the layers.

Corpus albicans A connective tissue scar left after the corpus luteum regresses.

Corpus luteum After ovulation, the follicular cell layer infolds and becomes this hormone-secreting organ. The corpus luteum makes progesterone and estrogen during the secretory stage of the menstrual cycle and during pregnancy.

Discharging follicle When a follicle has matured, it protrudes from the ovary surface. Enzymatic activity induced by a surge in luteinizing hormone causes the follicle to rupture.

Endocrine function The testes and ovaries have endocrine function. Each testis produces testosterone, which is responsible for male secondary sexual characteristics. Each ovary makes estrogen and progesterone as part of the menstrual cycle and during pregnancy.

Fallopian (uterine) tube A tube carrying sperm to the ovum and the fertilized ovum (zygote) to an implantation site in the uterine wall.

Mature ovum A mature follicle contains an ovum (egg), a fluid-filled space (antrum), and surrounding theca interna and theca externa layers.

Ovaries Each ovary produces female sex cells, secretes estrogen and progesterone, and regulates the postnatal growth of reproductive organs and the development of secondary sexual characteristics (e.g., pubic hair and female fat distribution).

Ovum An oocyte (egg). The "ovum" released at ovulation actually consists of an oocyte, as well as surrounding zona pellucida and corona radiata layers.

Scrotum See pp. 238–239.

Testes Testes are endocrine organs (producing testosterone) as well as the site of production of sperm cells (spermatogenesis).

Uterus The organ of gestation. The uterus has a fundus, body, isthmus, and cervix. The Fallopian (uterine) tubes enter at the junction of the fundus and body.

Vagina The female organ of copulation and the lower end of the birth canal. It is a highly dilatable tube extending from the cervix to the vaginal vestibule.

Testis

Scrotum

Testes—endocrine function

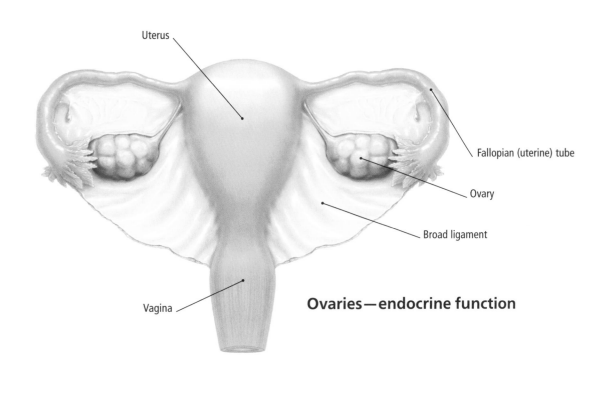

Uterus

Fallopian (uterine) tube

Ovary

Broad ligament

Vagina

Ovaries—endocrine function

Ovary—cross-sectional view

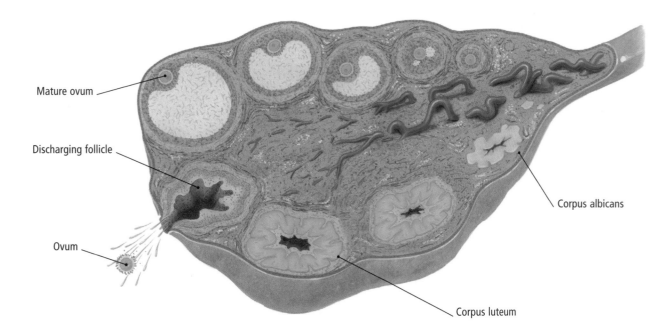

Mature ovum

Discharging follicle

Ovum

Corpus albicans

Corpus luteum

True or false?

1   *Luteinizing hormone and prolactin stimulate testosterone production by the testes.*

2   *Sertoli cells make androgen-binding protein under stimulation by follicle-stimulating hormone (FSH) from the anterior pituitary.*

3   *Leydig cells have very little lipid in their cytoplasm.*

4   *Feedback inhibition of FSH secretion by the anterior pituitary is by the hormone inhibin secreted from the Sertoli cells.*

5   *The prostate and seminal vesicles are directly stimulated by FSH from the anterior pituitary.*

6   *The enlargement of the larynx at puberty is a result of growth stimulated by testosterone.*

7   *Women have no circulating androgens.*

8   *FSH stimulates follicle development, ovulation, and production of estrogen.*

9   *Luteinizing hormone (LH) stimulates estrogen production by the corpus luteum.*

10   *Estrogen produced during the early menstrual cycle stimulates secretory activity by endometrial glands.*

Multiple choice

1 Functions of androgens include all of the following except

(A) sexual hair growth

(B) sebaceous gland secretion

(C) stimulation and maintenance of spermatogenesis

(D) maintenance of secretion by the prostate

(E) deposition of fat on the hips

2 Which of the following is an important precursor molecule for testosterone?

(A) cholesterol

(B) glycerol

(C) alanine

(D) guanine

(E) none of the above

3 Which of the following has a negative feedback effect on luteinizing hormone production by the anterior pituitary?

(A) inhibin

(B) activin

(C) testosterone

(D) thyroid hormone

(E) insulin

4 Which hormone secreted by Leydig cells has a positive effect on secretion of follicle-stimulating hormone by the anterior pituitary?

(A) inhibin

(B) activin

(C) testosterone

(D) thyroid hormone

(E) insulin

5 Which of the following produces androgens in the ovary?

(A) theca interna cell

(B) follicular cell

(C) zona pellucida

(D) oocyte

(E) none of the above

6 Which of the following most directly stimulates estrogen production in the ovary?

(A) luteinizing hormone

(B) follicle-stimulating hormone

(C) inhibin

(D) dehydroepiandrosterone

(E) activin

7 Which structure produces progesterone?

(A) uterine epithelium

(B) cervical epithelium

(C) vagina

(D) corpus luteum

(E) oocyte

Cyclical Changes in Ovary

The cyclical changes in the ovary are represented here in a series of illustrations of the developing ovarian follicle.

Secondary follicle

Primary follicle

Suspensory ligament of the ovary with blood vessels

Rupturing Graafian follicle

Oocyte

Corpus luteum produces progesterone

If no pregnancy occurs, the corpus luteum degenerates and progesterone levels fall.

Ovary

Fill in the blanks

1 *FSH stimulates production of _____ and _____ by the Sertoli cells of the testis.*

2 *Androgens play a key role in the differentiation of _____ in the male fetus.*

3 *The formation of female internal genitalia is inhibited in male fetuses by _____.*

4 *Excess female hormones in males leads to breast enlargement called _____.*

5 *_____ of the testis are responsible for production of testosterone.*

6 *The _____ phase of the menstrual cycle is mainly under the control of follicle-stimulating hormone.*

7 *The _____ cells of the ovarian follicle make androstenedione under the stimulation of luteinizing hormone.*

8 *The rupture of the ovarian follicle at ovulation is stimulated by _____.*

9 *The placenta produces _____ to maintain the corpus luteum.*

10 *The hormone _____ maintains the secretory function of endometrial glands.*

Ductus deferens

Testicular artery

Head of epididymis

Rete testis

Tunica albuginea

Septae

Body of epididymis

Efferent ductules

Tail of epididymis

Mediastinum testis

Seminiferous tubules

Lobules

Spermatozoa

Spermatid

Sertoli cell

Spermatocyte

Match the statement to the reason

1 Removal of the testes (castration) causes elevated levels of FSH and LH in the blood because…

a both FSH and LH are essential for sperm production.

2 Hyperprolactinemia inhibits male reproductive function because…

b the ovary produces estrogen, which has a negative feedback effect on the anterior pituitary.

3 Hypophysectomy (removal of the pituitary) stops sperm production because…

c this chemical can be converted to estradiol by the enzyme aromatase in the follicular cells.

4 Ovarian failure is characterized by elevated levels of FSH because…

d too much of that hormone leads to decreased secretion of FSH and LH by the anterior pituitary.

5 The production of androstenedione by theca interna cells of the ovary is a key step to estrogen production because…

e the negative feedback from the testes is lost.

Structure and Function of the Testis and Epididymis

Spermatozoa are produced in the walls of the seminiferous tubules by transformation of spermatogonia to spermatocytes, then spermatids, and finally spermatozoa. The spermatozoa then move through the rete testis (a network of tubules) and the efferent ductules to reach the tubules of the epididymis, before passing along the ductus deferens to the rest of the male reproductive tract.

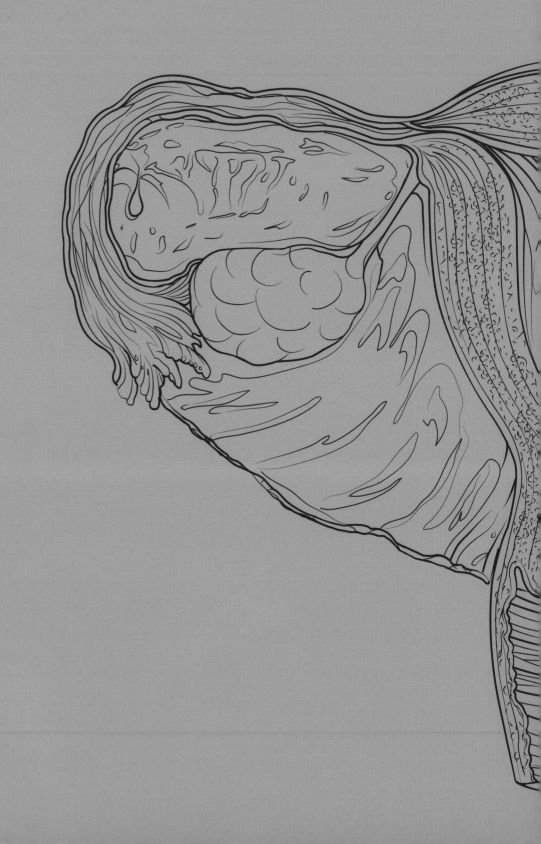

CHAPTER 9: THE REPRODUCTIVE SYSTEM

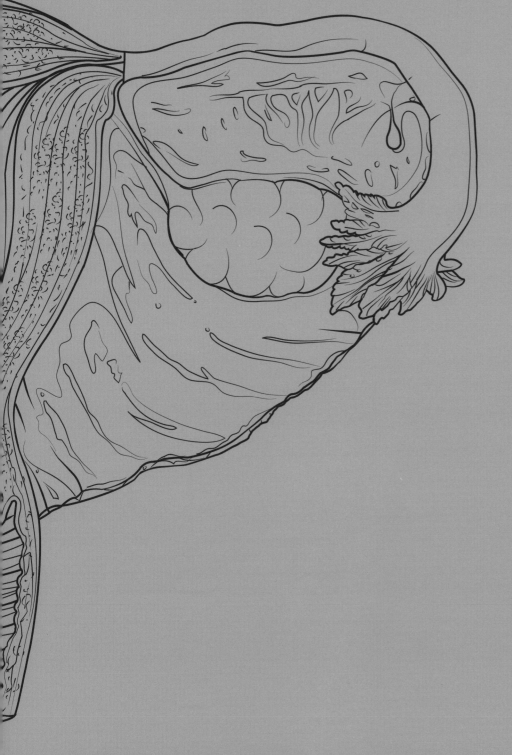

Male Reproductive System

The male reproductive system consists of the testes, which produce the sex cells called the spermatozoa and male reproductive hormones like testosterone; accessory glands (prostate, seminal vesicle, and bulbourethral gland), which produce the constituents of semen to keep sperm cells alive and functional following ejaculation into the vagina; and the male organ of copulation, the penis. The male reproductive system shares a common passage, the urethra, with the urinary system.

Key terms:

Ampulla of ductus deferens The dilated terminal portion of the ductus deferens immediately before it joins with the duct of the seminal vesicle to form the ejaculatory duct.

Bladder See pp. 218–219.

Bladder (detrusor) muscle Smooth muscle of the urinary bladder wall. The fibers are arranged in spiral and longitudinal layers.

Bulb of penis Part of the erectile tissue of the penis. It is continuous with the corpus spongiosum of the penile body. The spongy urethra passes through the center of the bulb.

Bulbourethral (Cowper's) gland This mucus-secreting gland drains into the intrabulbar part of the spongy (penile) urethra. It forms the initial part of the ejaculate.

Corpus cavernosum One of a pair of erectile tissue structures in the pendulous body of the penis. Each is surrounded by a tough tunica albuginea so that they develop great rigidity when filled with blood during erection.

Corpus spongiosum The continuation of the erectile tissue of the bulb into the body of the penis. It expands distally as the glans of the penis. The spongy urethra passes down its center.

Ductus deferens (vas deferens) A tube carrying sperm from the tail of the epididymis to the ejaculatory duct. It runs through the inguinal canal and curves over and behind the base of the bladder.

Ejaculatory duct The ductus deferens and duct of the seminal vesicle join to form an ejaculatory duct that opens into the prostatic urethra on the colliculus seminalis.

Epididymis The organ above and behind the testis that is responsible for making the sperm fertile. It receives efferent ductules from the testis and ends as the ductus deferens.

Fascia penis Connective tissue encircling the penis that has two layers: superficial fascia in continuity with the dartos of the scrotum, and deep fascia in continuity with the deep perineal fascia.

Glans penis The expanded end of the corpus spongiosum of the penis. It is pierced by the spongy urethra and is covered by the prepuce (foreskin) in uncircumcised males.

Membranous urethra The short segment of the urethra that passes through the pelvic diaphragm.

Penis The male organ of copulation. It has a root attached to the pelvis and a free-hanging body. The penis has three bodies of erectile tissue. The two upper erectile bodies are called crura where they are attached to the pelvis and become corpora cavernosa where they hang freely (in the body). Similarly, the bulb of the penis is continuous with the corpus spongiosum in the body. The crura/corpora cavernosa develop the highest pressures and greatest rigidity during erection, whereas the bulb/corpus spongiosum carries the penile urethra.

Prepuce (foreskin) A fold of skin that covers the glans penis. Beneath it lies the preputial sac. The prepuce attaches to the ventrum of the penis by a frenulum near the external urethral orifice.

Prostate gland The largest accessory genital gland. It produces prostate-specific acid phosphatase, prostate-specific antigen, amylase, and fibrinolysin. It is pierced by the prostatic urethra and the ejaculatory ducts, so both urine and sperm pass through this organ.

Prostatic urethra The segment of the urethra that is 1.25 inches (3 cm) long and passes through the prostate gland. It receives the openings of the ejaculatory ducts, the prostatic utricle, and the prostatic gland ducts.

Scrotum See pp. 238–239.

Seminal vesicle An accessory genital gland behind the base of the bladder. It secretes a viscous fluid rich in fructose (the major energy source for sperm) that makes up 50 to 70 percent of the ejaculate.

Superficial dorsal vein A large vein on the dorsum of the penis that drains into the great saphenous vein.

Testes Endocrine organs (producing testosterone), as well as the site of production of sperm cells (spermatogenesis).

Ureter The muscular tube about 10 inches (25 cm) long carrying urine from the renal pelvis to the urinary bladder. Its wall has an inner lining of stretchable transitional epithelium surrounded by smooth muscle and an adventitial connective tissue layer.

Urethra A tube carrying urine from the bladder to the external environment. In males, it has prostatic, membranous, and spongy (penile) parts.

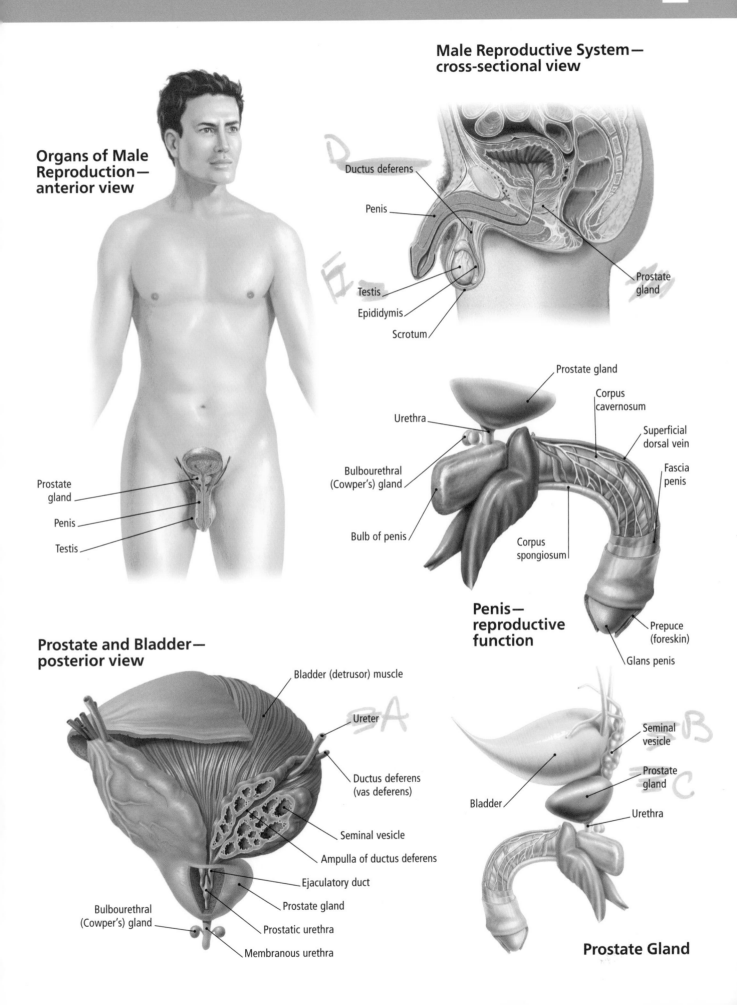

Organs of Male Reproduction—anterior view

Prostate gland

Penis

Testis

Male Reproductive System—cross-sectional view

Ductus deferens

Penis

Testis

Epididymis

Scrotum

Prostate gland

Penis—reproductive function

Prostate gland

Corpus cavernosum

Superficial dorsal vein

Fascia penis

Urethra

Bulbourethral (Cowper's) gland

Bulb of penis

Corpus spongiosum

Prepuce (foreskin)

Glans penis

Prostate and Bladder—posterior view

Bladder (detrusor) muscle

Ureter

Ductus deferens (vas deferens)

Seminal vesicle

Ampulla of ductus deferens

Ejaculatory duct

Prostate gland

Prostatic urethra

Bulbourethral (Cowper's) gland

Membranous urethra

Seminal vesicle

Prostate gland

Urethra

Bladder

Prostate Gland

True or false?

1   *The testis is covered by a fluid-filled sac called the tunica vaginalis.*

2   *The dense connective sheath of the testis is called the tunica albuginea.*

3   *The efferent ductules leave the testis at its inferior pole.*

4   *The ductus deferens enters the abdominal cavity through the femoral canal.*

5   *The ampulla of the ductus deferens is situated medial to the seminal vesicle.*

6   *The ejaculatory ducts pass through the seminal vesicles.*

7   *The ejaculatory ducts open into the prostatic urethra.*

8   *The bulbourethral glands have ducts that open into the prostatic urethra.*

9   *The bulb of the penis is surrounded by the ischiocavernosus muscle.*

10   *The prepuce of the penis covers the glans of the penis.*

Testes—posterior view

The testes are paired organs for the production of the male sex cells (spermatocytes). They also function as endocrine glands. The testes are suspended in the scrotum to achieve optimal temperature for spermatogenesis.

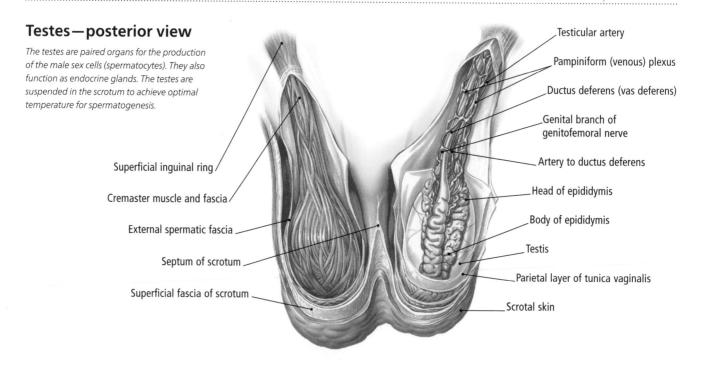

Superficial inguinal ring

Cremaster muscle and fascia

External spermatic fascia

Septum of scrotum

Superficial fascia of scrotum

Testicular artery

Pampiniform (venous) plexus

Ductus deferens (vas deferens)

Genital branch of genitofemoral nerve

Artery to ductus deferens

Head of epididymis

Body of epididymis

Testis

Parietal layer of tunica vaginalis

Scrotal skin

Multiple choice

1 *Which of the following sends connective tissue septa into the interior of the testis?*

Ⓐ tunica albuginea
Ⓑ parietal layer of tunica vaginalis
Ⓒ visceral layer of tunica vaginalis
Ⓓ internal spermatic fascia
Ⓔ cremasteric fascia

2 *Which cell type provides support, protection, and nourishment for spermatogenic cells?*

Ⓐ Leydig cells
Ⓑ spermatids
Ⓒ spermatocytes
Ⓓ Sertoli cells
Ⓔ myoid cells

3 *The process of differentiation of the spermatids is called*

Ⓐ spermatocytosis
Ⓑ androgenesis
Ⓒ meiosis
Ⓓ spermiogenesis
Ⓔ none of the above is correct

4 *Which of the following is made by the seminal vesicle?*

Ⓐ spermatozoa
Ⓑ fructose
Ⓒ alkaline phosphatase
Ⓓ mucus
Ⓔ acid phosphatase

5 *Which region most commonly gives rise to carcinoma of the prostate?*

Ⓐ periurethral zone
Ⓑ median lobe
Ⓒ anterior segment
Ⓓ peripheral zone
Ⓔ none of the above

6 *Which of the following is not found in the prostatic secretions?*

Ⓐ fibrinolysin
Ⓑ citric acid
Ⓒ acid phosphatase
Ⓓ zinc
Ⓔ manganese

7 *Which artery carries the bulk of blood for the corpus cavernosum?*

Ⓐ superior gluteal
Ⓑ inferior gluteal
Ⓒ internal pudendal
Ⓓ femoral
Ⓔ middle rectal

Torsion of testis

The testis and epididymis are suspended from the body by the spermatic cord, which carries the blood and nerve supply and lymphatic drainage of those organs. If the testis is rotated around the axis of the spermatic cord (testicular torsion), the venous drainage will be obstructed and the testis will undergo venous gangrene. This may occur with vigorous exercise or rapid flexion/extension movements of the hip—for example, when kick-starting a motorbike. If the testis is not rotated back to its original position within a few hours, the organ will become necrotic and need to be removed.

Color and label

i) Label each structure shown on the illustrations.

Male Reproductive System— anterior view

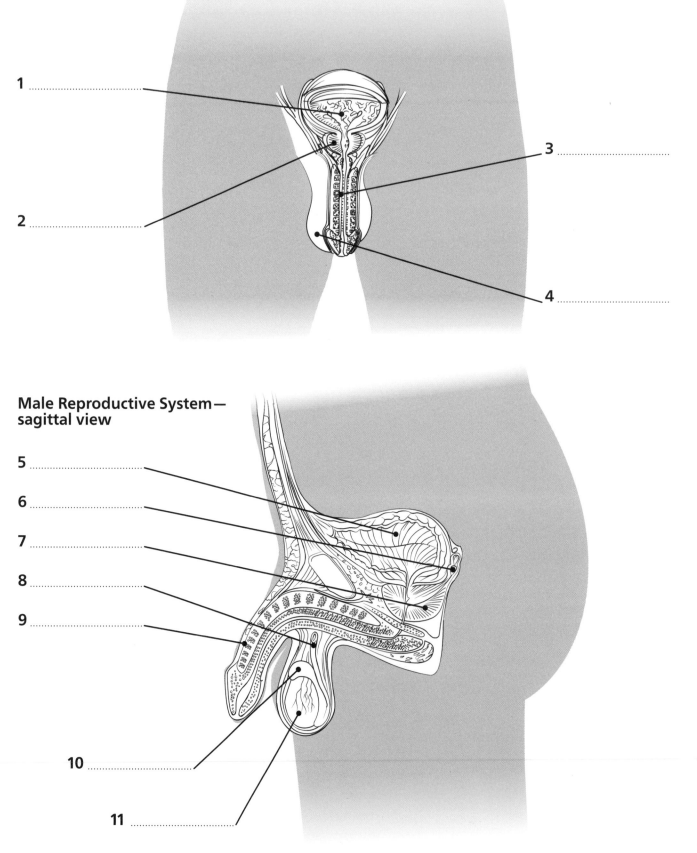

1 ....

2 ....

3 ....

4 ....

Male Reproductive System— sagittal view

5 ....

6 ....

7 ....

8 ....

9 ....

10 ....

11 ....

ii) Label each structure shown on the illustrations.

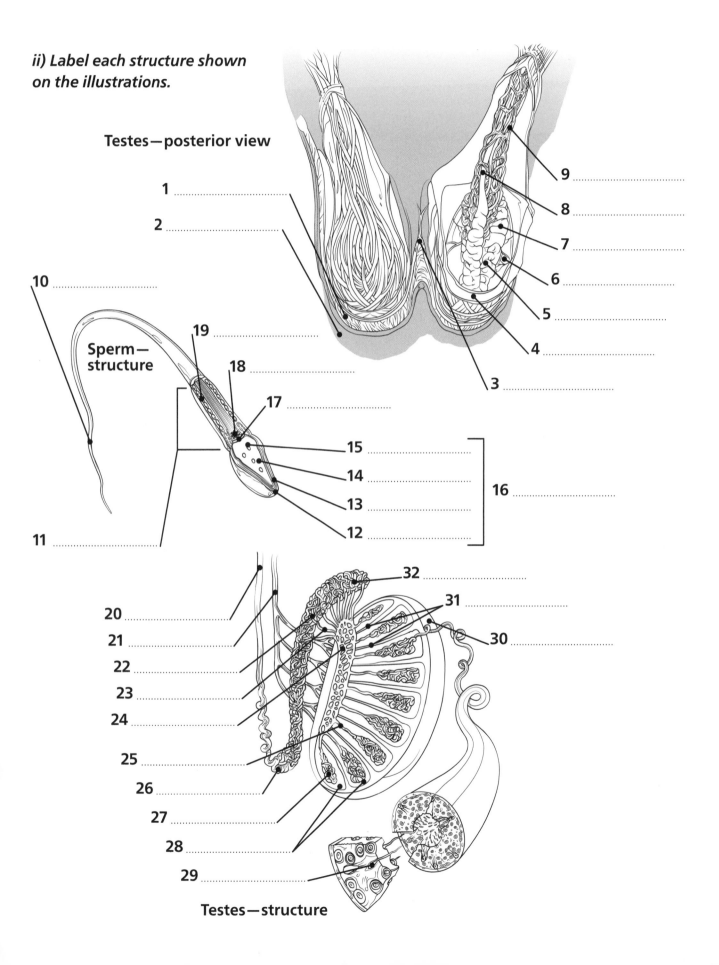

Testes—posterior view

1 ....

2 ....

9 ....

8 ....

7 ....

6 ....

5 ....

4 ....

3 ....

10 ....

Sperm—structure

19 ....

18 ....

17 ....

15 ....

14 ....

13 ....

12 ....

16 ....

11 ....

20 ....

21 ....

22 ....

23 ....

24 ....

25 ....

26 ....

27 ....

28 ....

29 ....

32 ....

31 ....

30 ....

Testes—structure

Fill in the blanks

1   The tail of a sperm is called a _____.

2   The venous drainage of the testis is specialized to form a structure called the
_____ plexus.

3   The cap-like structure covering the nucleus of the mature sperm is called the _____.

4   Optimal temperature for spermatogenesis is _____.

5   Sperm cells leaving the testis pass through the _____ on their way to the head of
the epididymis.

6   The three functions of the epididymis are _____, _____, and
_____.

7   The type of epithelium in the seminal vesicle is _____ to _____.

8   The lumen of the prostate gland may contain lumps of condensed glycoprotein and cell fragments
called _____.

9   The tough layer that encircles the corpora cavernosa of the penis is called the
_____.

10   The _____ produce a secretion rich in mucus to lubricate the urethra.

Match the statement to the reason

1   *Failure of descent of the testis causes infertility because…*

a   *enlargement of the central zone of the gland narrows the enclosed lumen.*

2   *Benign hypertrophy of the prostate can obstruct the urethra because…*

b   *the urethra makes a right-angle turn at this point.*

3   *Hypertrophy of the prostate can be assessed by a per rectal examination because…*

c   *the dartos muscle contracts to elevate the testis closer to the body.*

4   *Careless catheterization of the male urethra may rupture the intrabulbar fossa because…*

d   *the palpable posterior median groove is often lost when the gland enlarges.*

5   *The skin of the scrotum becomes wrinkled in cold weather because…*

e   *optimal temperature for spermatogenesis is below the body's core temperature.*

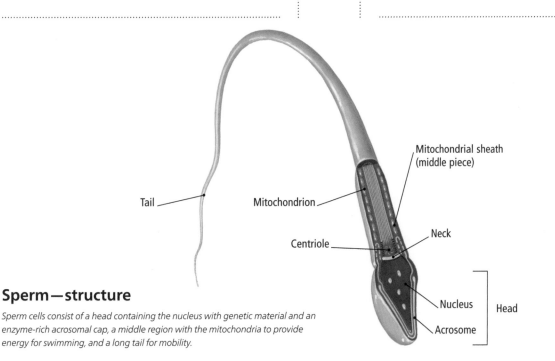

Sperm—structure

Sperm cells consist of a head containing the nucleus with genetic material and an enzyme-rich acrosomal cap, a middle region with the mitochondria to provide energy for swimming, and a long tail for mobility.

Female Reproductive System

The female gonads are the ovaries, which lie against the lateral pelvic wall and produce the ova and female sex steroid hormones (estrogen and progesterone). The uterine tubes carry the ovum to the body of the uterus, where the embryo can embed and develop into a fetus. Parturition, or birth, is the delivery of the full-term fetus through the birth canal (uterine cervix and vagina) to the external environment. The vagina also serves as the female organ of copulation, receiving the erect penis and the deposition of semen. The vagina opens at the external genitalia (vulva).

Key terms:

Adipose tissue This fat tissue includes yellow fat, which is found under the skin in adults and stores energy, and brown fat, which has many mitochondria and is found on the upper back in newborn infants.

Areola A circular zone of pigmented skin around the nipple. It contains sweat and sebaceous glands that keep the nipple moist.

Cervix The entrance to the uterus. It has intravaginal and supravaginal parts and has a cervical canal passing through it.

Clitoris A small erectile tissue structure. It has crura attached to the ischiopubic rami and a glans covered by a prepuce.

Fallopian (uterine) tube A tube carrying sperm to the ovum and the fertilized ovum (zygote) to an implantation site in the uterine wall.

Labium majus One of two elongated folds of skin over fat that run backward from the mons pubis. They flank the pudendal cleft with its openings of the urethra and vagina.

Labium minus One of two small folds of skin between the labia majora. They lie on each side of the vaginal opening and extend a fold anteriorly to form a hood (prepuce) over the clitoris.

Lactiferous duct The milk-carrying ducts from each mammary lobule. They dilate as lactiferous sinuses before opening at the nipple.

Lactiferous sinus The dilated portion of the lactiferous duct beneath the areola.

Lobules of mammary gland The breast has 15 to 20 lobules of glandular tissue separated by connective tissue.

Mammary glands Modified sweat glands that produce milk for neonates and infants. Breast tissue is located on the chest wall, with some tissue extending into the axilla (axillary tail).

Mons pubis The fatty elevation over the pubic body and symphysis. It is covered by hairs after puberty.

Nipple A prominence on the breast at which the lactiferous ducts open. It is made of smooth muscle that contracts to compress the ducts and causes the nipple to become erect.

Orifice of urethra The opening of the female urethra between the folds of the labia minora.

Ovary Each ovary produces female sex cells, secretes estrogen and progesterone, and regulates the postnatal growth of reproductive organs and the development of secondary sexual characteristics (e.g., female fat distribution and pubic hair).

Pectoralis major A muscle arising from the medial clavicle and upper six costal cartilages to insert into the crest of the greater tubercle of the humerus. It adducts, medially rotates, and flexes the arm.

Posterior labial commissure A projection of the tendinous center of the perineum into the pudendal cleft that gives the impression of a commissure joining the posterior ends of the labia majora.

Uterus The organ of gestation. The uterus has a fundus, body, isthmus, and cervix. The Fallopian (uterine) tubes enter at the junction of the fundus and body.

Vagina The female organ of copulation and the lower end of the birth canal. It is a highly dilatable tube extending from the cervix to the vaginal vestibule.

Vulva The female external genitalia. The region includes the mons pubis, labia (minora and majora), clitoris, and vaginal entrance.

Wall of vagina The anterior wall of the vagina is in contact with the base of the bladder.

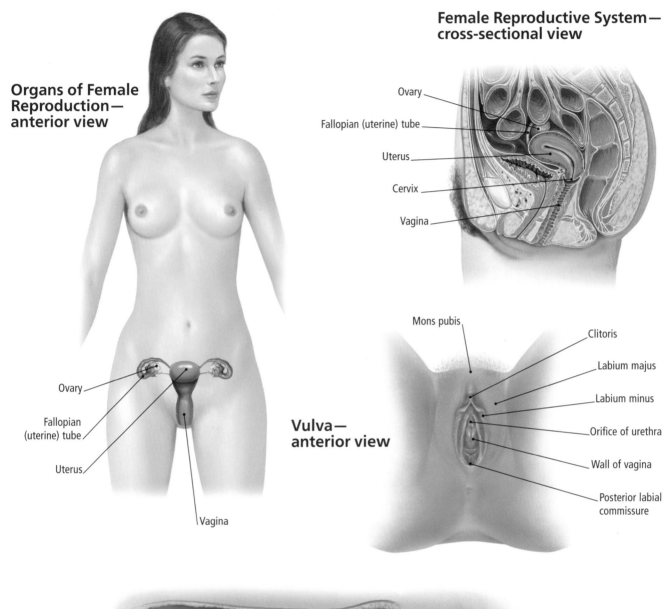

Organs of Female Reproduction— anterior view

Ovary

Fallopian (uterine) tube

Uterus

Vagina

Female Reproductive System— cross-sectional view

Ovary

Fallopian (uterine) tube

Uterus

Cervix

Vagina

Vulva— anterior view

Mons pubis

Clitoris

Labium majus

Labium minus

Orifice of urethra

Wall of vagina

Posterior labial commissure

Pectoralis major

Lobules of mammary gland

Lactiferous sinus

Areola

Nipple

Lactiferous duct

Adipose tissue

Mammary Glands

True or false?

1   The ovary is covered by a thin layer of squamous to low cuboidal epithelium, through which ova erupt.

2   The three phases of the ovarian cycle are follicular, ovulatory, and luteal.

3   The tough outer layer of the ovary is called the tunica vaginalis.

4   The mature ovarian follicle has a fluid-filled cavity called the corona radiata.

5   The glycoprotein-rich layer around the oocyte is called the zona pellucida.

6   The rupture of the mature follicle out of the ovary at ovulation is stimulated solely by a surge of follicle-stimulating hormone.

7   The narrowest part of the uterine tube is the ampulla.

8   The muscle of the uterus is called the myometrium.

9   The intravaginal cervix is surrounded by the fornix of the vagina.

10 The bulb of the vestibule provides lubrication to the vaginal entrance.

Polycystic ovarian syndrome

This is the most common endocrine disorder in women between the ages of 18 and 45. The condition refers to a group of symptoms experienced by women who have elevated androgens. The name of the condition comes from the multiple immature ovarian follicles that can be detected by ultrasonography. Symptoms and signs include excess body and facial hair, acne vulgaris, irregular or no periods, absent ovulation, and difficulty becoming pregnant. The disorder usually arises due to obesity, inadequate exercise, and family history. Treatment includes weight loss and exercise, oral contraceptives to assist with irregular periods, and in vitro fertilization to assist with conception.

Multiple choice

1 *What is the main source of arterial blood for the ovary?*
(A) internal iliac artery
(B) ovarian artery from aorta
(C) external iliac artery
(D) ovarian artery from superior vesical artery
(E) uterine artery from internal iliac artery

2 *Which structure is found at the very end of the uterine tube?*
(A) fimbriae
(B) isthmus
(C) ampulla
(D) infundibulum
(E) fundus

3 *Which part of the uterus is above the attachment of the uterine tubes?*
(A) cervix
(B) body
(C) fundus
(D) isthmus
(E) none of the above is correct

4 *The muscle layer of the uterus is called the*
(A) endometrium
(B) serosa
(C) muscularis mucosa
(D) muscularis adventitia
(E) myometrium

5 *Which internal feature of the cervical canal assists in keeping the canal closed?*
(A) palmate folds
(B) vaginal fornix
(C) myometrium
(D) isthmus
(E) internal os

6 *Which structure is related to the anterior wall of the lower vagina?*
(A) base of the bladder
(B) rectum
(C) anus
(D) urethra
(E) bulb of the vestibule

7 *Which structure(s) fill with blood during sexual arousal?*
(A) fundus of uterus
(B) endometrium
(C) bulb of the vestibule
(D) greater vestibular gland
(E) labia majora

Structure and Function of Ovary

The paired ovaries are responsible for the production and release of the oocyte (egg or ovum), for the production of estrogen to induce secondary sexual characteristics, and for the production of estrogen and progesterone to induce proliferation, secretion, and support of the uterine lining. They lie alongside the lateral pelvic wall adjacent to the opening of the Fallopian tube between the layers of the broad ligament.

Primary oocyte

Mature Graafian follicle

Discharging follicle

Corpus luteum

Ovum

Color and label

*i) Label each structure shown
on the illustrations.*

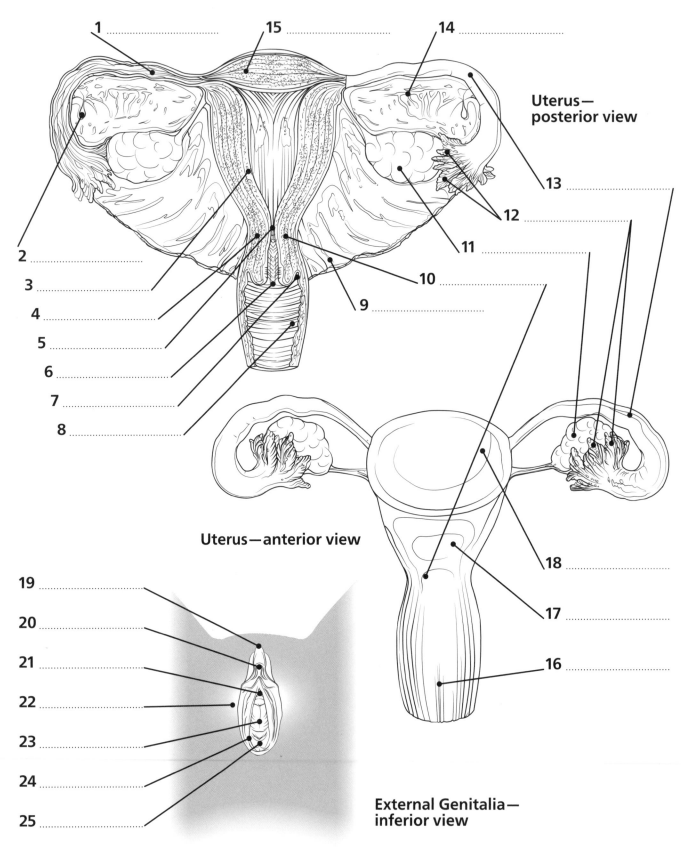

1 ....

15 ....

14 ....

Uterus—
posterior view

13 ....

12 ....

11 ....

10 ....

9 ....

2 ....

3 ....

4 ....

5 ....

6 ....

7 ....

8 ....

Uterus—anterior view

18 ....

17 ....

16 ....

19 ....

20 ....

21 ....

22 ....

23 ....

24 ....

25 ....

External Genitalia—
inferior view

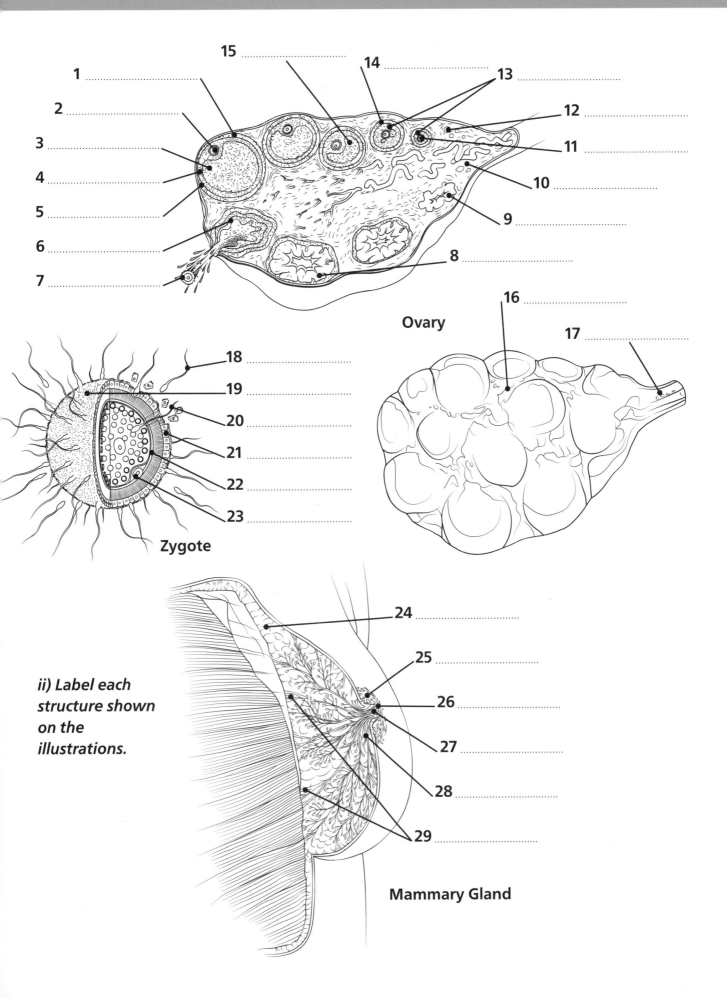

15
1
2
14
13
3
12
4
11
5
10
6
9
7
8

Ovary

16
17

18
19
20
21
22
23

Zygote

24
25
26
27
28
29

*ii) Label each
structure shown
on the
illustrations.*

Mammary Gland

Fill in the blanks

1   The fimbriae are attached to the rim of the _____ of the uterine tube.

2   The ovary has a mesenteric attachment called the _____.

3   The blood vessels of the ovary descend in the _____.

4   Fertilization usually occurs in the _____ of the uterine tube.

5   The myometrium is composed of _____.

6   The intravaginal cervix is surrounded by a space called the _____.

7   The _____ and _____ are the paired folds that surround the vaginal and urethral entrances.

8   The clitoris has a fold covering it called the _____.

9   Erectile tissue alongside the ischiopubic ramus in females is called the _____.

10   The fatty elevation of skin over the pubic symphysis in women is called the _____.

Structure of the Uterus

The uterus consists of an internal layer called the endometrium where the embryo implants to develop, a middle layer of smooth muscle called the myometrium, which will expel the fetus at childbirth, and an outer layer called the adventitia, or serosa.

Labels: Fallopian tube, Mesosalpinx of broad ligament, Ampulla, Infundibulum, Fimbriae, Mesovarium of broad ligament, Adventitia or serosa, Endometrium, Myometrium, Cervix, Vaginal fornix, Ovary, Mesometrium of broad ligament, Internal os, Vagina, External os

Match the statement to the reason

1   Blockage of the uterine tube due to pelvic inflammatory disease causes infertility because…

a   sperm cannot ascend the reproductive tract and oocytes cannot descend.

2   The uterus sheds its endometrium at the beginning of menstruation because…

b   the vaginal entrance will be exquisitely tender.

3   Retroversion of the uterus may lead to painful intercourse because…

c   the uterus prolapses, everting the vaginal mucosa as it descends.

4   Loss of pelvic floor support may lead to the cervix being visible at the vaginal entrance because…

d   the uterine fundus is pushed against the bony sacrum.

5   An abscess of the greater vestibular gland will lead to painful intercourse because…

e   hormone support by progesterone is lost at that time.

6   Uterine prolapse is often accompanied by urinary incontinence because…

f   the pelvic floor muscles contribute directly and indirectly to bladder and uterine support.

7   The shape of the external os of the cervix changes from a small circular profile to a transverse slit after vaginal delivery because…

g   small tears occur during parturition.

Conception and Pregnancy

Fertilization usually occurs in the ampulla of the uterine tube shortly after ovulation. Sperm must ascend the female reproductive tract by swimming. Once the sperm encounters the egg, it will release enzymes that dissolve the glycoprotein between the corona radiate cells that surround the egg. The sperm then penetrates the zona pellucida and fuses its cell membrane with that of the egg. The father's DNA is then transported into the egg to unite with that of the mother. The fertilized egg (zygote) will implant in the uterine wall about six to seven days after fertilization.

Key terms:

Amnion The innermost of the fetal membranes. It forms a protective amniotic fluid-filled sac around the embryo and fetus.

Blastocyst An embryonic stage consisting of a central fluid-filled cavity surrounded by a single layer of cells. It is the developmental stage that embeds in the wall of the uterus.

Chorionic villi A frond-like projection of the fetal circulation into the intervillous space where it is bathed in maternal blood.

Cotyledon A segment of the placenta separated from adjacent segments by connective tissue septa.

Embryo The developing human from the moment of conception to the end of the eighth week after fertilization.

Endometrium The mucosal lining of the uterus. It is invaded by the syncytial trophoblast to form the intervillous spaces.

Fertilization The union of male and female sex cells (gametes). It usually occurs in the ampulla of the Fallopian (uterine) tube.

Fetus The developing human from the beginning of the ninth week after fertilization to the moment of birth.

Maternal blood vessels These carry maternal blood into the intervillous spaces to provide nutrients and oxygen for diffusion into the fetal circulation.

Myometrium The smooth muscle layer of the uterine wall. It thickens during pregnancy to provide a means of expelling the fetus from the uterus at delivery.

Ovum An oocyte (egg). The "ovum" released at ovulation actually consists of an oocyte, as well as surrounding zona pellucida and corona radiata layers.

Perivitelline space A narrow space between the cell membrane of the oocyte and the zona pellucida.

Placenta An organ that attaches to the wall of the uterus and provides nutrients for the embryo and fetus.

Polar body A fragment containing excess genetic material as a result of completion of the first stage of meiosis (sex cell division) immediately before ovulation.

Sperm The male sex cell. It has a head, connecting piece, and tail. It must penetrate the corona radiata and zona glomerulosa layers to unite with the oocyte.

Syncytiotrophoblast One of the component layers of the placental membranes. During implantation of the embryo, it is highly invasive and erodes maternal tissue to form the intervillous spaces.

Umbilical artery One of two vessels carrying relatively deoxygenated and nutrient-poor blood from the fetus to the placenta. They are branches of the internal iliac arteries.

Umbilical cord A twisted cord joining the fetus to the placenta. It contains two umbilical arteries and a single umbilical vein suspended in embryonic connective tissue.

Umbilical vein The vessel carrying oxygenated and nutrient-rich blood from the placenta to the fetus. Originally there are two, but the right umbilical vein disappears during embryonic life.

Zona pellucida A translucent region surrounding the oocyte. It is composed of glycoproteins made by the oocyte. It may also be penetrated by processes of corona radiata cells.

Zygote The cell formed by the fusion of male and female sex cells before cleavage has begun.

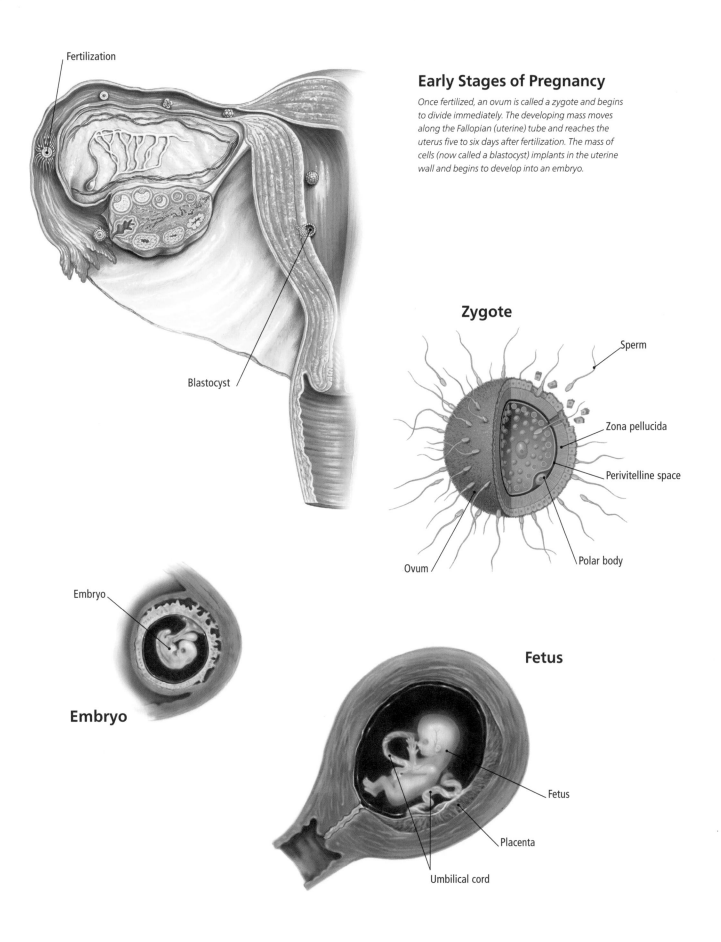

Fertilization

Blastocyst

Early Stages of Pregnancy

Once fertilized, an ovum is called a zygote and begins to divide immediately. The developing mass moves along the Fallopian (uterine) tube and reaches the uterus five to six days after fertilization. The mass of cells (now called a blastocyst) implants in the uterine wall and begins to develop into an embryo.

Zygote

Sperm

Zona pellucida

Perivitelline space

Polar body

Ovum

Embryo

Embryo

Fetus

Fetus

Placenta

Umbilical cord

True or false?

1 The zona pellucida is a thick layer of glycoprotein that surrounds the egg.

2 The zona pellucida stays around the zygote throughout embryonic life.

3 The embryo usually implants in the uterine wall during the follicular phase.

4 The early stage of the conceptus consists of a solid clump of cells called a morula.

5 The stage when the early embryo consists of a sphere with a hollow interior is called the blastocyst.

6 The placenta is formed from the inner cell mass of the blastocyst.

7 Implantation in the uterine wall is usually achieved by 10 days after fertilization.

8 The most invasive layer of the trophoblast is the cytotrophoblast.

9 A primary villus of the implantation site consists of a core of cytotrophoblast surrounded by the syncytiotrophoblast.

10 At about 17 days, the embryo develops three germ layers: ectoderm, mesoderm, and endoderm.

Umbilical cord

Umbilical vein

Umbilical arteries

Syncytiotrophoblast

Placenta

Chorionic villi

Endometrium

Maternal blood vessels

Myometrium

Placenta—cross-sectional view

The placenta is an organ that allows transfer of nutrients and oxygen from the maternal to the fetal circulation, and the removal of carbon dioxide and waste products. It is also an endocrine organ.

Multiple choice

1 *The fluid-filled sac surrounding the embryo from about 19 days is called the*
- (A) yolk sac
- (B) amniotic sac
- (C) decidua basalis
- (D) endometrium
- (E) intraembryonic coelom

2 *The gut tube of the embryo is lined with*
- (A) endoderm
- (B) intraembryonic mesoderm
- (C) ectoderm
- (D) amnion
- (E) extraembryonic mesoderm

3 *The nervous system is derived from the embryonic*
- (A) endoderm
- (B) intraembryonic mesoderm
- (C) ectoderm
- (D) amnion
- (E) extraembryonic mesoderm

4 *Which of the following gives rise to the peripheral nervous system cells?*
- (A) mesoderm
- (B) endoderm
- (C) amnion
- (D) neural crest
- (E) yolk sac

5 *How many chromosomes does a normal embryo have?*
- (A) 23 pairs
- (B) 46 pairs
- (C) 24 pairs
- (D) 48 pairs
- (E) none of the above

6 *Implantation of the blastocyst outside the uterine cavity is called*
- (A) molar pregnancy
- (B) placenta previa
- (C) hydatidiform mole
- (D) ectopic pregnancy
- (E) a neural tube defect

7 *Early splitting of the zygote into two embryos is likely to lead to*
- (A) nonidentical twins
- (B) Siamese twins
- (C) neural tube defect
- (D) fraternal twins
- (E) identical twins

Placenta—anterior view

The placenta is disc-shaped, with the fetal umbilical cord attached in the center. The umbilical cord contains two umbilical arteries carrying fetal deoxygenated blood, and a single umbilical vein carrying oxygenated blood back from the placenta.

Fetal Development

Once the embryo has developed all the organ systems (by about eight weeks after fertilization), the conceptus is called a fetus. Fetal life involves the maturation of the body systems in preparation for birth and survival in the external environment. During fetal life, the conceptus is dependent on the placenta for all its oxygen and nutrient needs, as well as the removal of toxins and waste products. The placenta is also an important endocrine organ, producing placental lactogen, chorionic gonadotropin, estrogen, and progesterone.

Key terms:

Anterior fontanelle The space between the paired frontal and parietal bones. It disappears by two years and will become the point called bregma in the adult.

Coronal suture The suture separating the frontal bone from the parietal bones.

External auditory (acoustic) meatus The external ear opening as seen in a skull. It lies within the temporal bone.

Frontal bone The bone underlying the forehead.

Lambdoid suture The suture between the single occipital and paired parietal bones. It is shaped like the Greek letter lambda.

Mandible The mandible develops from connective tissue in the mandibular process of the embryo.

Mastoid fontanelle A small fontanelle between the parietal, occipital, and temporal bones.

Mastoid part of temporal bone The mastoid process of the temporal bone develops as a separate ossification center in the skull.

Maxilla The bone of the anterior cheek. It develops from the maxillary process of the embryonic face.

Metopic suture The midline suture between the two frontal bones. It usually disappears by adult life.

Nasal bone A small bone that forms the superior and lateral margin of the piriform aperture of the bony nose.

Occipital bone The large bone on the underside of the skull. It provides attachment for the postvertebral muscles of the neck and bears occiptal condyles for articulation with the atlas (cervical vertebra 1).

Parietal bone A flat bone of the braincase (calvaria). It articulates with the frontal, temporal, sphenoid, and occipital bones.

Parietal tuber The central, slightly pointed region of the developing parietal bone.

Posterior fontanelle The space between the developing parietal and occipital bones.

Pterion The region where the frontal, parietal, sphenoid, and temporal bones meet.

Sagittal suture The suture running down the midline of the skull between the parietal bones of each side.

Squamous part of occipital bone The flattened part of the occipital bone. It will form the posterior cranial fossa, where the cerebellum sits.

Squamous part of temporal bone The flattened part of the temporal bone. It will form part of the calvaria surrounding the brain.

Styloid process An elongated spine projecting inferiorly from the temporal bone. It provides attachment for several muscles, including the stylohyoid and stylopharyngeus.

Temporal bone A bone made up of many small bones developing from discrete ossification centers in the fetal skull.

Tympanic ring of temporal bone A bony ring that supports the tympanic membrane.

Zygomatic bone The bone of the upper lateral cheek. It articulates with the maxilla, temporal, and frontal bones.

Zygomatic process of temporal bone An anterior projection of the temporal bone.

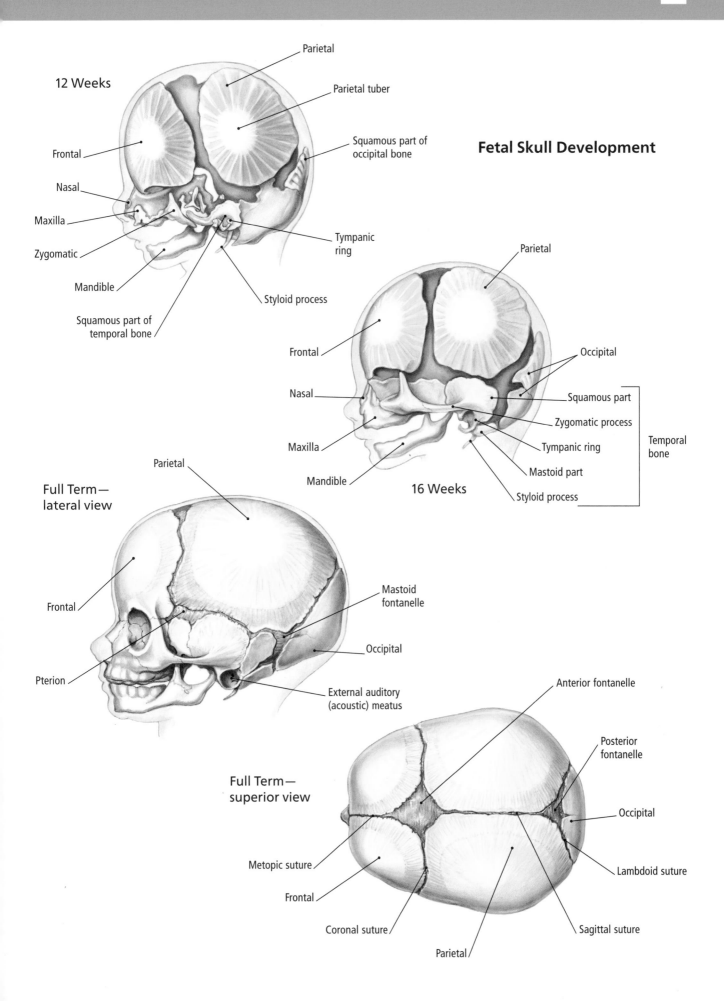

Fetal Skull Development

12 Weeks

- Parietal
- Parietal tuber
- Squamous part of occipital bone
- Frontal
- Nasal
- Maxilla
- Zygomatic
- Mandible
- Tympanic ring
- Styloid process
- Squamous part of temporal bone

16 Weeks

- Parietal
- Frontal
- Nasal
- Maxilla
- Mandible
- Occipital
- Squamous part
- Zygomatic process
- Tympanic ring
- Mastoid part
- Styloid process
- Temporal bone

Full Term—lateral view

- Parietal
- Frontal
- Pterion
- Mastoid fontanelle
- Occipital
- External auditory (acoustic) meatus

Full Term—superior view

- Anterior fontanelle
- Posterior fontanelle
- Occipital
- Lambdoid suture
- Sagittal suture
- Parietal
- Coronal suture
- Frontal
- Metopic suture

True or false?

1   *Most brain development occurs during the first trimester.*

2   *There are two umbilical veins in the umbilical cord.*

3   *The blood returning to the fetus via the umbilical vein bypasses the liver via the ductus venosus.*

4   *Blood entering the fetal right atrium is deoxygenated.*

5   *Pulmonary surfactant is made as early as 20 weeks gestation.*

6   *Maternal IgM molecules can cross the placenta.*

7   *Amniotic fluid can be sampled by needle biopsy in a procedure known as amniocentesis.*

8   *During pregnancy, prolactin, placental lactogen, estrogen, and progesterone all help to prepare the breast for lactation.*

9   *Exposure of the fetus to ethanol can cause facial abnormalities and intellectual disability.*

10   *Gases like oxygen and carbon dioxide must cross the placental barrier by active transport.*

Placenta previa

An abnormal extension of the placenta over, or close to, the internal opening of the cervical canal is called placenta previa. The hazard of placenta previa is that separation of the placenta from the uterine wall may occur due to mild uterine contractions, leading to vaginal hemorrhage and oxygen starvation and death of the fetus. If the placenta completely covers the cervical canal (total placenta previa), it would be impossible for the fetus to be delivered vaginally without it dying. In this case, the fetus must be delivered by caesarian section.

Multiple choice

1 *Which of the following feature(s) of the fetal circulation normally allows oxygenated blood to reach the systemic arteries?*

(A) ligamentum teres
(B) foramen ovale
(C) ventricular septal defect
(D) ductus arteriosus
(E) both B and D are correct

2 *Which part of the developing brain generates the most neurons?*

(A) ventricular germinal zone
(B) subventricular zone
(C) choroid plexus
(D) fourth ventricle
(E) basal ganglia

3 *Which organ in the fetal and neonatal chest makes the T lymphocytes?*

(A) thyroid gland
(B) aorta
(C) lymph nodes
(D) thymus
(E) lungs

4 *Which hormone maintains the corpus luteum during pregnancy?*

(A) chorionic gonadotropin
(B) maternal luteinizing hormone
(C) follicle-stimulating hormone
(D) androstenedione
(E) estrogen

5 *What are the contents of the fetal bowel that are excreted soon after birth called?*

(A) amnion
(B) meconium
(C) feculum
(D) excretum
(E) hepatulum

6 *Which of the following can result if the mother has been previously exposed to fetal rhesus-positive blood?*

(A) hydatidiform mole
(B) tetralogy of Fallot
(C) biliary atresia
(D) juvenile onset diabetes mellitus
(E) erythroblastosis fetalis

7 *Which hormone initiates uterine contractions at parturition?*

(A) estrogen
(B) progesterone
(C) oxytocin
(D) arginine vasopressin
(E) prolactin

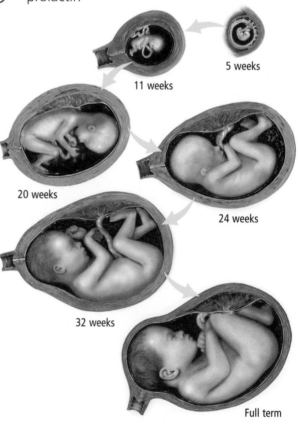

5 weeks
11 weeks
20 weeks
24 weeks
32 weeks
Full term

Changes in the Uterus and Fetus

The increase in maternal weight with pregnancy is not due only to the growth of the fetus, which usually amounts to only 5.5 to 6.6 lb (2.5 to 3 kg), but also to the increased mass of the uterine wall (more than 2.2 lb/1 kg) and the placenta, as shown here. By the end of pregnancy, the uterus and its contents usually weigh 22 lb (10 kg).

CHAPTER 10: THE BLOOD

Blood Cells

Blood cells make up approximately 46 percent of blood volume, with the rest being a fluid called plasma. Blood cells include red (erythrocyte) and white (leukocyte) types. Red blood cells carry gases (O_2, CO_2), whereas white blood cells are involved in immune function. White blood cells are further divided into granulocytes (have granules in their cytoplasm: neutrophils, eosinophils, and basophils) and agranulocytes (no granules in their cytoplasm: lymphocytes and monocytes). Platelets are fragments of cells that play a key role in hemostasis.

Key terms:

Basal lamina A layer of substance on which epithelial cells sit. It is usually produced by the epithelial cells themselves.

Basophil A type of leukocyte (white blood cell) with a bilobed nucleus and blue-staining granules. Basophils make up 1 percent of circulating leukocytes.

Blood cells Components of blood that are enclosed by a cell membrane. They include erythrocytes (red blood cells), lymphocytes, monocytes, eosinophils, neutrophils, basophils, and platelets.

Eosinophil A type of leukocyte (white blood cell) with a bilobed nucleus and red-staining granules. Eosinophils make up 2 to 4 percent of circulating leukocytes.

Erythrocytes (red blood cells) Cells that contain hemoglobin for the carriage of oxygen. They are biconcave discs with no nuclei.

Globin protein strand A folded chain of amino acids forming the protein (globin) part of hemoglobin.

Heme A protoporphyrin molecule enclosing an iron ion. The heme complex carries oxygen atoms and is found combined with globin protein chains in erythrocytes (red blood cells).

Iron ion A charged atom of iron held within the protoporphyrin heme molecule, thereby providing the oxygen binding site in hemoglobin.

Leukocytes (white blood cells) Cells within blood that do not contain hemoglobin and are mainly concerned with immune or defense functions. They include lymphocytes, monocytes, neutrophils, and eosinophils.

Lymphocyte A type of leukocyte (white blood cell) that is agranular but has a large, round, and slightly indented nucleus. Lymphocytes make up 20 to 45 percent of circulating leukocytes.

Macrophage A phagocytic cell derived from circulating monocytes. They are able to phagocytose (engulf) debris and microorganisms in tissues throughout the body.

Monocyte A large type of leukocyte (white blood cell) that is agranular but has an eccentrically placed kidney-shaped nucleus. Monocytes make up 2 to 8 percent of leukocytes.

Neutrophil A type of leukocyte (white blood cell) that makes up 50 to 75 percent of all leukocytes. Neutrophils have a multilobed nucleus, and their cytoplasm contains weakly staining granules.

Nucleus of endothelial cell The nucleus of the cell type that lines the internal surface of the blood vessel wall.

Platelets Tiny cellular fragments derived from megakaryocytes. They adhere to damaged endothelium and stop leakage of blood (hemostasis). They are formed in the bone marrow from the cytoplasm of megakaryocytes.

Components of Blood

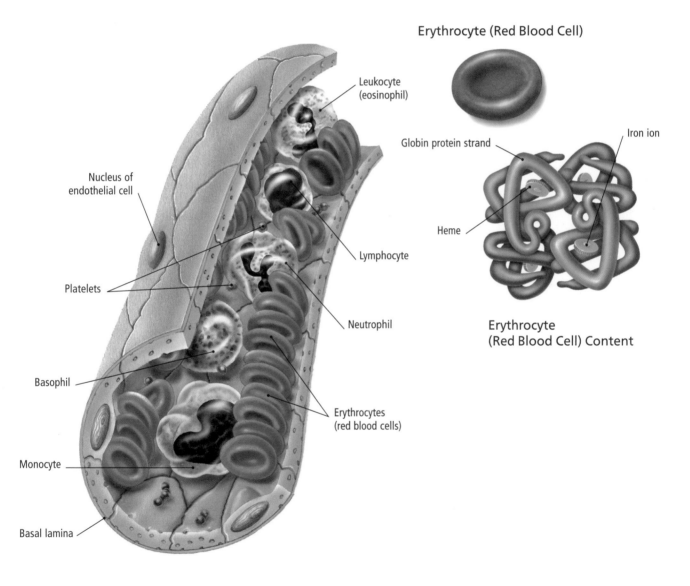

Erythrocyte (Red Blood Cell)

Leukocyte
(eosinophil)

Nucleus of
endothelial cell

Lymphocyte

Platelets

Neutrophil

Basophil

Erythrocytes
(red blood cells)

Monocyte

Basal lamina

Globin protein strand

Iron ion

Heme

Erythrocyte
(Red Blood Cell) Content

True or false?

1   *Mature mammalian red blood cells do not have nuclei.*

2   *Lymphocytes are the most common agranular white blood cell.*

3   *The granules of the neutrophil cytoplasm contain histamine.*

4   *Red blood cells have a biconcave disc shape.*

5   *Hemoglobin is composed of four globin chains, each with a central heme group.*

6   *Thalassemia is a group of conditions in which there is defective production of delta or epsilon chains of hemoglobin.*

7   *Lymphocytes make up between 20 and 40 percent of circulating leukocytes.*

8   *Monocytes circulate in the blood for about 100 days.*

9   *Platelet production is under the control of thrombopoietin released from the liver and kidney.*

10   *Platelets are fragments of neutrophil granulocytes.*

Monocyte

Macrophage

Neutrophil

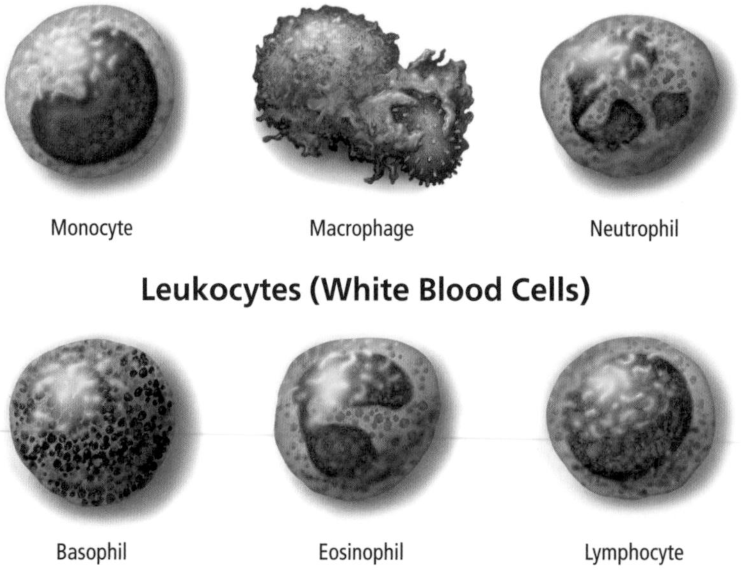

Leukocytes (White Blood Cells)

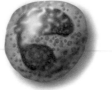
Basophil

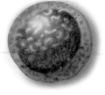
Eosinophil

Lymphocyte

Multiple choice

1 What is the diameter of a typical normal red blood cell?
- Ⓐ 2 to 3 μm
- Ⓑ 5 to 6 μm
- Ⓒ 7 to 8 μm
- Ⓓ 9 to 10 μm
- Ⓔ 13 to 14 μm

2 What is the usual lifetime of a red blood cell?
- Ⓐ 7 days
- Ⓑ 15 days
- Ⓒ 30 days
- Ⓓ 60 days
- Ⓔ 120 days

3 Red blood cells make up about what percentage of centrifuged blood volume (hematocrit)?
- Ⓐ 20 percent
- Ⓑ 35 percent
- Ⓒ 45 percent
- Ⓓ 55 percent
- Ⓔ 65 percent

4 Which is the most common type of granulocyte white blood cell?
- Ⓐ neutrophil
- Ⓑ eosinophil
- Ⓒ basophil
- Ⓓ mast cell
- Ⓔ monocyte

5 Which white blood cell is the first line of defense against parasites?
- Ⓐ neutrophil
- Ⓑ eosinophil
- Ⓒ basophil
- Ⓓ mast cell
- Ⓔ monocyte

6 Which white blood cell transforms into a tissue macrophage?
- Ⓐ neutrophil
- Ⓑ eosinophil
- Ⓒ basophil
- Ⓓ mast cell
- Ⓔ monocyte

7 Which white blood cell makes up the bulk of pus?
- Ⓐ neutrophil
- Ⓑ eosinophil
- Ⓒ basophil
- Ⓓ lymphocyte
- Ⓔ monocyte

8 Which cell gives rise to platelets?
- Ⓐ neutrophil
- Ⓑ eosinophil
- Ⓒ megakaryocyte
- Ⓓ lymphocyte
- Ⓔ macrophage

9 Which of the following blood cell types is most commonly invaded by intracellular bacteria?
- Ⓐ red blood cell
- Ⓑ neutrophil
- Ⓒ eosinophil
- Ⓓ lymphocyte
- Ⓔ basophil

Color and label

i) Label each structure shown on the illustrations.

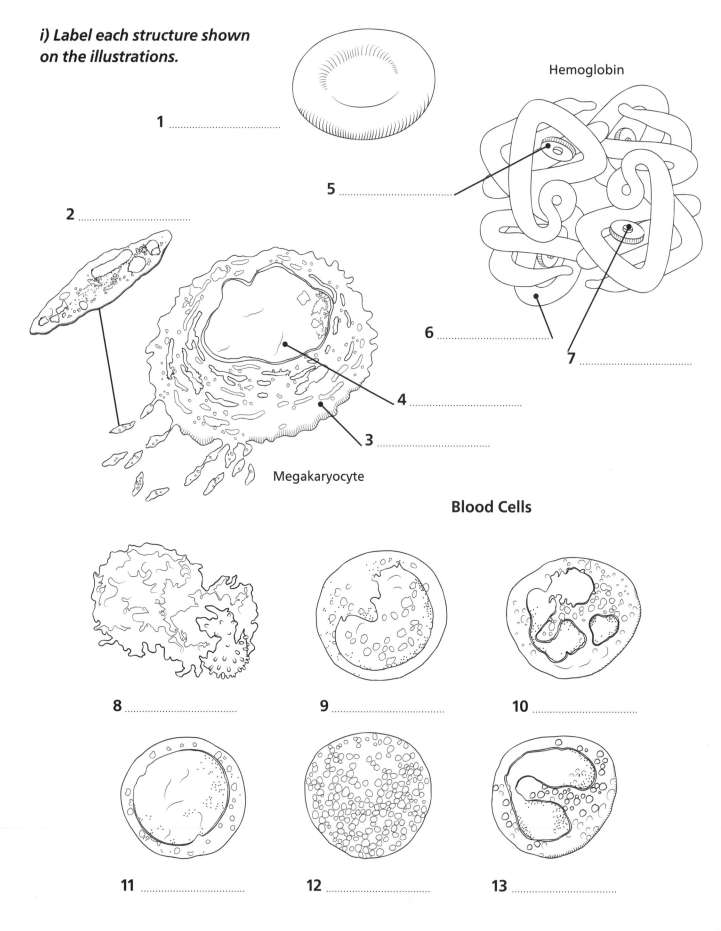

Hemoglobin

1 ....

2 ....

5 ....

6 ....

7 ....

4 ....

3 ....

Megakaryocyte

Blood Cells

8 ....

9 ....

10 ....

11 ....

12 ....

13 ....

*ii) Label each structure shown
on the illustrations.*

Blood—cellular level

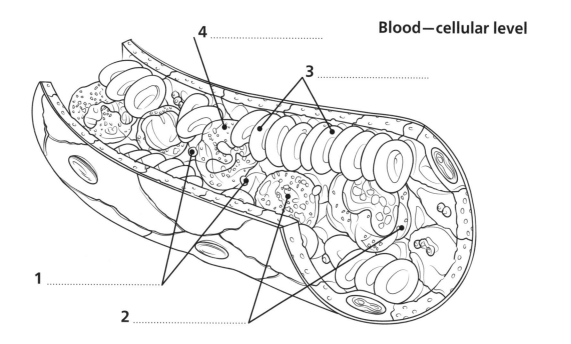

4 ....

3 ....

1 ....

2 ....

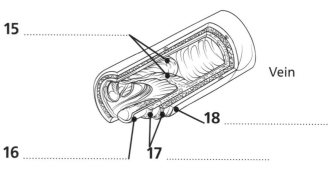

Artery

5 ....

6 ....

15 ....

Vein

16 ....

18 ....

17 ....

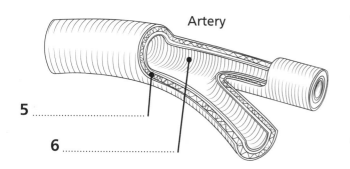

7 ....

Blood Vessels

8 ....

Capillaries

9 ....

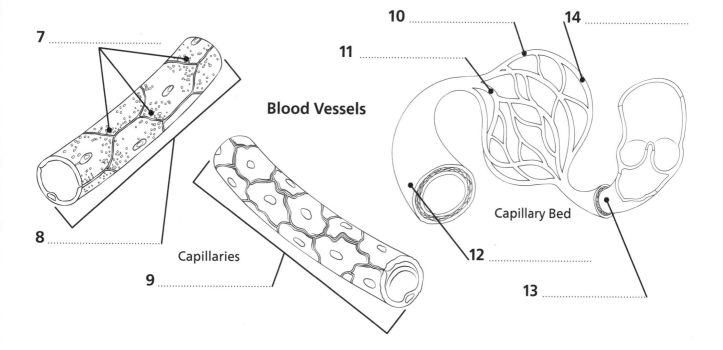

10 ....

14 ....

11 ....

Capillary Bed

12 ....

13 ....

Fill in the blanks

1   Immature red blood cells with nuclear fragments in their cytoplasm are called
    _____.

2   The most common type of agranular leukocyte is the _____.

3   The neutrophil granulocyte has a characteristic _____ nucleus.

4   The biconcave shape of red blood cells depends on a cytoskeletal protein called
    _____.

5   The major plasma proteins in blood are _____ and _____.

6   In a centrifuged blood sample, leukocytes make up a layer called the _____.

7   Basophil granulocytes resemble _____ cells of the tissue spaces.

8   The _____ removes old and defective red blood cells from circulation.

9   Leukocytes leave the bloodstream in a process called _____.

10 Red blood cells adopt a spherical shape in the condition known as _____.

Sickle cell blood

Normal blood

White blood cells

Abnormal red blood cells

Red blood cells

Platelets

Normal red blood cells

Sickle Cell Disease

In this blood disorder, red blood cells form curved or sickle shapes, resulting in the blockage of small blood vessels.

11 Each normal adult hemoglobin molecule has two _____ and two _____ globin chains.

12 Each hemoglobin chain contains a single _____ pigment molecule.

13 The oxygenated form of hemoglobin is called _____.

14 The binding of carbon dioxide with hemoglobin forms _____.

15 Old red blood cells may rupture in a process called _____.

16 The process of attraction of white blood cells to invading pathogens is called _____.

17 Movement of white blood cells by a sliding of their cytoplasm into cellular processes is called _____ movement.

18 The release of granules from the cytoplasm of white blood cells is called _____.

19 The process of determining how many of the different kinds of white blood cells are present in a blood film is called performing a _____ count.

20 The granules in the cytoplasm of eosinophils stain with acid dyes, so these cells may also be called _____.

Sickle cell disease (anemia)

Sickle cell disease is an inherited condition of abnormal hemoglobin in which red blood cells adopt a sickle shape and block small vessels under certain conditions. It is most common in people originating from sub-Saharan Africa, the Arabian Peninsula, and parts of India. The sickle-shaped cells are fragile and prone to rupture, so they have a short lifespan in circulation. Complications include anemia, swelling of the hands and feet, infection, and stroke. Patients may experience sickle cell crises, in which sickle cells block capillaries and cause acute ischemia (poor perfusion), pain, and death of tissue. However, carriers of the abnormal gene may actually benefit from a protective effect against malarial infection.

Match the statement to the reason

1   Erythrocytes have a biconcave disc shape because…

....

2   Iron deficiency anemia gives a pattern of small, faintly-colored red blood cells because…

....

3   Blood loss leads to an increased proportion of reticulocytes in the blood because…

....

4   Eosinophils are more common in the blood of patients with roundworm because…

....

5   Bacterial infection leads to neutrophilia because…

....

6   Spherical red cells (as in spherocytosis) may block small vessels because…

....

a   there is an insufficient amount of the ion to make the necessary hemoglobin.

....

b   it maximizes the exposure of hemoglobin for gas transfer.

....

c   this type of granulocyte contains enzymes that attack microbial cell membranes.

....

d   this cell type plays a key role in the immune response to parasites.

....

e   immature red blood cells are released prematurely into the bloodstream to maintain circulating hemoglobin.

....

f   they have insufficient flexibility to move through small capillaries.

....

1 Red blood cells in sickle cell anemia may lodge in capillaries because…

a histamine release from mast cells causes leakage of vascular fluid into tissue spaces.

2 Patients with abnormal red blood cells often have enlarged spleens because…

b they use these to destroy bacteria.

3 Neutrophils have lots of lysosomal enzymes because…

c there is a high concentration of carboxyhemoglobin there.

4 Anaphylactic shock involves loss of organ perfusion because…

d that is where defective cells are cleared from circulation.

5 Venous blood may have a bluish tinge because…

e the cells are both rigid and sticky.

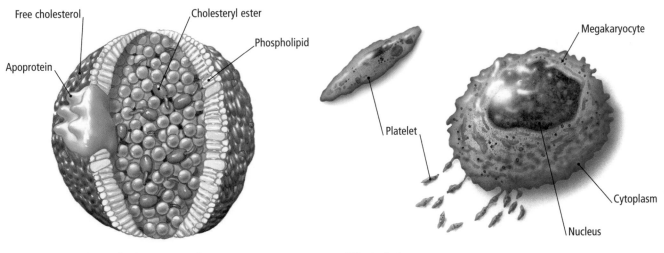

Cholesterol and Lipoprotein

Lipoproteins are protein molecules that carry fat within the blood. Types include HDL (high density lipoproteins), LDL (low density lipoproteins), IDL (intermediate density lipoproteins), and VLDL (very low density lipoproteins).

Platelets

Platelets are fragments of cells that circulate in the blood and have the ability to attach to areas of damaged blood vessels. By doing this, they can block small areas of hemorrhage.

Blood Production

After birth, red and white blood cells are produced in the red bone marrow in a process called hematopoiesis. Hematopoietic stem cells are capable of self-renewal. They give rise to committed myeloid or lymphoid progenitor cells that generate granulocyte/erythroid or lymphocytic blood cells, respectively. Myeloid stem cells can give rise to colony-forming units (CFUs) for the production of red blood cells or the different types of granulocyte or macrophage. Cells in the erythroid (red) cell line progressively develop hemoglobin and eventually lose their nuclei.

Key terms:

Basophilic erythroblast An immature type of cell in the erythroid lineage leading to mature red blood cells. It still contains a nucleus, and the cytoplasm absorbs basic or blue dyes during histological staining. This property of absorbing blue dyes gives the cell a blue color under the microscope. It has yet to make significant amounts of hemoglobin but contains the polyribosomes to manufacture globin protein.

Committed cell This cell (proerythroblast stage) is committed to becoming a red blood cell rather than some other type of blood cell (white blood cell or platelet).

Erythroblast An immature red blood cell. It goes through various stages (basophilic, polychromatophilic, orthochromatic) as hemoglobin increases in the cytoplasm and the nucleus degenerates and is ejected.

Erythrocyte A mature red blood cell. It has no nucleus, has a biconcave disc shape, and its cytoplasm is full of hemoglobin.

Erythroid colony-forming unit (CFU) A group of progenitor cells that will give rise to red blood cells and is able to produce a colony of differentiating erythroid line cells.

Erythropoietin A glycoprotein produced mainly in the peritubular interstitial cells of the inner cortex and outer medulla of the kidney. It regulates the production of red blood cells.

Hemoglobin (Hb) The oxygen- and carbon dioxide–carrying metalloprotein inside red blood cells. Adult hemoglobin usually consists of four globin chains (two alpha and two beta), each enclosing a heme molecule and a centrally placed iron atom.

Kidneys Paired organs on the posterior abdominal wall that regulate fluid and ionic balance in the body and also participate in the regulation of red blood cell production.

Orthochromatic erythroblast A type of immature red cell in the erythroid lineage that contains a shrunken nucleus and large amounts of hemoglobin. The nucleus may be in the process of ejection from the cell during this stage.

Polyribosomes Large molecules in the cytoplasm that manufacture protein from amino acids using the coded information on strands of messenger RNA (mRNA).

Polychromatophilic erythroblast A type of immature red cell in the erythroid lineage that is intermediate between the basophilic and orthochromatic erythroblast. Its cytoplasm stains with basic dyes (blue) because of the presence of polyribosomes but also with acidic dyes (pink) because of the hemoglobin that has been produced.

Proerythroblast A type of immature but committed cell that will follow the path of maturation to become a red blood cell. It is the first stage of the red blood cell lineage that can be recognized. Its development is induced by erythropoietin.

Red bone marrow The type of bone marrow where blood cells are produced. It contains a marrow stromal compartment (adipose cells, fibroblasts, stromal cells, endothelial cells, macrophages, and blood vessels) that produces support and regulatory factors to control hematopoiesis and a hematopoietic cell compartment to produce the blood cells.

Reticulocyte An almost-mature red blood cell with residual polyribosomes in its cytoplasm. Their presence in blood indicates recent stimulation of red blood cell production by erythropoietin.

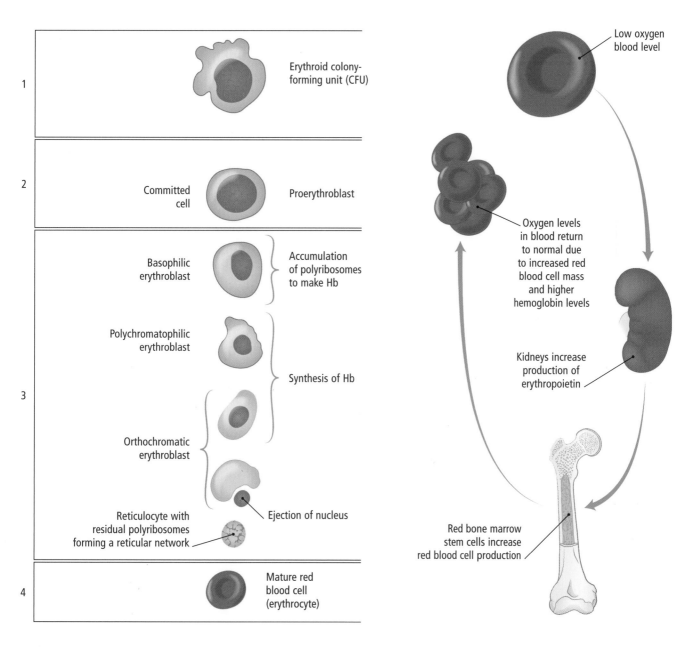

1   Erythroid colony-forming unit (CFU)

2   Committed cell   Proerythroblast

3   Basophilic erythroblast   Accumulation of polyribosomes to make Hb

Polychromatophilic erythroblast

Synthesis of Hb

Orthochromatic erythroblast

Reticulocyte with residual polyribosomes forming a reticular network   Ejection of nucleus

4   Mature red blood cell (erythrocyte)

Low oxygen blood level

Oxygen levels in blood return to normal due to increased red blood cell mass and higher hemoglobin levels

Kidneys increase production of erythropoietin

Red bone marrow stem cells increase red blood cell production

Production of Red Blood Cells

Red blood cells are produced in the red bone marrow from a hematopoietic stem cell. The process is regulated by the hormone erythropoietin, which is produced in the kidney. Differentiating red blood cells pass through proerythroblast, erythroblast, and reticulocyte stages.

True or false?

1   Most blood cell production during adult life is by the red bone marrow.

2   The erythroid colony-forming unit (CFU) gives rise to macrophages.

3   Blood cell production during fetal life can occur in the liver.

4   Megakaryocytes have a characteristic multilobed nucleus.

5   Megakaryocyte production is under the control of erythropoietin from the liver.

6   Erythropoietin production is increased by hypoxia.

7   Vitamin B6 is the most important vitamin for red blood cell production.

8   Iron deficiency anemia can be caused by chronic blood loss.

9   Allergic skin reactions involve neutrophil leukocytes.

10   Pus is composed mainly of the products of neutrophil granulocytes.

Acute lymphoblastic leukemia

Acute lymphoblastic leukemia is a cancer of white blood cells in which there is uncontrolled production of immature lymphocytes (lymphoblasts). The lymphoblasts crowd out the normal red bone marrow, leading to decreased production of red blood cells and normal white blood cells. The lymphoblasts may also lodge in other organs. The immune system is compromised, and patients will often present with bacterial infections such as pneumonia. The condition is most common in young children (2 to 5 years old), with another peak in old age. Symptoms include pallor, fatigue, fevers, shortness of breath, chest pain, coughing, nausea and vomiting, and bleeding (from low platelet levels). Treatments include chemotherapy, radiation therapy, and bone marrow and stem cell treatments.

Multiple choice

1 *The monoblast cell of the red bone marrow will eventually become a*

(A) macrophage
(B) red blood cell
(C) lymphocyte
(D) neutrophil granulocyte
(E) eosinophil granulocyte

2 *Which of the following is not a growth factor for hematopoiesis?*

(A) colony-stimulating factors
(B) erythropoietin
(C) thrombopoietin
(D) interleukin
(E) angiotensin

3 *Microcytic hypochromic anemia could be due to*

(A) vitamin B12 deficiency
(B) copper deficiency
(C) magnesium deficiency
(D) iron deficiency
(E) none of the above is correct

4 *Abnormally large red blood cells (megaloblasts) are seen in which type of anemia?*

(A) chronic blood loss
(B) iron deficiency anemia
(C) vitamin B12 deficiency
(D) acute blood loss
(E) sickle cell

5 *Which of the following cells is not capable of cell division?*

(A) pluripotent stem cell
(B) T cell progenitor
(C) B cell progenitor
(D) erythrocyte
(E) myeloid stem cell

6 *Which of the following gives rise to granulocytes?*

(A) metamyelocyte
(B) monoblast
(C) lymphoblast
(D) reticulocyte
(E) proerythroblast

7 *Cells leaving the red bone marrow must pass into the bloodstream through*

(A) marrow arteries
(B) central veins
(C) lymphatic channels
(D) capillary tight junctions
(E) sinusoid fenestrations

Red blood cells

Increased numbers of white blood cells (leukocytes)

Leukemia

This malignant disease involves the rapid and uncontrolled proliferation of leukocytes (white blood cells).

Fill in the blanks

1   The ability of a stem cell to give rise to many different types of mature cells is called
_____.

2   A deficiency of neutrophils in the bloodstream is called _____.

3   An increase in the white cell count in the blood is called _____.

4   A decrease in the white cell count is called _____.

5   The _____ colony-forming unit produces red blood cells.

6   The part of the red bone marrow that is not directly involved in hematopoiesis is called the
_____.

7   A reduced level of platelets in the blood is called _____.

8   The _____ colony-forming unit produces neutrophils and monocytes.

9   _____ can induce blood progenitor cells to produce cell clusters in vitro.

10   The cell called the _____ is the precursor of the lymphocyte.

11   Overproduction of red blood cells by the bone marrow is called _____.

12   A type of leukemia in which there is scarring of the bone marrow is called _____.

13   The process of blood cell production is called _____.

14   Nuclei extruded from developing red cells in the bone marrow are engulfed by
_____.

Match the statement to the reason

1 Red cell production is impaired in kidney disease because…

a the growth factor erythropoietin is produced in that organ.

2 Folic acid deficiency causes anemia because…

b the blood loss removes excess iron from the body.

3 Gastric mucosal atrophy causes megaloblastic anemia because…

c proliferation of myeloid stromal cells excludes normal hematopoietic tissue.

4 Patients with hemochromatosis can be treated by periodic blood withdrawal because…

d intrinsic factor from those cells is essential for vitamin B12 absorption.

5 The disease myelofibrosis causes anemia because…

e the vitamin is essential for DNA synthesis.

Pernicious anemia

Pernicious anemia is a condition in which there is inadequate production of intrinsic factor by the gastric mucosa. This factor is essential for the effective absorption of vitamin B12 by the terminal ileum. B12 is crucial for red blood cell production, so the presenting symptoms and signs are usually those of reduced red cell mass (tiredness, pallor, shortness of breath, and chest pain). The red blood cells are larger than normal (macrocytic) but have normal hemoglobin concentration. Neurological effects of reduced B12 may also be seen (numbness in hands and feet, poor balance, poor reflexes, depression, and confusion). Treatment is by injections or high oral doses of vitamin B12.

Blood Groups and Hemostasis

Red blood cells present specific molecules on their cell surface called antigens. The ABO system is the most important of blood groups and consists of four possible types (type A, type B, type AB, and type O), based on the presence or absence of two antigens (A antigen and B antigen) on the red blood cell surface. Another important system involves the rhesus factor. Individuals may have the rhesus factor on their red blood cells (Rh+) or not (Rh-). Hemostasis is the natural process of stopping blood loss from vessels.

Key terms:

Antibody A type of immune protein produced by plasma cells. Most antibodies circulate within the blood as immunoglobulins (IgG), but some are secreted to epithelial surfaces (IgA). Antibodies have binding sites to attach to specific molecules called antigens. These may be molecules on microorganisms, microbial toxins, or the body's own cells.

Antigen A molecule to which antibodies bind. Antigens may be foreign—for example, on invading bacteria, viruses, parasites, or fungi—or on the body's own cells.

Red blood cell A type of blood cell that has no nucleus and contains large amounts of hemoglobin in its cytoplasm. It carries oxygen and some carbon dioxide.

Rhesus negative Not having the rhesus factor antigen (D antigen) on one's red blood cells. Only about 15 percent of white Europeans and 6 percent of people of African descent are rhesus negative.

Rhesus positive Having the rhesus factor antigen (D antigen) on one's red blood cells. This is the common state in all races. A rhesus positive mother who has been previously exposed to rhesus positive blood and is carrying a rhesus negative fetus may produce antibodies that cross the placenta and harm the baby.

Type A Blood type in which the red blood cells display A antigen. The A antigen is composed of a specific carbohydrate chain containing one molecule each of N-acetyl galactosamine, N-acetyl glucosamine, and fucose and two molecules of galactose.

Type AB Type of blood in which the red blood cells display both A and B antigens.

Type B Blood type in which the red blood cells display the B antigen on their surface. The B antigen is composed of a specific carbohydrate chain containing one molecule each of N-acetyl glucosamine and fucose and three molecules of galactose.

Type O Blood type in which the red blood cells display neither the A nor B antigens on their surface.

Universal donor Individuals who are group O and rhesus (D antigen) negative are known as "universal donors" because their blood can be given to anyone without a transfusion reaction occurring. Their red blood cells do not have any of the A, B, or D (rhesus factor) antigens that could elicit an immune response.

Universal recipient Individuals who are group AB and rhesus (D antigen) positive are known as universal recipients because they can receive blood from anyone. Their blood does not have any antibodies against A, B, or D antigens.

ABO Blood Groups

There are several systems of blood antigens that must be considered in blood typing, but the most important is the ABO system. Individuals may have A or B antigens on their red blood cells (type A or B), both (type AB), or neither (type O). Someone from one blood group will make antibodies against red blood cells from someone with a different blood group antigen. If an individual belongs to one group, he or she can have blood from the same group, or from type O. Type AB individuals can receive blood from type A, B, or O individuals.

ANTIGEN (ON RED BLOOD CELL)	ANTIBODY (IN PLASMA)	BLOOD TYPE
A		Type A Cannot receive B or AB blood Can only receive A or O blood
B		Type B Cannot receive A or AB blood Can only receive B or O blood
A + B	Neither antibody	Type AB May receive any blood type Universal recipient
Neither A nor B	Both	Type O May receive only O blood Universal donor

True or false?

1   *The first event in hemostasis is vascular dilation.*

2   *The most significant blood group system is the ABO system.*

3   *People with AB blood group can accept blood from any type in the ABO system (universal recipient).*

4   *Platelets act to plug small points of leakage in the vessel wall.*

5   *Platelet deficiency gives rise to large-scale bruising.*

6   *The intrinsic pathway may be activated by local damage to the endothelium.*

7   *The clotting factors are made in the kidney.*

8   *The extrinsic pathway of blood clotting is usually activated inside the vessel.*

9   *The final common pathway of both intrinsic and extrinsic clotting pathways involves the conversion of fibrinogen to fibrin.*

10   *Plasmin is an enzyme that can degrade fibrin to dissolve clots.*

Hemostasis

Hemostasis is the process by which the body plugs holes in broken vessels. There are a series of events that contribute to this. First, vascular spasm constricts arterioles supplying the area where the vessels have been damaged. Second, a sticky platelet plug forms at the site of vessel damage. Third, the blood coagulates in a clotting factor cascade that ultimately converts fibrinogen to fibrin and binds the blood elements into a solid gel. The final event in hemostasis is clot retraction, in which actin and myosin in platelets pull the edges of wounds together and squeeze the liquid component of the coagulated blood out as serum.

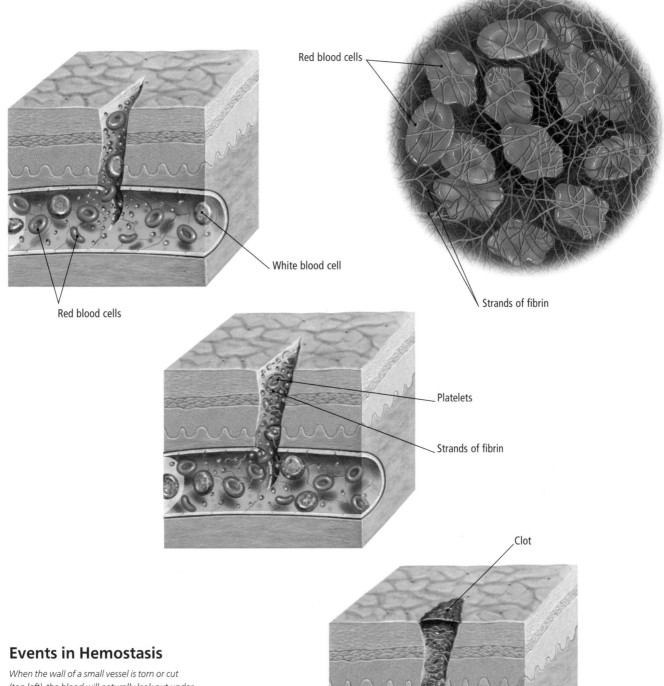

Red blood cells

White blood cell

Red blood cells

Strands of fibrin

Platelets

Strands of fibrin

Clot

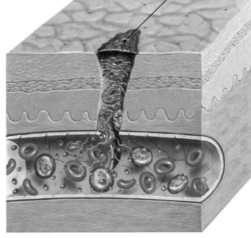

Events in Hemostasis

When the wall of a small vessel is torn or cut (top left), the blood will naturally leak out under pressure, but contact of platelets with damaged tissue and collagen will trigger aggregation of platelets to form a plug (middle). Coagulation of the extruded blood will also occur as the extrinsic pathway of coagulation is triggered, forming a clot (bottom). The end result is that red blood cells are trapped in a dense network of fibrin protein strands (top right).

Multiple choice

1 *Which type(s) of blood can be transfused into someone who is type A?*

(A) types A and B
(B) type A only
(C) types A and O
(D) type B
(E) types A and AB

2 *Which type of blood is called universal donor?*

(A) type AB
(B) type A
(C) type B
(D) type O
(E) none of the above is correct

3 *Erythroblastosis fetalis in a subsequent pregnancy can be prevented by giving the mother of a rhesus positive fetus an injection after delivery of*

(A) anti-D IgG
(B) fetal IgM
(C) antifetal IgE
(D) anti-G IgD
(E) none of the above is correct

4 *Which clotting factors are vitamin K dependent?*

(A) II, VII, IX, X
(B) I, VII, XI
(C) II, VIII, IX
(D) I, II, III, IV
(E) II, III, IV, X

5 *Which drug is most likely to interfere with hemostasis?*

(A) penicillin
(B) digoxin
(C) low-dose methotrexate
(D) warfarin
(E) vitamin D

6 *Which of the following has an anticoagulant effect by stabilizing antithrombin III clotting factors?*

(A) warfarin
(B) aspirin
(C) phenacetin
(D) plasmin
(E) heparin

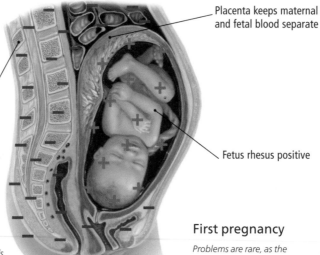

Placenta keeps maternal and fetal blood separate

Mother rhesus negative

Fetus rhesus positive

Rhesus (Rh) Factor
(see three images to the right)

In pregnancy, problems may occur if the mother is Rh negative and the fetus is Rh positive. The red blood cells of the fetus may be destroyed by the mother's Rh antibodies.

First pregnancy

Problems are rare, as the bloodstreams do not mix.

7 *The blood-clotting cascade begins in as little as*

(A) 100 milliseconds

(B) 1 second

(C) 3 seconds

(D) 15 seconds

(E) 3 minutes

8 *Which organ is the main site of clotting factor production?*

(A) spleen

(B) bone marrow

(C) liver

(D) gut wall

(E) none of the above

9 *Which of the following is an early event in hemostasis?*

(A) arteriolar vasoconstriction

(B) lysis of the fibrin clot

(C) contraction of fibroblasts

(D) conversion of fibrinogen to fibrin

(E) conversion of plasminogen to plasmin

10 *Activation of platelets leads to all of the following except*

(A) release of calcium

(B) release of ADP

(C) release of thromboxane A$_2$

(D) release of plasminogen

(E) attraction of other platelets

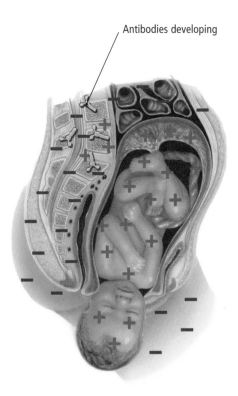

Antibodies developing

First pregnancy delivery

The baby's blood can leak into the maternal blood during delivery, causing the Rh-negative mother to develop antibodies that destroy Rh-positive blood cells.

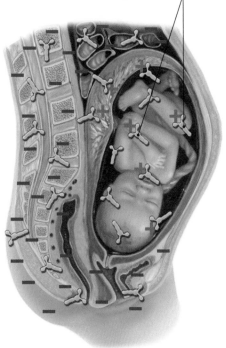

Rhesus positive antibodies produced as a result of exposure during first pregnancy attack the cells of the next Rh-positive fetus

Second pregnancy

If the mother does not receive an anti-D gamma globulin injection after the first pregnancy, she may develop antibodies that attack the blood cells of subsequent babies.

Fill in the blanks

1   *People with type O blood are called _____.*

2   *Rhesus incompatibility between fetus and mother can cause _____.*

3   *Lysis of fibrin clots is directly due to the action of _____.*

4   *Factor VIII in circulation is tightly bound to _____.*

5   *Abnormal aggregation of platelets in the vascular tree occurs in a condition called _____.*

6   *Exposure of the structural proteins _____ and _____ when the endothelium is damaged causes platelet adhesion.*

7   *Rapid treatment with _____ can dissolve a thrombus in an important artery.*

8   *The mechanical strength of a blood clot is mainly due to the _____.*

9   *A peptide hormone secreted by the endothelium to stimulate smooth muscle contraction and endothelial cell proliferation as part of vascular repair is called _____.*

10   *Clinically used vitamin K antagonists are _____ and _____.*

11   *Release of the _____ ion from platelets contributes to the clotting factor cascade.*

12   *Factor VIII is also called anti- _____ factor.*

13   *Sudden formation of a _____ in a cerebral artery is a common cause of stroke.*

14   *Sudden formation of a thrombus in a _____ is a common cause of myocardial infarction.*

Match the statement to the reason

1 Type O blood can be given to any ABO type because…

2 Warfarin is an anticoagulant because…

3 Heparin is a rapidly acting anticoagulant because…

4 Administration of tissue plasminogen activator shortly after thrombus formation in a coronary artery can clear the blockage rapidly because…

5 Liver disease causes defective blood coagulation because…

a it blocks formation of vitamin K–dependent clotting factors.

b type O red blood cells do not have A or B antigens on their cell surface.

c it converts plasminogen to plasmin to dissolve the fibrin clot.

d most clotting factors are made in that organ.

e it stabilizes antithrombin III clotting factors.

Hemophilia

Clotting factors are named according to a system of Roman numerals. The process of blood coagulation depends on a biochemical cascade in which one factor catalyzes the conversion of the next in series. Hemophilia is a mostly inherited condition in which one or more clotting factors are deficient. Hemophilia A is caused by a shortage of factor VIII and hemophilia B by inadequate factor IX. These are usually recessively inherited on the X chromosome and are thus much more common in males (who have only one copy of X). Hemophiliacs require transfusions of clotting factors to avoid internal bleeding, which is particularly damaging when it occurs in synovial joints.

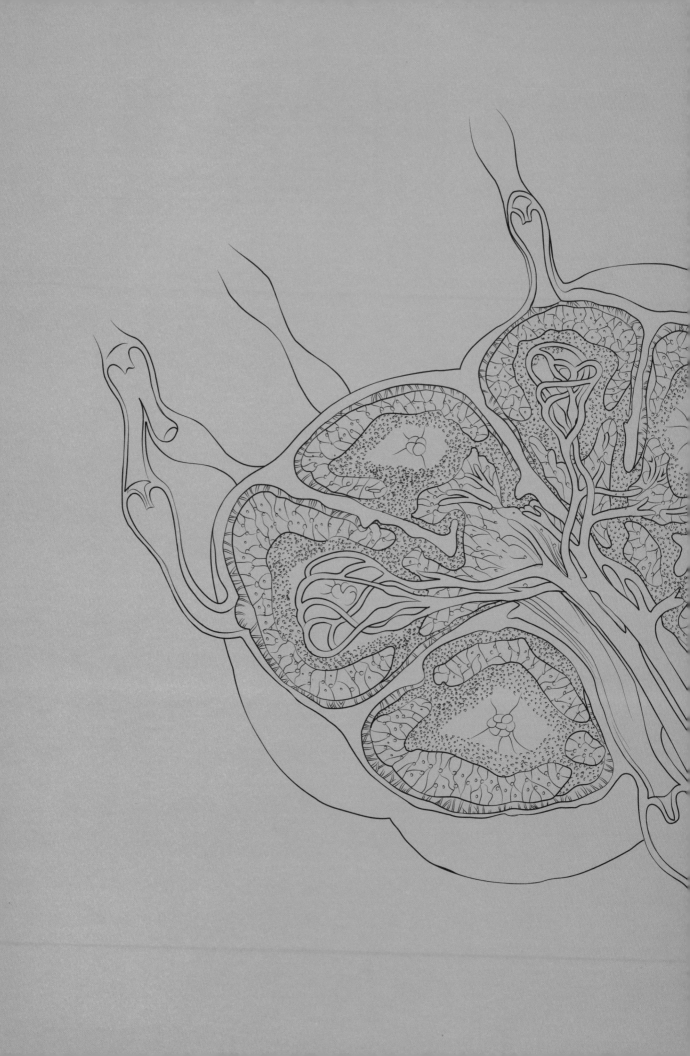

CHAPTER 11: THE IMMUNE SYSTEM

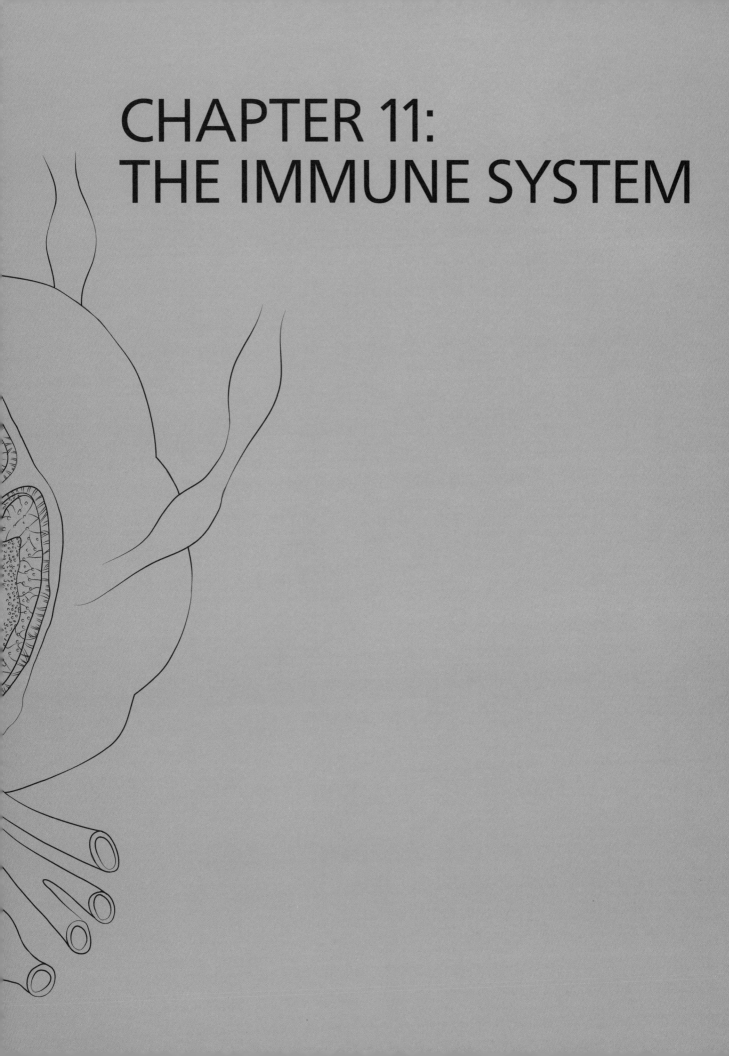

Lymph Nodes

Lymph nodes are collections of immune system cells located along the lymphatic vessels that carry excess tissue fluid back to the systemic venous side of circulation. Each lymph node consists of an outer cortex and an inner medulla, all surrounded by a capsule. Afferent lymphatic channels that carry lymph fluid to the node pierce the capsule. The node also has a hilum, where blood supply enters and leaves, and efferent lymphatic channels that carry lymph away from the node.

Key terms:

Anterior group Lymph nodes lying along the edge of the pectoralis minor muscle. The nodes drain the lateral upper chest and lateral breast.

Apical axillary nodes The final group of axillary lymph nodes at the apex of the axillary cavity before the lymph passes into the thoracic cavity.

Axillary nodes Groups of lymph nodes found around the walls of the axillary cavity. They drain to an apical group and into the subclavian lymph trunks.

Buccal nodes Superficial lymph nodes on the lateral wall of the oral cavity. They drain the cheek, lips, and lateral oral cavity.

Cervical nodes Lymph nodes in the neck are usually divided into deep and superficial groups. The deep group lies along the internal jugular vein, whereas the superficial group lies along the external jugular vein.

Cisterna chyli A dilated sac at the beginning of the thoracic duct near the aortic opening of the diaphragm. The cisterna chyli receives the intestinal, lumbar, and descending intercostal lymph trunks.

Common iliac nodes Deep lymph nodes lying along the common iliac vein. They receive lymph from the lower limb, pelvis, and lower abdominal wall.

Cubital nodes Superficial lymph nodes of the anterior elbow. They drain lymph from the hand and forearm.

External iliac nodes Deep lymph nodes lying along the external iliac vein. They drain lymph from the lower limb and inguinal region.

Intercostal nodes Deep lymph nodes of the posterior chest wall lying along the intercostal spaces. They drain the lateral chest wall.

Internal iliac nodes Deep lymph nodes of the lateral pelvic cavity. They drain lymph from the pelvic organs.

Lateral group Lymph nodes lying along the lateral border of the axillary cavity. They drain lymph from the upper limb.

Palmar and dorsal plexus Plexuses of fine lymph channels on the respective sides of the hand.

Parasternal nodes Deep lymph nodes in the anterior chest immediately behind the sternum and costal cartilages. They drain lymph from the anterior chest wall.

Parotid nodes Superficial lymph nodes palpable over the parotid salivary gland immediately anterior to the external ear. They drain lymph from the temporal fossa and cheek.

Plantar and dorsal plexus Plexuses of fine lymph channels on the respective sides of the foot.

Popliteal nodes Superficial lymph nodes palpable behind the knee. They drain lymph from the leg and foot.

Posterior mediastinal nodes Deep lymph nodes of the posterior chest. They drain the esophagus, great vessels, and posterior chest wall.

Retroauricular nodes Superficial lymph nodes palpable behind the external ear. They drain the posterior scalp and external ear.

Sacral nodes Lymph nodes anterior to the sacrum that drain the posterior pelvic organs.

Superficial inguinal nodes Lymph nodes palpable in the inguinal region. They are usually arranged in a "T" shape immediately below the inguinal ligament and drain lymph from the lower limb.

Thoracic duct The largest and longest lymphatic channel in the body. It drains lymph from the lower body and limbs, the left chest, left upper limb, and left head and neck, all into the venous circulation at the junction of the left brachiocephalic and left subclavian veins.

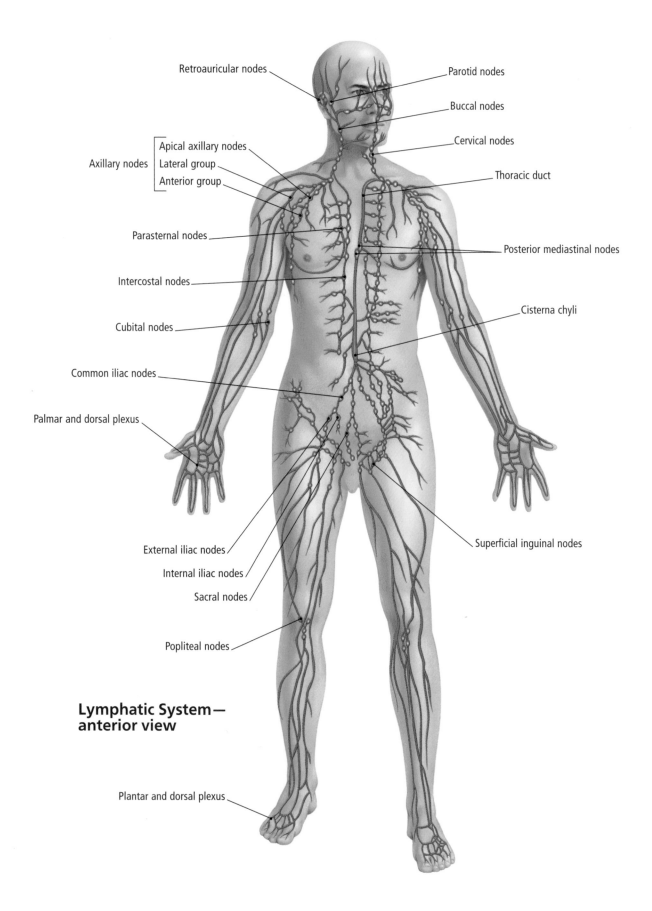

Retroauricular nodes

Parotid nodes

Buccal nodes

Cervical nodes

Axillary nodes — Apical axillary nodes / Lateral group / Anterior group

Thoracic duct

Parasternal nodes

Posterior mediastinal nodes

Intercostal nodes

Cisterna chyli

Cubital nodes

Common iliac nodes

Palmar and dorsal plexus

Superficial inguinal nodes

External iliac nodes

Internal iliac nodes

Sacral nodes

Popliteal nodes

**Lymphatic System—
anterior view**

Plantar and dorsal plexus

True or false?

1   *Most lymph nodes are located in the limbs and superficial tissues.*

2   *The largest lymphatic channel in the body is the thoracic duct.*

3   *The highest concentration of lymph nodes in the chest is in the middle mediastinum.*

4   *Lymph nodes of the intestine are found within the layers of the attaching mesenteric folds.*

5   *The brain is drained by lymph nodes inside the skull.*

6   *The germinal centers of lymph nodes develop in response to antigen stimulation.*

7   *The lymphatic duct drains the entire lower half of the body.*

8   *An enlarged rubbery lymph node is usually involved in bacterial infection.*

9   *Inguinal lymph nodes are aligned parallel to the inguinal ligament.*

10   *Lymph nodes involved with cancer will become red and tender.*

Lymph Nodes

Lymph nodes are collections of lymphoid cells along lymphatic channels. They filter the lymph fluid and provide surveillance for foreign molecules, such as those on bacteria, viruses, or fungi.

Afferent lymphatic vessels

Trabecula

Follicle of cortex

Capsule

Capillary

Vein

Artery

Efferent lymphatic vessel

Multiple choice

1 *Lymph from the lower limbs drains mainly through nodes in which region?*

Ⓐ gluteal
Ⓑ internal iliac
Ⓒ inguinal
Ⓓ umbilical
Ⓔ trochlear

2 *Venom from a snake bite on the finger would be carried by the*

Ⓐ arteries
Ⓑ lymphatics
Ⓒ venules
Ⓓ fascial planes
Ⓔ muscle membranes

3 *Which of the following is a potential site of the start of the thoracic duct?*

Ⓐ cisterna chyli
Ⓑ inguinal nodes
Ⓒ internal iliac nodes
Ⓓ left internal jugular vein
Ⓔ popliteal nodes

4 *A tumor on the tip of the tongue is likely to spread first to the*

Ⓐ submental node
Ⓑ jugulodigastric node
Ⓒ juguloomohyoid node
Ⓓ submandibular node
Ⓔ supraclavicular node

5 *Which body region does not usually drain to the axillary nodes?*

Ⓐ finger
Ⓑ elbow
Ⓒ medial breast
Ⓓ lateral breast
Ⓔ pectoralis major muscle

6 *Where are most B lymphocytes located in a lymph node?*

Ⓐ capsule
Ⓑ subcapsular sinus
Ⓒ medullary cord
Ⓓ medullary sinus
Ⓔ lymphatic nodule

7 *Where are most T lymphocytes located in a lymph node?*

Ⓐ deep cortex
Ⓑ subcapsular sinus
Ⓒ medullary cord
Ⓓ medullary sinus
Ⓔ lymphatic nodule

Lymphedema

Lymphedema is swelling of tissue due to an accumulation of extracellular fluid arising from blockage of lymphatic channels. Normally, a small proportion of the fluid carried to the peripheral tissues in the arteries is carried back to the body core by lymphatic channels. When these channels are blocked due to invasion by cancer cells or parasitic infestation, the fluid accumulates in the tissue. Lymphedema is commonly seen in Western countries in patients who have had clearance of lymph nodes and channels as a result of surgical management of cancer (e.g., breast, melanoma). In developing countries, it can also be caused by filariasis (an infestation with nematodes).

Fill in the blanks

1   The space below the lymph node capsule is called the _____.

2   The lymphatic duct opens into the junction of the _____ and the _____.

3   The _____ drains the right upper limb and right side of the head.

4   The _____ is a dilated sac-like structure that may give rise to the thoracic duct in about 20 percent of people.

5   The _____ is the hypothetical first lymph node draining a malignant neoplasm.

6   The _____ lymph nodes are found around the bifurcation of the trachea.

7   The _____ is the part of the lymph node where blood vessels enter and leave.

8   The medullary cords of lymph nodes contain mainly _____ and _____ cells.

9   B lymphocytes in the lymph nodes differentiate into _____ cells.

10   The lymph node capsule sends connective tissue _____ into the interior of the lymph nodes.

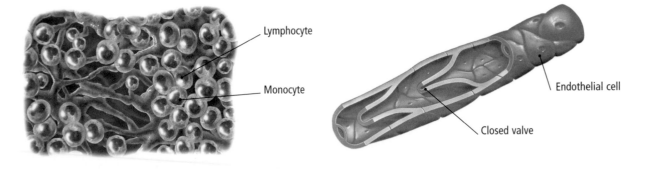

Immune System Tissue

Immune system tissue is concerned with the defense of the body against attack by microorganisms. It also removes abnormal cells, such as those that have become cancers.

Lymphatic Vessel

Lymphatic vessels carry excess tissue fluid back from the periphery of the body to the venous side of the systemic circulation. They have internal valves and may have a small amount of smooth muscle in their walls, but are very low–pressure vessels.

Match the statement to the reason

1 *Lymph from the testes drains to the abdominal lymph nodes because…*

a *these organs are in the upper abdomen during fetal life.*

2 *Breast cancer surgery is often accompanied by axillary lymph node clearance because…*

b *lymph channels and their nodes often develop alongside veins.*

3 *Removal of axillary lymph nodes can cause lymphedema because…*

c *the medial part of the chest drains directly to parasternal lymph nodes.*

4 *Tumor from the medial side of the breast can easily spread to the chest because…*

d *lymph node clearance also removes the associated lymph channels.*

5 *Lymph nodes are often found alongside venous channels because…*

e *most of this organ drains laterally to the armpit.*

Lymphoma

Lymphoma is a malignant neoplasm arising from cells of the lymph nodes. Symptoms include fevers, night sweats, weight loss, loss of appetite, and enlarged lymph nodes. Enlarged lymph nodes around the airways can make breathing difficult. Lymphoma may also infiltrate other lymphoid organs such as the spleen, causing enlargement. Types of lymphoma include Hodgkin's and non-Hodgkin's (which accounts for about 90 percent of cases). Infection with the Epstein-Barr virus is a risk factor for Hodgkin's lymphoma. Risk factors for non-Hodgkin's lymphoma include autoimmune disease, infection with HIV, immunosuppressant medications, and exposure to some pesticides. Treatment is by chemotherapy, radiation therapy, targeted therapy, and surgery.

Color and label

*i) Label each structure shown
on the illustration.*

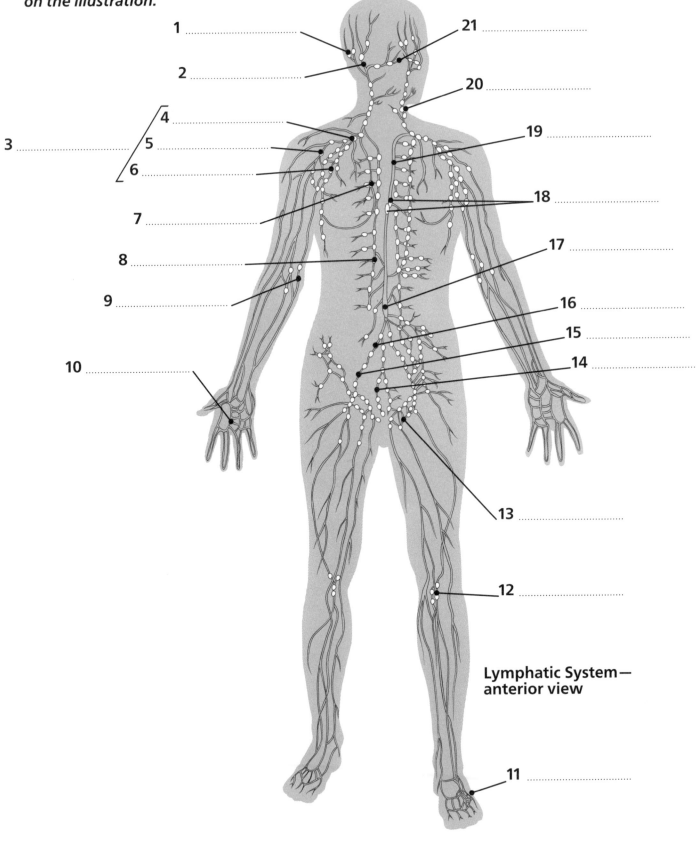

1 ....

2 ....

3 ....

4 ....

5 ....

6 ....

7 ....

8 ....

9 ....

10 ....

11 ....

12 ....

13 ....

14 ....

15 ....

16 ....

17 ....

18 ....

19 ....

20 ....

21 ....

**Lymphatic System—
anterior view**

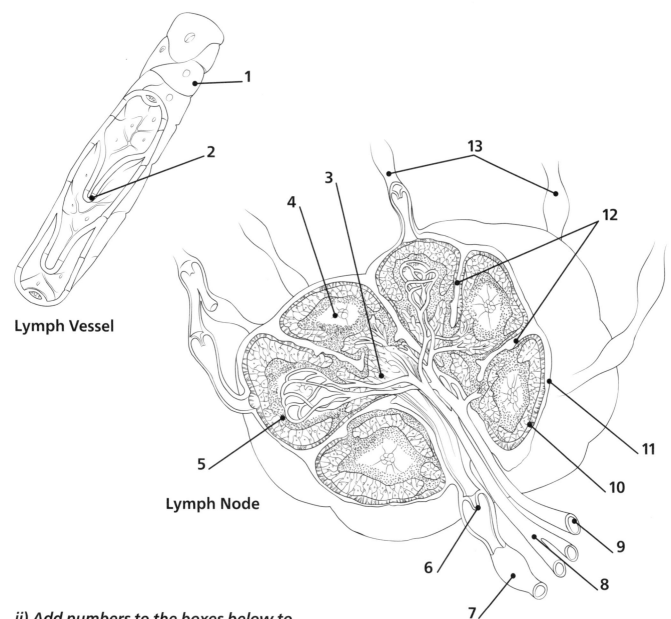

Lymph Vessel

Lymph Node

ii) Add numbers to the boxes below to match each label to the correct part of the illustrations.

Valve	☐
Closed valve	☐
Trabeculae	☐
Follicle of cortex	☐
Afferent lymphatic vessels	☐
Subcapsular sinus	☐
Efferent lymphatic vessel	☐

Artery	☐
Vein	☐
Lymphatic endothelial cell	☐
Capsule	☐
Medullary sinus	☐
Blood capillary	☐

Spleen and Thymus

The spleen is a lymphoid organ that is associated with the gut and is found in the upper left side of the abdomen. It is surrounded by a soft connective tissue capsule and contains red and white pulp. The spleen serves as a blood filter, removing old and defective red blood cells, as well as bacteria and viruses. The thymus is an organ in the chest that produces T (thymus-dependent) lymphocytes. The thymus is most active in early childhood and involutes around puberty.

Key terms:

Artery A high-pressure vessel carrying blood away from the heart. Arteries have a thick middle layer of elastic tissue and smooth muscle, called the tunica media.

Basal lamina A layer of substance on which epithelial cells sit. It is usually produced by the epithelial cells themselves.

Capsular vein Part of the venous network draining the subcapsular region of the thymic cortex.

Capsule The connective covering of the spleen. It sends sheets of connective tissue (trabeculae) into the interior of the spleen.

Connective tissue septum Extensions of the capsule of the thymus into the interior of the organ to separate the lobules.

Cortex The outer part of the thymus gland where T lymphocytes are produced.

First capillary venule A vessel located at the corticomedullary junction of the thymus. Capillary venules are where mature T-cells leave the organ.

Hassall's corpuscle A collection of aged epithelial cells in the medulla of the thymus.

Impression of the colon (left colic flexure) The region of the visceral surface of the spleen where it makes contact with the left colic (splenic) flexure.

Impression of the kidney The region of the visceral surface of the spleen where it is in contact with the upper pole of the left kidney and left adrenal gland.

Impression of the stomach A depression on the visceral surface of the spleen from its contact with the fundus and greater curvature of the stomach.

Left lobe The thymus consists of left and right lobes divided into incomplete lobules. Each lobule has an outer cortex and inner medulla.

Lymphoid organs Organs where cells responsible for the immune defense of the body are located. Examples include the lymph nodes, tonsils (lingual, palatine, and pharyngeal), spleen, and thymus.

Medulla The medulla is the core of the thymus where residual, aged epithelial cells form Hassall's corpuscles.

Notch in superior border The superior border of the spleen has this notching as a remnant of lobation during fetal life.

Red pulp The red pulp of the spleen monitors and recycles old and defective red blood cells and cellular debris. It contains an interconnected network of splenic sinusoids with discontinuous walls that allow transfer of cells between the red pulp and the blood.

Right lobe The thymus consists of left and right lobes divided into incomplete lobules. Each lobule has an outer cortex and inner medulla.

Spleen A lymphoid organ in the upper left abdomen. It has a diaphragmatic surface in contact with the diaphragm and a visceral surface in contact with the stomach, tail of pancreas, left kidney, and left colic flexure. It recycles old and damaged red blood cells and clears the blood of debris.

Splenic artery A large and tortuous branch of the celiac trunk that runs behind the stomach and along the superior border of the pancreas to the hilum of the spleen.

Splenic vein A vein draining from the entrance to the spleen (splenic hilum). From the spleen, it runs behind the tail and body of the pancreas.

Superior border The superior border of the spleen is in line with the ninth rib and is notched due to the remnants of a division into lobes during fetal life.

T lymphocyte T lymphocytes are white blood cells that develop in the cortex of the thymus. A selection process ensures that T lymphocytes produced by the spleen do not attack the body's own tissues.

Thymus A lymphoid organ in the anterior mediastinum of the chest. It is active before puberty to produce T lymphocytes but becomes a fatty-fibrous remnant in adult life.

Trabecular arteries Arteries in the connective tissue walls (trabeculae) that penetrate the interior of the spleen from the capsule.

Venous sinusoids Splenic sinusoids have discontinuous walls, allowing exchange of blood cells with the red pulp spaces.

White pulp nodule The immune component of the spleen. The white pulp nodules contain T and B lymphocytes, antigen-presenting cells, and plasma cells.

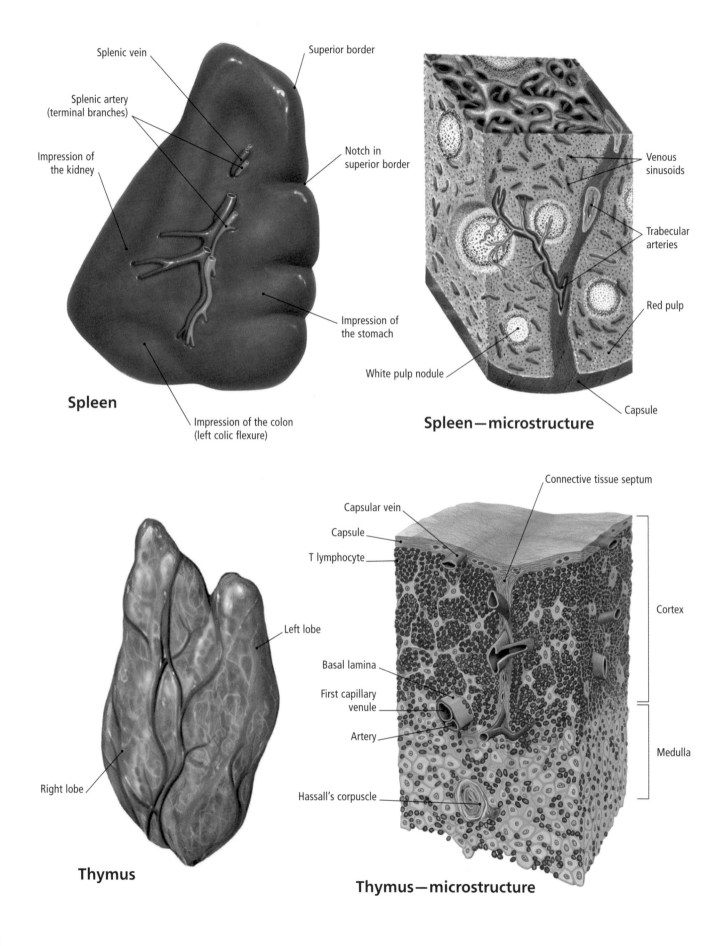

Spleen

Splenic vein
Splenic artery (terminal branches)
Impression of the kidney
Superior border
Notch in superior border
Impression of the stomach
Impression of the colon (left colic flexure)

Spleen—microstructure

Venous sinusoids
Trabecular arteries
Red pulp
White pulp nodule
Capsule

Thymus

Left lobe
Right lobe

Thymus—microstructure

Connective tissue septum
Capsular vein
Capsule
T lymphocyte
Basal lamina
First capillary venule
Artery
Hassall's corpuscle
Cortex
Medulla

True or false?

1   *The spleen has a firm, rigid capsule.*

2   *The spleen has red and white pulp components.*

3   *Splenic sinusoids terminate as open-ended vessels within the red pulp.*

4   *The white pulp is the blood filter component of the spleen.*

5   *Macrophages of the red pulp remove bacteria in the blood that have been coated with complement proteins.*

6   *The thymus consists of an outer cortex and an inner medulla.*

7   *Most T lymphocyte development takes place in the medulla of the thymus.*

8   *Lymphocytes that come to reside in the thymus originally come from the fetal liver.*

9   *The thymus is closely related to the left atrium.*

10   *The thymus develops from both endodermal and ectodermal embryonic layers.*

Thymus

Spleen

Location of Spleen and Thymus

The spleen and thymus are the two largest lymphoid organs. The thymus is located in the anterior mediastinum, in front of the great vessels. The spleen lies in the upper left side of the abdominal cavity.

Multiple choice

1 Fracture of which rib is most likely to damage the spleen?

(A) rib 6
(B) rib 7
(C) rib 8
(D) rib 10
(E) rib 12

2 Which organ is not in contact with the spleen?

(A) stomach
(B) liver
(C) left kidney
(D) left colic flexure
(E) tail of the pancreas

3 The spleen is attached to the stomach by the

(A) lesser omentum
(B) transverse mesocolon
(C) phrenicolienal ligament
(D) left crus of the diaphragm
(E) gastrosplenic ligament

4 Which of the following is most likely to cause an enlarged spleen?

(A) sickle cell anemia
(B) bowel cancer
(C) breast cancer
(D) congenital heart disease
(E) septicemia

5 At which stage of life does the thymus involute (degenerate)?

(A) embryonic life
(B) fetal life
(C) two years after birth
(D) puberty
(E) old age

6 Where is the thymus located?

(A) lower neck
(B) anterior mediastinum
(C) posterior mediastinum
(D) middle mediastinum
(E) pleural cavity

7 Which structure in the thymus is composed of keratinized epithelial cells?

(A) red pulp
(B) trabeculae
(C) Hassall's corpuscle
(D) capsule
(E) germinal zone

Ruptured spleen

The spleen is an abdominal lymphoid organ lying beneath ribs 9 to 11 on the left side. Although it has a connective tissue capsule, the splenic tissue is quite soft and friable, so blunt force to the left side of the lower chest (e.g., from a motor vehicle accident) can rupture the spleen. The spleen then bleeds profusely, potentially leading to life-threatening blood loss. Patients may have local tenderness at the lower left chest or left shoulder-tip pain due to irritation by blood of the diaphragmatic parietal peritoneum. The spleen may also be prone to rupture in malaria, leukemia, and lymphoma.

Fill in the blanks

1   The spleen is joined to the kidney by a peritoneal fold called the _____.

2   The splenic artery runs a tortuous course along the _____ of the
_____.

3   The hilum of the spleen is directly in contact with the tail of the _____.

4   The _____ surface of the spleen faces the rib cage.

5   The _____ of the spleen serves the role of clearance of old and defective red
blood cells.

6   The _____ of the spleen serves the immune function of the organ.

7   The embryonic origin of the thymus is from both _____ and _____
tissues of the region of the third pharyngeal pouch.

8   Aged epithelial cells accumulate in the thymus as _____.

9   The _____ of the thymus is the site where negative selection deletes unsuitable
T lymphocytes.

10   The process of programmed cell death of unsuitable T lymphocytes is called _____.

11   Fractures of ribs _____ to _____ on the left side could damage
the spleen.

12   The stroma of the spleen is composed of _____ connective fibers.

13   The thymus is mainly located in the _____ and _____
mediastinum.

14   Lobules of the thymus are separated from each other by connective tissue _____.

Match the statement to the reason

1 Malaria often leads to splenic enlargement because…

a the organ enlarges to clear large numbers of infected red blood cells.

2 A spleen must be enlarged at least three times to be palpable because…

b blood irritates the diaphragmatic parietal peritoneum.

3 In DiGeorge syndrome there is a deficiency of T lymphocytes because…

c the cortical epithelial cells of the thymus fail to develop.

4 Rupture of the spleen can cause shoulder-tip pain because…

d these cells play a key role in cell-mediated immunity.

5 Congenital deficiency of T lymphocytes makes the individual susceptible to tumors because…

e the organ is usually found posterior to the left midaxillary line.

Congenital T cell deficiency (DiGeorge syndrome)

DiGeorge syndrome is a genetic problem in which there is deletion of about 40 genes from chromosome 22. The syndrome includes an abnormal face, congenital heart disease, parathyroid deficiency, cleft palate, and developmental delay. It also includes immunodeficiency because the cortex of the thymus fails to develop. The failure of the thymus to organize means that T cells cannot differentiate. Consequently, the cell-mediated immune response to viruses and cancer cells is defective, but the nonspecific and humoral immune response is less affected. Children may also develop autoimmune disease such as rheumatoid arthritis and Graves' disease.

Color and label

*i) Label each structure shown
on the illustration.*

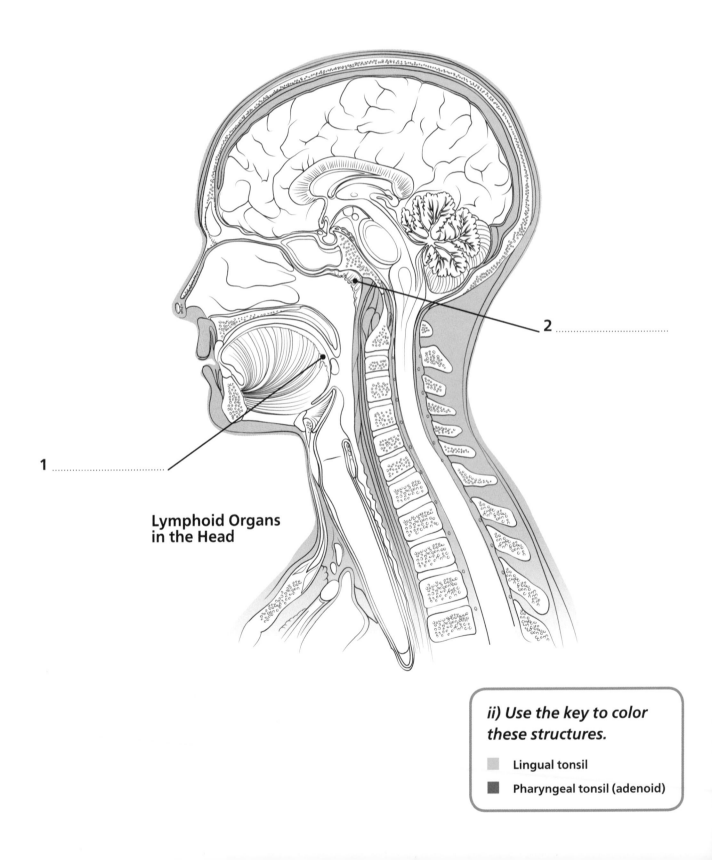

**Lymphoid Organs
in the Head**

1 ....

2 ....

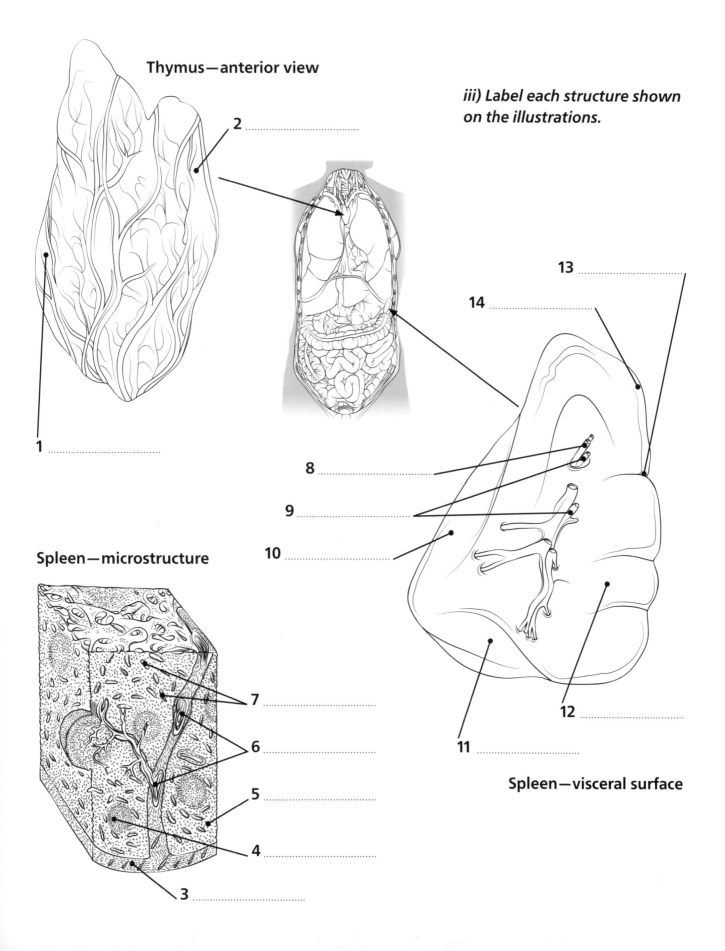

Thymus—anterior view

2 ....

iii) Label each structure shown on the illustrations.

13 ....

14 ....

8 ....

9 ....

10 ....

1 ....

Spleen—microstructure

7 ....

6 ....

5 ....

4 ....

3 ....

12 ....

11 ....

Spleen—visceral surface

Cellular/Humoral Response

The humoral immune response relies on antibodies that attach to foreign antigen molecules. If B lymphocytes encounter the specific antigen they are intended for, they become activated, dividing repeatedly to produce daughter cells that mature into plasma cells and memory B cells. Plasma cells secrete antibodies immediately against a specific antigen, whereas memory B cells will respond to the antigen on a subsequent exposure. Cell-mediated immunity involves several different classes of T cells. Cell-mediated immunity responds mainly to cells infected with intracellular pathogens such as viruses and intracellular bacteria (e.g., tuberculosis).

Key terms:

Antibody See pp. 324–325.

Antigen See pp. 324–325.

B lymphocyte A type of lymphocyte that arises in the bone marrow from a lymphoid stem cell. Mature B cells enter the circulation and become plasma cells to produce antibodies.

Blood vessel Tubular structures that carry blood through the body. High-pressure vessels carrying blood away from the heart are called arteries, those carrying blood to the heart are called veins, and those serving nutrient and gas exchange in the tissues are called capillaries.

Helper T cell A type of T lymphocyte that interacts with B cells in the presence of an antigen-presenting cell (e.g., dendritic cell or macrophage). This interaction ensures that B cells have access to new antigens and also stimulates the proliferation of the B cells to make antibodies against that new antigen.

Killer (or cytotoxic) T cell A type of T lymphocyte that binds with infected antigen-presenting cells such as macrophages. When this binding occurs, the killer cell releases perforin to rupture the cell membrane of the infected macrophage and destroy the cell and its resident viruses or mycobacteria.

Macrophage A phagocytic cell derived from circulating monocytes. They are able to phagocytose (engulf) debris and microorganisms in tissues throughout the body.

Memory B cell A type of B lymphocyte that retains a memory of exposure to a specific antigen. When re-exposure occurs, there is a rapid and efficient response to produce antibodies against the antigen.

Memory T cell A type of T lymphocyte that retains a memory of exposure to a specific antigen. When re-exposure occurs, there is a rapid and efficient response to produce T-cell-mediated types of immunity (e.g., T helper and T killer function).

Plasma B cell The stage of B lymphocyte differentiation in which the cells become antibody-production factories, manufacturing antibodies against antigens on microorganisms or other foreign tissues.

Proteins Long chains of amino acids with complex folding that produces specific three-dimensional shapes. Immune proteins such as antibodies, complement, and cytokines or interleukins are used by cells of the immune system.

Suppressor T cell A type of T lymphocyte that acts on helper T cells to moderate or inhibit helper T cell activity. Suppressor T cells also help regulate the progression of B cells to become plasma cells.

T cell See T lymphocyte.

T lymphocyte A type of lymphocyte that develops in the thymus during early postnatal life. T cells should only respond to foreign antigens bound to their own MHC (major histocompatibility complex) molecules. T cells exhibit self-tolerance, meaning they will not respond to self-antigens.

Virus A life form that consists of DNA or RNA strands (the viral genetic code) enclosed in a protein shell. Viruses rely on the genetic machinery of cellular organisms to produce copies of themselves.

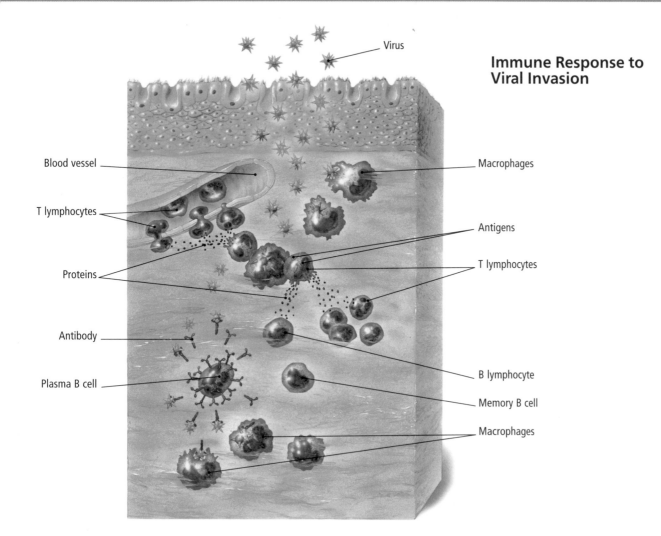

Virus

Immune Response to Viral Invasion

Blood vessel

Macrophages

T lymphocytes

Antigens

T lymphocytes

Proteins

Antibody

B lymphocyte

Plasma B cell

Memory B cell

Macrophages

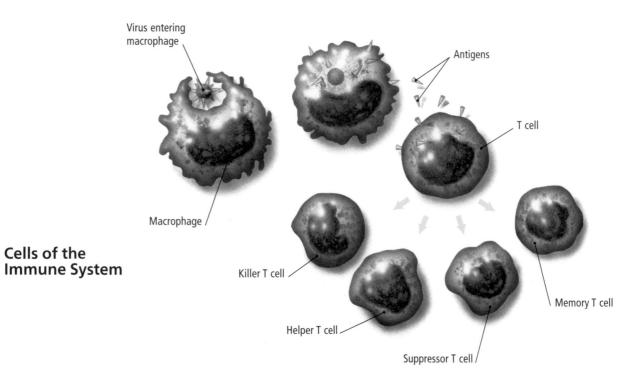

Virus entering macrophage

Antigens

T cell

Macrophage

Killer T cell

Memory T cell

Cells of the Immune System

Helper T cell

Suppressor T cell

True or false?

1   Neutrophils and macrophages are part of the innate immune system.

2   Acquired immunity often develops in response to exposure to an infectious pathogen.

3   Complement system molecules are synthesized in the spleen.

4   Plasma cells usually circulate in the bloodstream.

5   T lymphocytes give rise to plasma cells.

6   Autoimmune diseases are due to a failure of self-tolerance.

7   The thymus develops in the third pharyngeal pouch and cleft.

8   The thymus increases in size steadily during adult life.

9   Splenectomy would have no adverse health effects.

10   The human immunodeficiency virus (HIV) mainly infects B lymphocytes.

Thymus

The Thymus

The thymus is a lymphoid organ in the anterior part of the thoracic cavity, which produces T lymphocytes.

Multiple choice

1 *Which of the following is not an immunoglobulin?*
- (A) IgG
- (B) IgM
- (C) IgE
- (D) IgZ
- (E) IgA

2 *Which type of immunoglobulin is secreted onto an epithelial surface?*
- (A) IgG
- (B) IgM
- (C) IgE
- (D) Igkappa
- (E) IgA

3 *Which cell produces immunoglobulins?*
- (A) plasma cell
- (B) T lymphocyte
- (C) macrophage
- (D) neutrophil granulocyte
- (E) mast cell

4 *Which cell type is most effective against intracellular pathogens?*
- (A) plasma cell
- (B) T lymphocyte
- (C) macrophage
- (D) neutrophil granulocyte
- (E) mast cells

5 *Which cell differentiates into a plasma cell?*
- (A) mast cell
- (B) T lymphocyte
- (C) macrophage
- (D) neutrophil granulocyte
- (E) B lymphocyte

6 *Which cell type engulfs inert foreign material?*
- (A) mast cell
- (B) T lymphocyte
- (C) macrophage
- (D) neutrophil granulocyte
- (E) B lymphocyte

7 *Which cell type in the brain is part of the monocyte/macrophage lineage?*
- (A) neurone
- (B) oligodendrocyte
- (C) astrocyte
- (D) endothelium
- (E) microglia

Autoimmune disease

Autoimmune disease arises when the body's immune system mounts an attack on its own tissues. It is, therefore, a failure of the natural self-tolerance that should arise in lymphocytes during development. Examples include rheumatoid arthritis, systemic lupus erythematosus (SLE), juvenile onset diabetes mellitus, rheumatic heart disease, and ankylosing spondylitis. Symptoms and signs will vary according to the actual tissues affected, but autoimmune diseases are usually more common in women than men. Some diseases are triggered by an infection with a microorganism (virus or bacterium) that has an antigen with a similar structure to molecules on the body's own cells.

Fill in the blanks

1   The largest immunoglobulin molecule is _____.

2   Lymphocytes that remember exposure to an antigen and can initiate a response on re-exposure are called _____ cells.

3   The molecules on pathogenic microorganisms that initiate an immune response are called _____.

4   Anaphylaxis is life-threatening because of the drop in _____ and constriction of the _____.

5   The immediate response during anaphylaxis is release of _____ from _____.

6   The main cell types defending against cancer are the _____ and _____.

7   Tonsillar tissue behind the nose is located in the wall of the _____.

8   The tonsillar tissue in the pharynx and posterior third of the tongue is collectively called _____.

9   Enlargement of the tonsillar tissue occurs in a common condition called _____.

10   The cause of AIDS is infection with the _____.

11   Lymph nodes in the groin region are called _____ nodes.

12   Lymph nodes are usually located alongside vessels of the _____ side of the systemic circulation.

13   The deep lymph nodes of the neck are located alongside the _____ vein.

Match the statement to the reason

1 Abnormal proliferation of plasma cell precursors increases immunoglobulin levels in the blood because…

a that cell line is responsible for making antibodies.

2 The complement system gets its name because…

b the virus infects macrophages, dendritic cells, and CD-4-bearing helper T cells.

3 Infection with HIV causes opportunistic infections and cancers because…

c these cells are able to prevent invasion of body tissue by gut flora.

4 Immune system cells are concentrated in tonsillar tissue near the mouth and nasal passages because…

d the constituent molecules enhance the effectiveness of antibodies.

5 Immune system cells are concentrated in Peyer's patches in the mucosa and submucosa of the intestine because…

e they can protect the entrances to the digestive and respiratory systems.

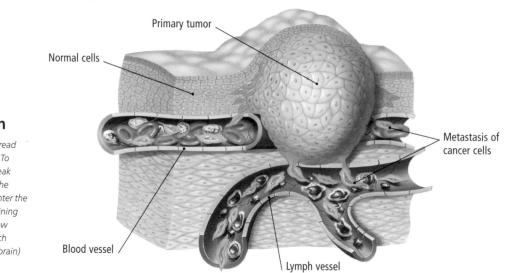

Metastasis by a Malignant Neoplasm

Malignant neoplasms (cancers) can spread locally or to distant sites (metastasize). To reach distant organs, the cells must break through the basement membrane of the epithelium they arise from and then enter the capillaries and lymphatic channels draining their site of origin. Cancer cells then flow passively with the blood until they reach another capillary bed (e.g., lung, liver, brain) where they lodge and grow.

Primary tumor

Normal cells

Metastasis of cancer cells

Blood vessel

Lymph vessel

Answers

pp. 22–23: 1. True, 2. True, 3. False—The diaphragm separates the thoracic and abdominal cavities, 4. True, 5. False—The diaphragm separates the thoracic and abdominal cavities, 6. False—The dorsal body cavity contains the central nervous system, 7. False—Most of the lymphatic system is a diffuse system of channels and small nodes, 8. True, 9. True, 10. False—The nucleus contains DNA, 11. False—The nucleolus is within the nucleus, 12. True, 13. False—The mineral in bone is hydroxyapatite (calcium phosphate), 14. False—They are covered by hyaline cartilage, 15. True, 16. False—Tubular structure provides maximal mechanical strength for weight, 17. False—The outer membrane of bone is fibrous, not muscular, 18. False—The nervous system acts over milliseconds to minutes, 19. True, 20. True.

pp. 24–25: 1—E, 2—B, 3—D, 4—E, 5—C, 6—D, 7—B, 8—A, 9—C, 10—D, 11—A, 12—D, 13—B, 14—E.

pp. 26–27: *i) Label each structure shown on the illustrations:* 1. Nucleus, 2. Golgi apparatus, 3. Cilium, 4. Microvilli, 5. Location of chromatin, 6. Mitochondrion, 7. Endoplasmic reticulum, 8. Ribosome, 9. Peroxisome, 10. Nucleolus, 11. Cytoplasm, 12. Nuclear pores, 13. Lysosome, 14. Centriole, 15. Synaptic knob, 16. Axon terminal, 17. Axon, 18. Myelin sheath, 19. Dendrite, 20. Mitochondrion, 21. Nucleolus, 22. Nuclear membrane, 23. Golgi apparatus, 24. Cell body.

ii) Use the key to color these structures: refer to i), above, for answers.

iii) Add numbers to the boxes to match each label to the correct part of the illustration: 1. Red blood cell, 2. Platelet, 3. Cytoplasm, 4. Nucleus, 5. Heme, 6. Globin protein strand, 7. Iron ion, 8. Macrophage, 9. Monocyte, 10. Neutrophil, 11. Lymphocyte, 12. Basophil, 13. Eosinophil.

pp. 28–29: 1—D, 2—C, 3—E, 4—C, 5—A, 6—E, 7—A, 8—D, 9—E, 10—B, 11—A, 12—D, 13—C, 14—E.

pp. 30–31: *i) Label each structure shown on the illustrations:* 1. Epidermis, 2. Dermis, 3. Subcutaneous fat, 4. Sweat gland, 5. Ruffini endings, 6. Free nerve ending, 7. Nerve ending, 8. Krause bulb, 9. Pacinian corpuscle, 10. Dermal papilla, 11. Deep fascia, 12. Hair follicle, 13. Sebaceous gland, 14. Stratum spinosum and granulosum, 15. Stratum granulosum, 16. Stratum corneum, 17. Hair, 18. Meissner corpuscles, 19. Root of nail, 20. Cuticle, 21. Lunula, 22. Nail.

ii) Add numbers to the boxes to match each label to the correct part of the illustrations: 1. Epidermis, 2. Precuticular epithelium, 3. Internal root sheath, 4. External root sheath, 5. Dermal hair papilla, 6. Nerve, 7. Follicle sheath, 8. Melanocyte, 9. Hair bulb, 10. Internal root sheath, 11. External root sheath, 12. Follicle sheath, 13. Erector pili muscle (arrector pili), 14. Sebaceous gland, 15. Medulla, 16. Cortex, 17. Cuticle, 18. Hair shaft, 19. Skull bone, 20. Pericranium, 21. Loose areolar tissue, 22. Aponeurosis, 23. Skin, 24. Hair follicle, 25. Hair.

pp. 32–33: 1. brachium or arm, 2. umbilicus, 3. inguinal, 4. antebrachium or forearm, 5. anatomical position, 6. crus or leg, 7. axilla, 8. cervical, 9. popliteal, 10. pollex, 11. pericardium, 12. endocrine, 13. plantar, 14. skin, 15. cubital, 16. ribosomes, 17. axon, 18. chromatin, 19. elastin, 20. collagen.

p. 34: 1—C, 2—A, 3—E, 4—D, 5—B.

p. 35: 1—D, 2—B, 3—A, 4—E, 5—C.

pp. 44–45: 1. True, 2. False—The hyoid has no direct contact with the vertebral column. The hyoid bone is supported by muscles and long ligaments only, 3. False—There are just 2 pairs of floating ribs, 4. False—The shape of the lumbar vertebrae allows only forward flexion and extension and some lateral flexion, but no trunk rotation, 5. True, 6. False—Only the scapula and clavicle belong to the pectoral girdle, 7. True, 8. True, 9. True, 10. False—The bones of the palm are the metacarpals, 11. True, 12. False—The three components of the hip bone meet at the acetabulum, 13. False—We sit on our ischial tuberosities, 14. True, 15. True, 16. True, 17. True.

pp. 46–47: 1—C, 2—A, 3—D, 4—B, 5—C, 6—A, 7—E, 8—B, 9—B, 10—A, 11—D, 12—B, 13—C, 14—E.

pp. 48–49: *i) Add numbers to the boxes to match each label to the correct part of the illustration:* 1. Spongy bone, 2. Muscle, 3. Tendon, 4. Epiphyseal plate, 5. Marrow cavity, 6. Bone marrow, 7. Cortical bone, 8. Periosteum, 9. Endosteum, 10. Inner circumferential lamella, 11. Volkmann's canal, 12. Interstitial lamellae, 13. Outer circumferential lamellae, 14. Concentric lamellae, 15. Haversian canals, 16. Trabecula of spongy bone.

ii) Label each structure shown on the illustrations: 1. Parietal bone, 2. Maxilla, 3. Upper (maxillary) teeth, 4. Lower (mandibular) teeth, 5. Mandible , 6. Mental foramen, 7. Nasal septum, 8. Zygomatic bone, 9. Greater wing of sphenoid bone, 10. Temporal bone, 11. Lesser wing of sphenoid bone, 12. Nasal bone, 13. Frontal bone, 14. Lambdoid suture, 15. Mastoid process, 16. Mandible, 17. External occipital protuberance, 18. Occipital bone, 19. Temporal bone, 20. Lambda, 21. Parietal bone, 22. Sagittal suture.

pp. 50–51: 1. Volkmann's canals, 2. Haversian canals, 3. periosteum, 4. periosteum; endosteum, 5. cranial vault or brain case; facial skeleton, 6. nutrient foramina, 7. seven; twelve; five; five; three to five, 8. hyoid bone, 9. pelvic or ventral, dorsal, 10. greater; lesser tubercle, 11. trochlear, 12. proximal and distal; proximal, middle and distal, 13. obturator foramen; ilium; ischium, 14. sciatic, 15. iliac, 16. linea aspera, 17. intercondylar fossa, 18. medial malleolus, 19. talus; cuneiforms, 20. talus.

p. 52: 1—D, 2—A, 3—B, 4—E, 5—C.

p. 53: 1—D, 2—A, 3—E, 4—B, 5—C.

pp. 54–55: *i) Label each structure shown on the illustrations:* 1. Cervical vertebrae (C1–C7), 2. Thoracic vertebrae (T1–T12), 3. Lumbar vertebrae (L1–L5), 4. Sacrum, 5. Coccyx, 6. Cervical region (C1–C7), 7. Thoracic region (T1–T12), 8. Lumbar region (L1–L5), 9. Sacral region (S1–S5), 10. Coccygeal region, 11. Transverse processes, 12. Spinous processes, 13. Axis (C2), 14. Atlas (C1).

ii) Use the key to color these structures: refer to i), above, for answers.

iii) Label each structure shown on the illustrations: 1. Body, 2. Transverse foramen, 3. Pedicle, 4. Lamina, 5. Spinous process, 6. Vertebral foramen, 7. Superior articular process (facet), 8. Posterior tubercle, 9. Anterior tubercle, 10. Inferior articular process (facet), 11. Sulcus for ventral ramus of spinal nerve, 12. Body, 13. Transverse process, 14. Pedicle, superior notch, 15. Spinous process, 16. Lamina, 17. Transverse costal facet, 18. Superior articular process (facet), 19. Superior costal demifacet, 20. Body, 21. Inferior articular process (facet), 22. Inferior costal demifacet, 23. Superior articular process (facet), 24. Lamina, 25. Spinous process, 26. Transverse process, 27. Mammillary process, 28. Pedicle, 29. Vertebral foramen, 30. Body, 31. Inferior articular process (facet), 32. Pedicle.

p. 56: 1. False—Fibrous joints are the most stable, 2. True, 3. True, 4. False—Ellipsoidal joints allow movement in two axes, 5. True, 6. False—The nucleus pulposus is within the annulus fibrosus, 7. False— It is a saddle-shaped joint, 8. False—The knee joint is bicondylar and allows rotation around the long axis of the femur as well as flexion/extension, 9. True, 10. True.

p. 57: 1—A, 2—E, 3—E, 4—C, 5—E, 6—C, 7—D.

p. 58: 1. coronal; sagittal, 2. condyle of the mandible; temporal, 3. atlantoaxial or C1/C2 intervertebral joint, 4. atlanto-occipital, 5. parasagittal; flexion and extension; rotation, 6. pronation; supination, 7. flexion; extension; rotation, 8. flexion; extension; medial and lateral rotation, 9. inversion; eversion, 10. talofibular; calcaneofibular.

p. 59: 1—E, 2—B, 3—A, 4—C, 5—D.

pp. 66–67: 1. True, 2. False—It is the interaction between actin and myosin that achieves the contraction, 3. True, 4. True, 5. False—The correct order is external oblique, internal oblique, and transversus abdominis, from outside to inside, 6. True, 7. True, 8. True, 9. True, 10. False—The triceps brachii attaches to the infraglenoid tubercle, 11. False—The biceps brachii also supinates the forearm and flexes the shoulder, 12. False—The triceps brachii attaches to the olecranon of the ulna, 13. True, 14. True, 15. False—Intrinsic muscles are also found in the palm, 16. False—The gluteus maximus is principally a hip extensor. The gluteus medius and minimus are the main muscles for supporting the hip during stance, 17. True, 18. True, 19. False—The gastrocnemius muscle (medial and lateral heads) also flexes the knee, 20. False—There are four layers of muscles in the foot.

pp. 68–69: *i) Label each structure shown on the illustration:* 1. Sarcoplasmic reticulum, 2. Transverse tubules, 3. Sarcomere, 4. Myofibril, 5. Myosin, 6. Actin, 7. Myosin head, 8. Myosin crossbridge, 9. Actin, 10. Myosin tail, 11. Muscle fiber, 12. Myofibril, 13. Nuclei.

ii) Add numbers to the boxes to match each label to the correct illustration: 1. Unipennate, 2. Bipennate, 3. Multipennate, 4. Spiral, 5. Radial, 6. Quadrate, 7. Strap, 8. Cruciate, 9. Triangular, 10. Multicaudal, 11. Digastric, 12. Circular, 13. Fusiform, 14. Bicipital, 15. Tricipital, 16. Quadricipital.

pp. 70–71: 1—D, 2—B, 3—A, 4—C, 5—E, 6—A, 7—C, 8—D, 9—B, 10—E, 11—E, 12—C, 13—A, 14—C.

pp. 72–73: 1. masseter; temporalis; medial pterygoid; lateral pterygoid, 2. neck flexion; to the opposite side, 3. sternum; ribs; lumbar vertebrae, 4. pubis; ilium; ischium; coccyx, 5. pubococcygeus; iliococcygeus; puborectalis (ischio)coccygeus, 6. pectoralis major; triceps brachii, 7. pronator quadratus; pronator teres, 8. flexion; flexion; supination, 9. biceps brachii; supinator, 10. intrinsic, 11. superficial and deep flexor, 12. hip abductors, 13. adductors; hamstrings, 14. ischial tuberosity, 15. rectus femoris, 16. biceps femoris; semimembranosus; semitendinosus, 17. piriformis; gemelli; obturators; quadratus, 18. adductor magnus; adductor longus; adductor brevis; pectineus; gracilis; obturator externus.

pp. 74–75: *i) Add numbers to the boxes to match each label to the correct part of the illustration:* 1. Temporalis, 2. Masseter, 3. Scalenus anterior, 4. Scalenus medius, 5. Levator scapulae, 6. Trapezius (cut), 7. Trapezius, 8. Sternocleidomastoid, 9. Sternohyoid, 10. Frontalis.

ii) Label each structure shown on the illustration: 1. Trapezius, 2. Latissimus dorsi, 3. External oblique, 4. Thoracolumbar fascia.

iii) Color the trapezius red and the latissimus dorsi orange: refer to i) above, for answers.

p. 76: 1—C, 2—B, 3—A.

p. 77: 1—C, 2—E, 3—A, 4—B, 5—D.

p. 84: 1. False—The ascending and descending tracts are in the white matter, 2. True, 3. True, 4. False—Dorsal root ganglion cells are purely sensory in function, 5. True, 6. True, 7. False—The sciatic nerve is the largest nerve in the body, 8. True, 9. False—The ulnar nerve is vulnerable to injury where it runs posterior to the medial epicondyle, 10. True.

p. 85: 1—A, 2—C, 3—C, 4—B, 5—D, 6—A, 7—E.

p. 86: 1. central canal, 2. gray matter, 3. C5; T1, 4. deltoid; shoulder tip, 5. musculocutaneous, 6. thoracic (T) 1; lumbar (L) 1 or 2, 7. pelvic splanchnic, 8. obturator, 9. saphenous; femoral, 10. tibial; sciatic.

p. 87: 1—D, 2—B, 3—A, 4—C.

pp. 88–89: *i) Add numbers to the boxes to match each label to the correct part of the illustrations:* 1. Radial nerve, 2. Median nerve, 3. Ulnar nerve, 4. Musculocutaneous nerve, 5. Anterior interosseous nerve, 6. Digital branches of radial nerve, 7. Superficial branch of radial nerve, 8. Deep branch of radial nerve, 9. Axillary nerve, 10. Common palmar digital branches of median nerve, 11. Superficial branch of ulnar nerve, 12. Flexor retinaculum, 13. Median nerve, 14. Superficial branch of radial nerve, 15. Ulnar nerve.

ii) Add numbers to the boxes to match each label to the correct part of the illustrations: 1. Lateral femoral cutaneous nerve, 2. Femoral nerve, 3. Obturator nerve, 4. Sciatic nerve, 5. Saphenous nerve, 6. Common fibular nerve, 7. Superficial fibular nerve, 8. Deep fibular nerve, 9. Lateral plantar nerve, 10. Medial plantar nerve, 11. Lateral sural cutaneous nerve, 12. Medial sural cutaneous nerve, 13. Tibial nerve, 14. Branches from femoral nerve, 15. Posterior femoral cutaneous nerve.

pp. 94–95: 1. True, 2. False—The axon transmits information to other nerve cells, 3. True, 4. False—The main neuron of the cerebral cortex is the pyramidal neuron, 5. False—The primary motor cortex is on the precentral gyrus, 6. True, 7. True, 8. True, 9. True, 10. True, 11. False—The corticospinal tract controls the spinal cord motor neurons, 12. False—The caudate and putamen are mainly concerned with motor function and language. The nucleus accumbens and septal area are concerned with reward and addiction, 13. True, 14. False—The sense of smell is processed on the medial temporal lobe, 15. False—Usually language areas are in the left hemisphere, 16. True, 17. True, 18. False—The sympathetic nervous system carries the autonomic motor signals for pupillary dilation, 19. True—this is called prosopagnosia, 20. True.

pp. 96–97: 1—A, 2—D, 3—E, 4—B, 5—A, 6—E, 7—C, 8—A, 9—A, 10—E, 11—B, 12—C, 13—D, 14—A.

pp. 98–99: *i) Add numbers to the boxes to match each label to the correct part of the illustrations:* 1. Frontal lobe, 2. Temporal lobe, 3. Occipital lobe, 4. Parietal lobe, 5. Primary somatosensory cortex, 6. Primary motor cortex, 7. Motor speech (Broca's) area, 8. Auditory association area, 9. Auditory cortex, 10. Wernicke's sensory speech area, 11. Reading comprehension area, 12. Visual cortex, 13. Visual association area, 14. Somatic sensory association area, 15. Central sulcus.

ii) Label each structure shown on the illustration: 1. Olfactory bulb (I), 2. Optic nerve (II), 3. Oculomotor nerve (III), 4. Trochlear nerve (IV), 5. Trigeminal nerve (V), 6. Abducens nerve (VI), 7. Facial nerve (VII), 8. Vestibulocochlear nerve (VIII), 9. Glossopharyngeal nerve (IX), 10. Vagus nerve (X), 11. Spinal accessory nerve (XI), 12. Hypoglossal nerve (XII).

pp. 100–101: 1. choroid plexus, 2. lateral, 3. cerebrospinal, 4. dural venous sinuses, 5. fornix, 6. corpus callosum, 7. internal capsule, 8. optic chiasm, 9. hypoglossal, 10. oculomotor; trochlear; abducens, 11. lateral geniculate, 12. medial geniculate, 13. middle cerebellar, 14. trigeminal; facial, 15. glossopharyngeal; vagus, 16. central, 17. calcarine, 18. lateral, 19. fourth.

p. 102: 1—D, 2—A, 3—B, 4—E, 5—C.

pp. 104–105: *i) Label each structure shown on the illustration:* 1. Thalamus, 2. Choroid plexus of lateral ventricle, 3. Pineal body, 4. Cerebral peduncle, 5. Trochlear nerve (IV), 6. Cerebellar peduncles, 7. Atlas (C1), 8. Second cervical nerve, 9. Spinal accessory nerve (XI), 10. Sulcus limitans, 11. Facial colliculus, 12. Dorsal median sulcus, 13. Pons, 14. Inferior colliculus, 15. Superior colliculus, 16. Lateral geniculate body, 17. Medial geniculate body, 18. Pulvinar, 19. Habenula.

ii) Use the key to color these structures: refer to i), above, for answers.

iii) Label each structure shown on the illustrations: 1. Spinal nerves C1–C8, 2. Spinal nerves T1–T12, 3. Spinal nerves L1–L5, 4. Spinal nerves S1–S5, 5. Coccygeal spinal nerve, 6. Celiac (solar) plexus, 7. Peripheral nerves, 8. Spinal cord, 9. Sympathetic ganglia, 10. Aortic arch.

pp. 110–111: 1. True, 2. True, 3. False—The olfactory tract carries sensory information to the brain, 4. True, 5. False—The taste buds lie within the surrounding trench of each circumvallate papilla, 6. True, 7. False—Olfaction is important for the subtleties of taste, because taste receptors have a limited range of responses (sweet, sour, bitter, salt, umami), 8. False—The lacrimal gland is in the upper lateral part of the orbit, 9. True, 10. True, 11. True, 12. False—The vitreous humor fills the posterior bulb of the eye, 13. True, 14. False—The fovea is lateral to the optic disk, 15. True, 16. False—The middle ear is filled with air, 17. False—The pharyngotympanic tube connects the middle ear with the nasopharynx, 18. True, 19. True, 20. False—The stapedius is important in dampening auditory ossicle movement during loud noise.

pp. 112–113: 1—A, 2—D, 3—E, 4—A, 5—C, 6—B, 7—D, 8—C, 9—E, 10—E, 11—D, 12—B, 13—D, 14—D.

pp. 114–115: *i) Add numbers to the boxes to match each label to the correct part of the illustration:* 1. Helix, 2. Pinna, 3. Cartilage, 4. Lobule, 5. External auditory canal (meatus), 6. Temporal bone, 7. Middle ear (tympanic cavity), 8. Eustachian (auditory) tube, 9. Cochlea, 10. Cochlear nerve branch, 11. Vestibular nerve branches, 12. Semicircular canals, 13. Ossicles (malleus, incus, and stapes), 14. Tympanic membrane (eardrum).

ii) Label each structure shown on the illustrations: 1. Frontal sinus, 2. Nasal bone, 3. Nasal cartilage, 4. Alar cartilage, 5. Nasal vestibule (nostril), 6. Maxilla (hard palate), 7. Soft palate, 8. Inferior nasal concha, 9. Middle nasal concha, 10. Superior nasal concha, 11. Sphenoidal sinus, 12. Olfactory centers in the brain, 13. Olfactory receptors, 14. Sphenoid bone, 15. Nasal cavity, 16. Olfactory bulb, 17. Frontal lobe of the brain, 18. Olfactory mucosa, 19. Bowman's gland (olfactory gland), 20. Cilia, 21. Olfactory cell, 22. Cribriform plate of the ethmoid bone, 23. Olfactory tract.

pp. 116–117: 1. parasympathetic; sympathetic, 2. iris, 3. parasympathetic; accommodation, 4. choroid capillary bed/choriocapillaris, 5. rod; cone, 6. photoreceptors; retinal ganglion cells, 7. fovea, 8. vestibule; bony labyrinth, 9. inner hair cells; outer hair cells, 10. endolymph, 11. auditory tube; acute otitis media, 12. stria vascularis, 13. semicircular; ampullae, 14. high frequency, 15. vestibular; vestibulocochlear, 16. basilar, 17. knob-like endings; cilia, 18. olfactory part of the temporal, 19. nucleus of the solitary tract; insula, 20. chorda tympani; facial.

p. 118: 1—C, 2 –D, 3—A, 4—E, 5—B.

p. 119: 1—E, 2—C, 3—A, 4—D, 5—B.

pp. 126–127: 1. False—The heart is located centrally behind the sternum, but with the left ventricle projecting down and to the left, 2. False—The fibrous pericardium is outside the serous pericardium, 3. True, 4. False—The left side of the heart (left atrium) receives blood from the pulmonary veins, 5. True, 6. True, 7. False—There are two pulmonary veins on each side, 8. False—The muscular wall of the left ventricle is thicker than that of the atria (left or right) because the left ventricle must attain high pressures during systole, 9. True, 10. False—The septum between the left and right ventricles is the thickest wall, 11. True, 12. False—Relatively deoxygenated blood flows through the pulmonary artery on its way to the lungs, 13. False—The mitral valve controls the orifice between the left atrium and left ventricle. There is no valve between the left and right atria, 14 True, 15. False—The heart chambers are lined with endocardium. The pericardium is on the outside of the heart, 16. False—The anterior part of the right atrium has muscular elevations on its interior called the musculi pectinati, 17. True, 18. True, 19. False—The left atrium lies at the heart base, 20. True.

pp. 128–129: 1—C, 2—A, 3—B, 4—C, 5—E, 6—D, 7—D, 8—E, 9—C, 10—D, 11—B, 12—A, 13—D, 14—E, 15—A.

pp. 130–131: *i) Color the arteries in red and the veins in blue: refer to ii), below, for answers.*

ii) Add numbers to the boxes to match each label to the correct part of the illustration: 1. Brachiocephalic artery, 2. Right brachiocephalic vein, 3. Superior vena cava, 4. Ascending aorta, 5. Right pulmonary artery, 6. Right superior pulmonary vein, 7. Right inferior pulmonary vein, 8. Right atrium, 9. Cusp of tricuspid valve, 10. Right ventricle, 11. Papillary muscles, 12. Inferior vena cava, 13. Descending thoracic aorta, 14. Chordae tendineae, 15. Cusp of mitral valve, 16. Pulmonary valve, 17. Left atrium, 18. Left inferior pulmonary vein, 19. Left superior pulmonary vein, 20. Ligamentum arteriosum, 21. Left pulmonary artery, 22. Aortic arch, 23. Left brachiocephalic vein, 24. Left subclavian artery, 25. Left common carotid artery.

iii) Label each structure shown on the illustration: 1. Left brachiocephalic vein, 2. Brachiocephalic trunk, 3. Right common carotid artery, 4. Right subclavian artery, 5. Right brachiocephalic vein, 6. Right lung (upper lobe), 7. Ascending aorta, 8. Superior vena cava, 9. Pericardium, 10. Right atrium,

11. Right ventricle, 12. Pleura, 13. Right lung (lower lobe), 14. Diaphragm, 15. Left ventricle, 16. Pulmonary trunk, 17. Left pulmonary artery, 18. Left lung (upper lobe), 19. Aortic arch, 20. Left subclavian artery, 21. Left common carotid artery.

p. 132: 1. coronary sinus, 2. aorta, 3. apex, 4. muscular; membranous, 5. tricuspid, 6. pulmonary artery, 7. right atrium, 8. ascending aorta, 9. trabeculae carneae, 10. musculi pectinati, 11. right ventricle, 12. left ventricle.

p. 133: 1—C, 2—E, 3—A, 4—D, 5—B.

pp. 138–139: 1. False—Oxygenated blood to the brain passes mainly through internal carotid arteries, 2. True, 3. True, 4. False—The external iliac artery supplies the lower limb. The internal iliac artery supplies pelvic organs and the buttock, 5. True, 6. False—The celiac artery usually supplies the stomach, spleen, liver, gall bladder, and part of the duodenum, 7. True, 8. False—The ulnar artery is found on the medial palmar side of the wrist, 9. False—The external iliac artery becomes the femoral artery at the base of the thigh, 10. False—The femoral artery accompanies the femoral nerve, 11. True, 12. True, 13. True, 14. True, 15. False—Renal arteries are branches of the abdominal aorta, below the diaphragm, 16. True, 17. False—The femoral artery becomes the popliteal artery at the adductor canal. The popliteal artery lies posterior to the knee joint, 18. True, 19. True, 20. False—It is the dorsalis pedis artery that is palpable on the dorsum of the foot. The posterior tibial artery can only be felt behind the medial malleolus.

pp. 140–141: 1—D, 2—B, 3—C, 4—A, 5—E, 6—B, 7—D, 8—B, 9—A, 10—B, 11—E, 12—A, 13—D, 14—D.

pp. 142–143: *i) Label each structure shown on the illustrations:* 1. Inferior vena cava, 2. Liver, 3. Portal vein, 4. Duodenum, 5. Pancreaticoduodenal vein, 6. Superior mesenteric vein, 7. Right colic vein, 8. Appendicular vein, 9. Colon, 10. Rectum, 11. Small intestine, 12. Left colic veins, 13. Inferior mesenteric vein, 14. Pancreas, 15. Splenic vein, 16. Spleen, 17. Left gastric vein, 18. Stomach, 19. Thoracic aorta, 20. Pancreaticoduodenal artery, 21. Right colic artery, 22. Ileocolic artery, 23. Appendicular artery, 24. Rectal artery, 25. Sigmoidal artery, 26. Jejunal arteries, 27. Left colic artery, 28. Inferior mesenteric artery, 29. Superior mesenteric artery, 30. Splenic artery, 31. Gastroduodenal artery, 32. Common hepatic artery, 33. Celiac trunk, 34. Hepatic artery proper, 35. Thoracic aorta, 36. Inferior vena cava.

ii) Use the key to color these structures: refer to i), above, for answers.

iii) Add numbers to the boxes to match each label to the correct part of the illustrations: 1. Celiac trunk, 2. Superior mesenteric artery, 3. Segmental artery, 4. Right renal artery, 5. Left renal artery, 6. Ureteric branch of the left renal artery, 7. Right gonadal artery, 8. Ureter, 9. Abdominal aorta, 10. Inferior mesenteric artery, 11. Arcuate artery, 12. Interlobar artery, 13. Renal pyramid (medulla), 14. Cortex, 15. Left adrenal gland, 16. Common iliac artery, 17. Internal iliac artery, 18. External iliac artery, 19. Obturator artery, 20. Obliterated umbilical artery, 21. Superior vesical artery, 22. Uterine artery, 23. Vaginal artery, 24. Middle rectal artery, 25. Internal pudendal artery, 26. Inferior gluteal artery, 27. Superior gluteal artery, 28. Lateral sacral artery, 29. Iliolumbar artery.

p. 144: 1. aorta, 2. internal carotid; vertebral, 3. external carotid, 4. brachiocephalic trunk, 5. intercostal, 6. lumbar, 7. internal iliac, 8. popliteal, 9. radial and ulnar, 10. celiac; superior mesenteric; inferior mesenteric.

p. 145: 1—E, 2—D, 3—A, 4—C, 5—B.

p. 148: 1. True, 2. True, 3. True, 4. False—The long or great saphenous vein runs up the inner side of the thigh. The short saphenous vein ends at the popliteal fossa, 5. False—The superior vena cava drains the head, neck, upper limbs, and upper chest, 6. True, 7. True, 8. False—The left renal vein is longer than the right because the inferior vena cava (into which they both drain) is on the right side of the abdomen, 9. True, 10. True.

p. 149: 1—B, 2—C, 3—A, 4—E, 5—C, 6—B, 7—D.

p. 150: 1. superior vena cava; inferior vena cava, 2. external jugular vein; internal jugular vein, 3. basilic; cephalic, 4. renal, 5. small/short saphenous; great saphenous, 6. femoral, 7. brachial, 8. posterior tibial; anterior tibial, 9. portal, 10. median cubital.

p. 151: 1—A, 2—B, 3—E, 4—D, 5—C, 6—G, 7—F.

pp. 152–153: *i) Label each structure shown on the illustration:* 1. Common carotid artery, 2. External jugular vein, 3. Axillary vein, 4. Brachial artery, 5. Superior vena cava, 6. Common iliac vein, 7. Radial artery, 8. Ulnar artery, 9. Posterior tibial artery, 10. Great saphenous vein, 11. Anterior tibial artery, 12. Fibular artery, 13. Small/short saphenous vein, 14. Popliteal vein, 15. Femoral artery, 16. Obturator artery, 17. Obturator vein, 18. Common iliac artery, 19. Basilic vein, 20. Inferior vena cava, 21. Cephalic vein, 22. Heart, 23. Arch of aorta, 24. Subclavian vein.

ii) Use the key to color these structures: refer to i), above, for answers.

iii) Add numbers to the boxes to match each label to the correct part of the illustrations: 1. Supraorbital artery, 2. Posterior branch of superficial temporal artery, 3. Supratrochlear artery, 4. Anterior branch of superficial temporal artery, 5. Occipital artery, 6. Superficial temporal artery, 7. Transverse facial artery, 8. Facial artery, 9. Transverse cervical artery, 10. External carotid artery, 11. Anterior branch of superficial temporal vein, 12. Supraorbital vein, 13. Supratrochlear vein, 14. Facial vein, 15. Submental vein, 16. Internal jugular vein, 17. Brachiocephalic vein, 18. Subclavian vein, 19. External jugular vein, 20. Retromandibular vein, 21. Posterior auricular vein, 22. Occipital vein, 23. Posterior branch of superficial temporal vein.

p. 154: 1. False—Arteries have the thickest walls of all vessels, 2. True, 3. False—Although there are semilunar valves where the pulmonary trunk and aorta arise from the heart, there are no valves in arteries. The high pressure and velocity of flow in arteries make valves unnecessary, 4. False—Not all capillaries have fenestrations. Fenestrations are found where large molecules must cross the vessel wall, e.g. in endocrine glands, intestines, pancreas, 5. True, 6. False—Pressure is much higher in the femoral artery, 7. True, 8. True, 9. False—Smooth muscle is mainly found in the tunica media, 10. True.

p. 155: 1—D, 2—E, 3—A, 4—D, 5—B, 6—E, 7—C.

p. 156: 1. endothelial, 2. media, 3. intima, 4. media, 5. pericytes, 6. adventitia, 7. fenestrations, 8. lumen, 9. internal elastic lamina, 10. external elastic lamina.

p. 157: 1—D, 2—A, 3—E, 4—B, 5—C.

pp. 162–163: 1. False—There are usually three nasal conchae or turbinates on each side, 2. True, 3. True, 4. False—The sphenoid sinus opens into the posterior roof region, 5. True, 6. True, 7. False—The respiratory part of the nasal cavity is lined with ciliated pseudostratified columnar epithelium with goblet cells, 8. False—The maxillary sinus opens into the middle meatus beneath the middle concha, 9. True, 10. False—The component hard parts of the larynx are cartilaginous, 11. True, 12. False—The laryngeal entrance is strengthened by the epiglottis and other small cartilages, 13. True, 14. False—The vestibular fold is superior to the vocal fold, 15. False—The voice is produced by vibration of the vocal fold, 16. False—The arytenoid cartilages rotate and slide to move the vocal folds, 17. False—The laryngeal saccule provides lubrication for the vocal fold, 18. False—The epiglottis is superior and anterior to the superior larynx, 19. True, 20. False—The laryngeal muscles are supplied by the vagus nerve.

pp. 164–165: 1—E, 2—D, 3—C, 4—A, 5—C, 6—B, 7—D, 8—E, 9—D, 10—B, 11—C, 12—A, 13—E.

pp. 166–167: *i) Label each structure shown on the illustrations:* 1. Frontal sinus, 2. Sphenoid sinus, 3. Ethmoid sinuses, 4. Maxillary sinus, 5. Lacrimal gland, 6. Lacrimal canals, 7. Nasolacrimal duct.

ii) Add numbers to the boxes to match each label to the correct part of the illustration: 1. Base of skull, 2. Middle nasal concha, 3. Inferior nasal concha, 4. Soft palate, 5. Uvula, 6. Palatine tonsil, 7. Dorsum of tongue, 8. Epiglottis, 9. Aryepiglottic fold, 10. Esophagus, 11. Parathyroid glands, 12. Thyroid gland

(lateral lobe), 13. Inferior constrictor muscle, 14. Middle constrictor muscle, 15. End of greater horn of hyoid bone, 16. Stylopharyngeus muscle, 17. Angle of mandible, 18. Stylohyoid muscle, 19. Superior constrictor muscle, 20. Parotid gland.

p. 168: 1. erectile, 2. middle nasal meatus, 3. sphenoid, 4. middle nasal concha, 5. vestibule, 6. vestibule, 7. vocal folds, 8. protection of the airway; phonation (voice production), 9. epiglottis; corniculate; cuneiform, 10. hyoid.

p. 169: 1—D, 2—C, 3—A, 4—B, 5—E.

pp. 172–173: 1. False—Tracheal cartilages are not complete rings, they are U-shaped, 2. False—The posterior wall of the trachea is smooth muscle (the trachealis), 3. True, 4. True, 5. False—The visceral or pulmonary pleura covers each lung, 6. False—The branching of the trachea occurs at the level of the sternal body (thoracic vertebrae 5 to 6), 7. True, 8. False—Normally the pleural fluid is less than 20 ml, 9. True, 10. False—The division of lungs into lobes is variable. The left lung may have a partial horizontal fissure, and the right lung may have an incomplete horizontal fissure in some people, 11. False—The two lobes of the left lung are separated by the oblique fissure, 12. False—The sternal angle is level with the second costal cartilage, and the horizontal fissure is usually at the level of the fourth costal cartilage, 13. True, 14. True, 15. False—The anterior border of the left lung has the notch, and it is called the cardiac notch, 16. True, 17. True, 18. True, 19. True, 20. True.

pp. 174–175: 1—A, 2—B, 3—E, 4—A, 5—E, 6—D, 7—D, 8—A, 9—C, 10—B, 11—A, 12—B, 13—E, 14—B.

pp. 176–177: *i) Color the trachea red and the inferior lobar bronchi blue: refer to ii), below, for the correct answers.*

ii) Add numbers to the boxes to match each label to the correct part of the illustration: 1. Trachea, 2. Left primary bronchus, 3. Right primary bronchus, 4. Superior lobar bronchi, 5. Middle lobar bronchus, 6. Inferior lobar bronchi, 7. Right lung, 8. Left lung.

iii) Label each structure shown on the illustrations: 1. Trachea, 2. Right primary bronchus, 3. Left primary bronchus, 4. Submucosal gland, 5. Trachealis muscle, 6. Annular ligament, 7. Cartilage, 8. Respiratory epithelium, 9. Submucosa, 10. Smooth muscle, 11. Respiratory epithelium, 12. Cartilage, 13. Submucosal gland.

pp. 178–179: 1. trachealis, 2. esophagus, 3. carina, 4. recurrent laryngeal, 5. asthma, 6. bronchial, 7. cervical; costal; mediastinal; diaphragmatic, 8. inferior or lower, 9. lingula, 10. base, 11. cardiac notch, 12. cupola, 13. 4th, 14. right atrium, 15. costal; diaphragmatic, 16. terminal bronchioles, 17. type 1 pneumocyte/alveolar epithelium; basement membrane; alveolar capillary endothelium, 18. respiratory, 19. pulmonary; bronchopulmonary (hilar); tracheobronchial; paratracheal, 20. type 1 pneumocyte or alveolar cell.

p. 180: 1—E, 2—C, 3—A, 4—B, 5—D.

p. 181: 1—A, 2—E, 3—D, 4—C, 5—B.

pp. 182–183: *i) Add numbers to the boxes to match each label to the correct part of the illustration:* 1. Pharynx, 2. Trachea, 3. Right primary bronchus, 4. Superior lobar bronchus, 5. Middle lobar bronchus, 6. Diaphragm, 7. Left primary bronchus, 8. Nasal cavity.

ii) Label each structure shown on the illustrations: 1. Manubrium, 2. Body of sternum, 3. Costal cartilage, 4. Xiphoid process, 5. Rectus abdominus muscle, 6. External intercostal muscles, 7. Internal intercostal muscles, 8. Ribs 8–10 ("floating" ribs), 9. Ribs 1–7 ("true" ribs), 10. Nasal cavity, 11. Nasopharynx, 12. Oropharynx, 13. Laryngopharynx, 14. Larynx, 15. Trachea, 16. Body of sternum, 17. Xiphoid process, 18. Esophagus, 19. Inferior vena cava, 20. Central tendon, 21. Twelfth rib, 22. Celiac trunk, 23. Abdominal aorta, 24. Right crus of diaphragm, 25. Vertebral column, 26. Left crus of diaphragm, 27. Quadratus lumborum muscle.

iii) Color the rectus abdominus muscle in red: refer to ii), above, for correct answer.

pp. 188–189: 1. False—The parotid duct opens opposite the crown of the second upper molar, 2. True, 3. True, 4. False—Enamel is the most superficial layer of the crown of human teeth, 5. True, 6. False—Wisdom teeth usually erupt at about 18 years, 7. True, 8. True, 9. False—The lingual tonsil is found on the posterior one third of the tongue, 10. False—The hyoglossus muscle is a depressor of the tongue, 11. True, 12. True, 13. True, 14. True, 15. False—Filiform papillae are conical in shape, 16. False—The parotid gland lies anterior to the external ear, 17. True, 18. False—The wall of the lower part of the esophagus contains exclusively smooth muscle, 19. False—The esophagus lies posterior to the left atrium, not the left ventricle.

pp. 190–191: 1—D, 2—E, 3—A, 4—C, 5—D, 6—A, 7—C, 8—A, 9—B, 10—E, 11—D, 12—B, 13—A, 14—B, 15—C, 16—E.

p. 192: 1. canine, 2. periodontal ligaments, 3. premolar or bicuspid, 4. maxillary, 5. mandibular, 6. accessory, 7. mylohyoid, 8. parasympathetic; facial. 9. sulcus terminalis, 10. styloglossus, 11. palatoglossal, 12. palatoglossal; palatopharyngeal, 13. pharyngeal constrictors, 14. vagus.

p. 193: 1—D, 2—A, 3—B, 4—G, 5—F, 6—C, 7—E.

pp. 194–195: i) Label each structure shown on the illustrations:
1. Palatopharyngeal arch and muscle, 2. Lingual tonsil, 3. Palatine tonsil, 4. Palatoglossal arch and muscle, 5. Terminal sulcus, 6. Vallate papillae, 7. Median sulcus, 8. Filiform papillae, 9. Fungiform papilla, 10. Central incisor, 11. Lateral incisor, 12. Canine, 13. First premolar, 14. Second premolar, 15. First molar, 16. Second molar, 17. Third molar, 18. Capillary plexus, 19. Dentin, 20. Enamel, 21. Crown of tooth, 22. Neck of tooth, 23. Root of tooth, 24. Alveolar vein, 25. Alveolar artery, 26. Alveolar nerve, 27. Apical foramen, 28. Alveolar process, 29. Root canal, 30. Pulp cavity, 31. Gingiva.

ii) Label each structure shown on the illustrations: 1. Sublingual folds (with openings of sublingual ducts), 2. Sublingual caruncle (with opening of submandibular duct), 3. Frenulum, 4. Submandibular duct, 5. Sublingual gland, 6. Lingual nerve, 7. Deep lingual vein, 8. Accessory parotid gland, 9. Parotid duct, 10. Tongue, 11. Frenulum, 12. Sublingual gland, 13. Submandibular gland, 14. Facial vein, 15. Internal jugular vein, 16. Masseter muscle, 17. Parotid gland.

pp. 198–199: 1. False—The fundus lies under the left dome of the diaphragm, 2. True, 3. False—The left kidney is a posterior relation of the body of the stomach. The spleen is a relation of the fundus, 4. True, 5. False—The zymogenic or chief cells make the proteolytic enzyme precursor pepsinogen, 6. True, 7. False—The celiac trunk branches provide most of the blood supply to the stomach, 8. False—The greater omentum attaches along the greater curvature of the stomach, 9. False—The pyloric sphincter is smooth muscle, 10. True, 11. False—The bile duct opens into the second part of the duodenum, 12. True, 13. True, 14. True, 15. False—Tenia coli are derived from the external longitudinal layer of the muscularis externa, 16. False—Appendices epiploicae are most commonly seen on the transverse and sigmoid colon, 17. True, 18. True, 19. False—Internal hemorrhoids are derived from dilation of the internal venous plexus at the anorectal junction, 20. True.

pp. 200–201: i) Label each structure shown on the illustrations: 1. Serosa (mesothelium), 2. Serosa (connective tissue), 3. Plica circularis, 4. Submucosa, 5. Muscularis externa (outer longitudinal fibers), 6. Mucosa, 7. Muscularis mucosae, 8. Nerves of the mesenteric plexus, 9. Muscularis externa (inner circular layer), 10. Villi, 11. Mesentery, 12. Duodenum, 13. Transverse colon (reflected down), 14. Ascending colon, 15. Cecum, 16. Appendix, 17. Rectum, 18. Anus, 19. Ileum, 20. Sigmoid colon, 21. Jejunum, 22. Pylorus (of stomach), 23. Descending colon, 24. Stomach.

ii) Label each structure shown on the illustrations: 1. Plica circularis, 2. Circular muscular layer, 3. Longitudinal muscular layer, 4. Wall of small intestine, 5. Lamina propria, 6. Basal lamina, 7. Intestinal epithelium, 8. Lymphatic lacteal, 9. Lymphocytes, 10. Rectum, 11. Anal column, 12. Anus, 13. Anal valve, 14. External anal sphincter, 15. Internal anal sphincter, 16. Internal anal sphincter, 17. External anal sphincter, 18. Rectum, 19. Sigmoid colon.

pp. 202–203: 1—A, 2—D, 3—D, 4—E, 5—A, 6—A, 7—B, 8—C, 9—B, 10—D, 11—E, 12—D, 13—B, 14—D.

pp. 204–205: 1. cardia, 2. cardiac notch, 3. rugae, 4. gastrohepatic; lesser omentum, 5. intrinsic factor, 6. inner oblique; middle circular; outer longitudinal, 7. gastrin, 8. simple columnar, 9. duodenal cap, 10. goblet, 11. secretin, 12. cholecystokinin, 13. Peyer's patches, 14. fat absorption, 15. gut associated lymphoid tissue, 16. vermiform appendix, 17. ileocecal; appendicular, 18. transverse mesocolon; pancreas, 19. common iliac vessels, 20. transverse rectal folds.

p. 206: 1—E, 2—B, 3—D, 4—C, 5—A.

p. 207: 1—A, 2—C, 3—B, 4—E, 5—D.

p. 210: 1. False—Glycogen is a molecule that stores carbohydrate, 2. False—The left and right hepatic ducts join to form the common hepatic duct, 3. True, 4. False—The hepatic veins enter the inferior vena cava as it passes through the liver substance, 5. True, 6. True, 7. True, 8. False—The bile duct is accompanied by the gastroduodenal and posterior superior pancreaticoduodenal arteries, 9. True, 10. False—The islets of Langerhans are solely endocrine tissue.

p. 211: 1—A, 2—E, 3—A, 4—E, 5—C, 6—D, 7—B.

pp. 212–213: i) Label each structure shown on the illustrations: 1. Cystic artery, 2. Fundus (of gallbladder), 3. Body (of gallbladder), 4. Neck (of gallbladder), 5. Cystic duct, 6. Common hepatic duct, 7. Left hepatic duct, 8. Right hepatic duct, 9. Cystic duct, 10. Gallbladder, 11. Common bile duct, 12. Inferior vena cava, 13. Liver, 14. Portal vein, 15. Duodenum, 16. Pancreaticoduodenal vein, 17. Superior mesenteric vein, 18. Ascending colon, 19. Right colic vein, 20. Appendicular vein, 21. Cecum, 22. Rectum, 23. Jejunum, 24. Left colic vein, 25. Inferior mesenteric vein, 26. Pancreas (cut), 27. Splenic vein, 28. Left gastric vein, 29. Stomach, 30. Spleen.

ii) Label each structure shown on the illustrations: 1. Neck (of pancreas), 2. Head (of pancreas), 3. Accessory pancreatic duct, 4. Main pancreatic duct, 5. Uncinate process, 6. Tail (of pancreas), 7. Body (of pancreas), 8. Fundus (of stomach), 9. Cardia, 10. Pyloric region, 11. Body (of stomach), 12. Longitudinal muscle layer, 13. Cardiac muscle, 14. Circular muscle layer, 15. Submucosal neurovascular layer, 16. Lesser curvature, 17. Duodenum (first part), 18. Pyloric sphincter, 19. Pylorus (of stomach), 20. Mucosa and submucosa, 21. Greater curvature.

iii) Use the key to color these structures: refer to ii), above, for answers.

p. 214: 1. caudate, 2. coronary, 3. subphrenic, 4. hepatorenal recess, 5. fundus, 6. portal vein, 7. bile duct; main pancreatic duct; hepatopancreatic ampulla, 8. spleen, 9. splenic artery; splenic vein, 10. uncinate.

p. 215: 1—C, 2—A, 3—E, 4—B, 5—D.

p. 220: 1. False—The right kidney is situated slightly lower than the left because of the liver, 2. True, 3. True, 4. False—The renal arteries are approximately the same length. It is the renal veins that are asymmetrical, 5. True, 6. True, 7. True, 8. False—The ureter is entirely a retroperitoneal structure, 9. False—The urinary bladder is usually at the level of sacral vertebra 3 or below, 10. False—The male urethra is much longer than the female urethra.

p. 221: 1—E, 2—D, 3—A, 4—A, 5—D, 6—B, 7—E.

p. 222: 1. perinephric (perirenal) space, 2. sinus, 3. renal pelvis, 4. cortical columns, 5. papilla, 6. renal vascular, 7. seminal vesicles; ampullae of ductus deferens, 8. pubococcygeus, 9. urethra, 10. uterus.

p. 223: 1—B, 2—E, 3—D, 4—A, 5—C.

pp. 228–229: 1. False—Glomeruli are located solely in the renal cortex, 2. True, 3. False—There are about 1 million in both kidneys combined,

4. False—The glomerulus is the main site of ultrafiltration, 5. True, 6. True, 7. False—The vessel leaving each renal corpuscle is an efferent arteriole, 8. True, 9. False—The proximal convoluted tubule attaches to the urinary pole, 10. True, 11. True, 12. False—The ascending limb of the loop of Henle has both thin and thick segments, 13. False—The loop of Henle descends into the renal medulla in the juxtamedullary nephron, 14. True, 15. False—Thiazide diuretics mainly act at the distal convoluted tubule to inhibit the Na+, K+ ATPase pump to reduce NaCl and water reabsorption, 16. False—ADH mainly acts at the collecting tubule to promote water absorption, 17. True, 18. True, 19. True, 20. True.

pp. 230–231: *i) Label each structure shown on the illustrations:* 1. Inferior vena cava, 2. Right adrenal gland, 3. Right kidney, 4. Right renal artery, 5. Right renal vein, 6. Perirenal fat, 7. Right ureter, 8. Abdominal aorta, 9. Left ureter, 10. Left renal vein, 11. Left renal artery, 12. Segmental artery, 13. Left kidney, 14. Left adrenal gland, 15. Inferior vena cava, 16. Right adrenal gland, 17. Renal pyramid (medulla), 18. Tip of renal papilla, 19. Cortex, 20. Renal pelvis, 21. Renal column, 22. Major calyx, 23. Minor calyx, 24. Right ureter, 25. Abdominal aorta.

ii) Label each structure shown on the illustrations: 1. Adrenal glands, 2. Right kidney, 3. Left kidney, 4. Inferior vena cava, 5. Abdominal aorta, 6. Right ureter, 7. Left ureter, 8. Urinary bladder, 9. Afferent arteriole, 10. Glomerular capillaries, 11. Bowman's capsule around a glomerulus, 12. Glomerulus, 13. Artery, 14. Loop of Henle, 15. Capillary network, 16. Proximal convoluted tubule, 17. Efferent arteriole.

pp. 232–233: 1—A, 2—A, 3—C, 4—D, 5—E, 6—A, 7—E, 8—C, 9—A, 10—B, 11—A, 12—D, 13—B, 14—C.

pp. 234–235: 1. podocytes, 2. parietal layer of Bowman's capsule, 3. mesangial cells, 4. proximal convoluted tubule, 5. tight junctions, 6. basolateral domain; plasma membrane folds, 7. lysosomes, 8. thick segment of the descending limb of the loop of Henle, 9. simple squamous, 10. low simple cuboidal, 11. macula densa, 12. afferent; efferent, 13. juxtaglomerular, 14. Cortical, 15. Vasa recta, 16. atrial natriuretic factor, 17. glomerular ultrafiltrate, 18. collecting tubule, 19. Interlobar, 20. arcuate.

p. 236: 1—A, 2—C, 3—E, 4—B, 5—D.

p. 237: 1—B, 2—E, 3—D, 4—A, 5—C.

p. 240: 1. False—Ureters run in front of the transverse processes of the lumbar vertebrae, 2. False—The ureter has an internal diameter of less than 1 mm, so even small stones are difficult to pass, 3. True, 4. True, 5. False—When it is full, the urinary bladder rises out of the pelvic cavity, 6. True, 7. False—There is no gland around the neck of the female bladder, 8. True, 9. False—The male urethra passes down the bulb and corpus spongiosum of the penis, 10. True.

p241: 1—A, 2—B, 3—D, 4—B, 5—A, 6—E, 7—C.

pp. 242–243: *i) Add numbers to the boxes to match each label to the correct part of the illustrations:* 1. Bladder, 2. Region of internal urethral sphincter, 3. Symphysis pubis, 4. Prostate gland, 5. External urethral sphincter, 6. Penis, 7. Urethra, 8. Urethral meatus (opening), 9. Uterus, 10. Bladder, 11. Internal urethral sphincter, 12. Symphysis pubis, 13. Vagina, 14. External urethral sphincter, 15. Urethra.

ii) Label each structure shown on the illustrations: 1. Right inferior vena cava, 2. Right testicular vein, 3. Right testicular artery, 4. Right ureter, 5. Right internal iliac artery, 6. Bladder, 7. Opening (meatus) of right ureter, 8. Prostate gland, 9. Penis, 10. Scrotum, 11. Urethral meatus (opening), 12. Spongy urethra, 13. Bulb of penis, 14. Bulbourethral (Cowper's) gland, 15. Urethra, 16. Neck of bladder, 17. Left external iliac vein, 18. Left external iliac artery, 19. Opening (meatus) of left ureter, 20. Left internal iliac vein, 21. Left common iliac vein, 22. Left common iliac artery, 23. Left ureter, 24. Abdominal aorta, 25. Right inferior vena cava, 26. Right ovarian vein, 27. Right ovarian artery, 28. Right ureter, 29. Right internal iliac artery, 30. Opening (meatus) of right ureter, 31. Urethra, 32. Trigone, 33. Opening (meatus) of left ureter, 34. Left internal iliac vein, 35. Left external iliac artery, 36. Left external iliac vein, 37. Left common iliac vein, 38. Left common iliac artery, 39. Left ureter, 40. Abdominal aorta.

pp. 244–245: 1. pelviureteric junction, 2. pelviureteric junction, 3. peristalsis, 4. ischial spine, 5. renal artery, 6. common iliac vessels, 7. cervix, 8. uterine artery, 9. vesicoureteric junction, 10. trigone, 11. transitional, 12. urachus/median umbilical ligament, 13. vesical, 14. splanchnic, 15. internal urethral sphincter, 16. urogenital, 17. intrabulbar fossa, 18. intrabulbar fossa, 19. lacunae, 20. external urethral orifice.

p. 246: 1—D, 2—A, 3—E, 4—C, 5—B.

p. 247: 1—B, 2—E, 3—D, 4—C, 5—A.

p. 252: 1. False—Many hormones are small peptide molecules, 2. False—Steroid hormones bind to nuclear receptors to exert their action, 3. False—The posterior pituitary is an outgrowth of the hypothalamus of the brain, 4. True, 5. False—Stomach acid and enzymes can easily break down peptide hormones, 6. True, 7. False—Hormones mainly exercise their effects over hours to years, 8. True, 9. True, 10. False—Some endocrine glands (e.g. the adrenal cortex) have a mesodermal origin, 11. False—Some hormones can be carried by other body fluids (e.g. in body cavities) or diffuse locally in the extracellular space, 12. False—Some peptide hormones (e.g. insulin) are more than 50 amino acids long, 13. False—The negative feedback loop is the most important physiological mechanism regulating homeostasis, in that secretion of a hormone is turned off when a sufficient effect is brought about by the hormone, 14. True, 15. True.

p. 253: 1—C, 2—E, 3—C, 4—A, 5—D, 6—C, 7—E.

p. 254: 1. nasal, 2. parathyroid glands, 3. pituitary; pineal, 4. lobes, 5. lower larynx; trachea, 6. pancreas, 7. adrenal gland, 8. adrenal, 9. medulla, 10. cortex.

p. 255: 1—E, 2—B, 3—D, 4—A, 5—C, 6—G, 7—F.

p. 258: 1. False—ACTH is normally made only by the anterior pituitary gland, 2. False—Luteinizing hormone is made by basophil cells of the anterior pituitary, 3. True, 4. True, 5. True, 6. False—The pinealocytes are the secretory cells, 7. False—Thyroid hormones are synthesized from iodine and tyrosine, 8. True, 9. False—The thyroid gland contains follicles and colloid. The parathyroids have a follicular-like arrangement, but no colloid, 10. False—The chief cells of the parathyroid gland secrete parathyroid hormone.

p. 259: 1—B, 2—D, 3—E, 4—A, 5—B, 6—D, 7—A.

pp. 260–261: *i) Label each structure shown on the illustrations:* 1. Cerebrum, 2. Pineal gland, 3. Brainstem, 4. Cerebellum, 5. Neurosecretory cell, 6. Optic chiasm, 7. Infundibulum, 8. Hypophyseal artery, 9. Anterior pituitary (adenohypophysis), 10. Posterior pituitary (neurohypophysis), 11. Hypophyseal portal system, 12. Axon, 13. Mammillary body, 14. Hypothalamus.

ii) Label each structure shown on the illustrations: 1. Lumen filled with colloid, 2. Thyroid epithelium, 3. Capsule of parathyroid, 4. Chief cell, 5. Capillary, 6. Oxyphil cells, 7. Thyroid cartilage (of larynx), 8. Left lobe of thyroid gland, 9. Isthmus of thyroid gland, 10. Right lobe of thyroid gland, 11. Parathyroid glands, 12. Thyroid, 13. Esophagus.

pp. 262–263: 1. somatostatin, 2. hypothalamo-hypophyseal portal, 3. supraoptic; paraventricular, 4. acidophil; basophil; chromophobe, 5. morning; awakening, 6. acromegaly, 7. pituitary dwarfism, 8. testosterone; Leydig, 9. follicle-stimulating hormone, 10. cortisol, 11. Herring bodies, 12. antidiuretic hormone/vasopressin, 13. superior cervical ganglion, 14. neural crest, 15. goiter, 16. triiodothyronine, 17. thyrotropin (thyroid-stimulating hormone), 18. calcitriol (vitamin D), 19. oxyphil, 20. calcitonin.

p. 264: 1—B, 2—E, 3—D, 4—A, 5—C.

p. 265: 1—D, 2—B, 3—E, 4—C, 5—A, 6 –G, 7—F.

p. 268: 1. True, 2. False—The outermost layer of the adrenal cortex is the zona glomerulosa, 3. True, 4. False—Adrenal sex steroids are made by the zone reticularis of the cortex, 5. True, 6. True, 7. False—Most chromaffin cells secrete adrenaline (epinephrine), 8. False—Beta cells produce insulin, 9. True, 10. True.

p. 269: 1—B, 2—D, 3—B, 4—B, 5—E, 6—D, 7—A.

pp. 270–271: *i) Add numbers to the boxes to match each label to the correct part of the illustrations:* 1. Head (of pancreas), 2. Accessory pancreatic duct, 3. Uncinate process, 4. Main pancreatic duct, 5. Tail (of pancreas), 6. Body (of pancreas), 7. Neck (of pancreas), 8. Exocrine pancreas, 9. Insuloacinar portal vessels.

ii) Label each structure shown on the illustrations: 1. Left suprarenal artery, 2. Left adrenal medulla, 3. Left kidney, 4. Left adrenal gland, 5. Left adrenal cortex, 6. Zona glomerulosa, 7. Zona fasciculata, 8. Zona reticularis, 9. Medullary plexus of veins, 10. Medullary vein, 11. Medulla, 12. Deep plexus of veins, 13. Sinusoidal vessels, 14. Subcapsular plexus of veins, 15. Capsular artery, 16. Capsule.

p. 272: 1. ACTH (adrenocorticotropic hormone), 2. hirsutism; clitoris, 3. gluconeogenesis, 4. alpha; beta, 5. vasodilatory, 6. central vein, 7. somatostatin, 8. pancreatic islets, 9. beta, 10. glycogenolysis (breakdown of glycogen).

p. 273: 1—D, 2—C, 3—A, 4—B, 5—E.

p. 276: 1. True, 2. True, 3. False—Like all cells making steroid hormones, Leydig cells have abundant lipid in their cytoplasm, 4. True, 5. False—These accessory reproductive glands are stimulated mainly by testosterone, 6. True, 7. False—Androgens are secreted by the adrenal cortex and some ovarian cells in women. They play a key role in regulating libido in women, 8. True, 9. False—LH mainly stimulates progesterone production by the corpus luteum, 10. False—Progesterone secretion during the later half of the menstrual cycle stimulates endometrial gland secretion.

p. 277: 1—E, 2—A, 3—C, 4—B, 5—A, 6—A, 7—D.

p. 278: 1. inhibin; activin, 2. male genitalia, 3. Müllerian inhibiting substance, 4. gynecomastia, 5. Leydig cells, 6. follicular, 7. theca interna, 8. luteinizing hormone, 9. human chorionic gonadotropin, 10. progesterone.

p. 279: 1—E, 2—D, 3—A, 4—B, 5—C.

p. 284: 1. True, 2. True, 3. False—The efferent ductules leave at the posterosuperior pole, 4. False—The ductus deferens enters the abdomen through the inguinal canal, 5. True, 6. False—The ejaculatory ducts pass through the central area of the prostate gland, 7. True, 8. False—The bulbourethral glands have ducts that open into the intrabulbar fossa of the spongy urethra, 9. False—The bulb of the penis is surrounded by the bulbospongiosus muscle, 10. True.

p. 285: 1—A, 2—D, 3—D, 4—B, 5—D, 6—E, 7—C.

p. 286–287: *i) Label each structure shown on the illustrations:* 1. Urinary bladder, 2. Prostate gland, 3. Penis, 4. Scrotum, 5. Urinary bladder, 6. Seminal vesicle, 7. Prostate gland, 8. Ductus deferens, 9. Penis, 10. Epididymis, 11. Testis.

ii) Label each structure shown on the illustrations: 1. Superficial fascia of scrotum, 2. Scrotal skin, 3. Septum of scrotum, 4. Parietal layer of tunica vaginalis, 5. Body of epididymis, 6. Testis (covered by visceral layer of tunica vaginalis), 7. Head of epididymis, 8. Ductus deferens, 9. Pampiniform plexus of veins, 10. Tail, 11. Mitochondrial sheath (middle piece), 12. Cell membrane, 13. Acrosome, 14. Nucleus, 15. Nuclear vacuole, 16. Head, 17. Neck, 18. Centriole, 19. Mitochondrion, 20. Ductus deferens, 21. Testicular artery, 22. Body of epididymis, 23. Efferent ductules, 24. Rete testis, 25. Mediastinum testis, 26. Tail of epididymis, 27. Seminiferous tubules,

28. Lobules, 29. Spermatozoon, 30. Tunica albuginea, 31. Septae, 32. Head of epididymis.

p. 288: 1. flagellum, 2. pampiniform, 3. acrosome, 4. 35ºC, 5. efferent ductules, 6. sperm maturation; sperm storage; sperm transport, 7. simple columnar; pseudostratified, 8. corpora amylacea, 9. tunica albuginea, 10. bulbourethral glands.

p. 289: 1—E, 2—A, 3—D, 4—B, 5—C.

p. 292: 1. True, 2. True, 3. False—The tough outer layer of the ovary is called the tunica albuginea, 4. False—The corona radiata is the layer of cells that surrounds the primary oocyte in the mature follicle, 5. True, 6. False—Luteinizing hormone stimulates ovulation, 7. False—The isthmus is the narrowest part. The ampulla is the widest, 8. True, 9. True, 10. False—The bulb of the vestibule is erectile tissue. The greater vestibular gland lubricates the vaginal entrance.

p. 293: 1—B, 2—A, 3—C, 4—E, 5—A, 6—D, 7—C.

pp. 294–295: *i) Label each structure shown on the illustrations:* 1. Uterine tube (cut open), 2. Hydatid of Morgagni, 3. Endometrium, 4. Myometrium, 5. Internal os of cervix, 6. External os of cervix, 7. Vaginal fornix (lateral), 8. Vagina, 9. Broad ligament, 10. Cervix, 11. Ovary, 12. Fimbriae, 13. Uterine tube, 14. Mesosalpinx (of broad ligament), 15. Fundus (of uterus), 16. Vagina, 17. Body (of uterus), 18. Fundus (of uterus), 19. Mons pubis, 20. Clitoris, 21. Orifice of urethra, 22. Labium majorum, 23. Orifice of vagina, 24. Labium minorum, 25. Hymen.

ii) Label each structure shown on the illustrations: 1. Mature (Graafian) follicle, 2. Mature ovum, 3. Follicular fluid, 4. Theca interna, 5. Theca externa, 6. Discharging follicle (ovulation), 7. Ovum, 8. Corpus luteum, 9. Corpus albicans, 10. Ovarian stroma, 11. Primary oocyte, 12. Primary follicle, 13. Thecal cells, 14. Primordial follicle, 15. Antrum, 16. Ovarian surface epithelium, 17. Ovarian hilum, 18. Sperm, 19. Ovum, 20. Sperm entering ovum, 21. Zona pellucida, 22. Perivitelline space, 23. Polar body, 24. Adipose tissue, 25. Areola, 26. Nipple, 27. Lactiferous sinus, 28. Lactiferous duct, 29. Fibrocollagenous septa (Cooper's suspensory ligament).

p. 296: 1. infundibulum, 2. mesovarium, 3. suspensory ligament of the ovary, 4. ampulla, 5. smooth muscle, 6. fornix, 7. labia minora; labia majora, 8. prepuce, 9. crus of the clitoris, 10. mons pubis.

p. 297: 1—A, 2—E, 3—D, 4—C, 5—B, 6—F, 7—G.

p. 300: 1. True, 2. False—The zygote emerges from the zona pellucida at about 72 hours after entering the uterine cavity, 3. False—The embryo implants during the secretory phase of the menstrual cycle, 4. True, 5. True, 6. False—The inner cell mass gives rise to the proper embryo. The outer cell mass gives rise to the placenta and membranes, 7. True, 8. The syncytiotrophoblast is the most invasive, 9. True, 10. True.

p. 301: 1—B, 2—A, 3—C, 4—D, 5—A, 6—D, 7—E.

p. 304: 1. False—Brain development extends across all of gestation and into early postnatal life, 2. False—There is only 1 umbilical vein, but two umbilical arteries in the cord, 3. True, 4. False—This blood is oxygenated because it has come almost directly from the placenta, 5. False—Pulmonary surfactant is not made in large amounts until at least 35 weeks gestation, 6. False—Only maternal IgG molecules are small enough to cross the placenta to the fetus, 7. True, 8. True, 9. True, 10. These gases cross the placental membranes by simple diffusion.

p. 305: 1—E, 2—B, 3—D, 4—A, 5—B, 6—E, 7—C.

p. 310: 1. True, 2. True, 3. False—Neutrophil granulocytes have granules of enzymes (myeloperoxidase, elastase, lysozyme) in their cytoplasm, 4. True, 5. True, 6. False—Thalassemia involves defective synthesis of alpha or beta chains, 7. True, 8. False—Monocytes usually leave the circulation within 100 hours, 9. True, 10. False—Platelets are fragments of megakaryocytes.

p. 311: 1—C, 2—E, 3—C, 4—A, 5—B, 6—E, 7—A, 8—C, 9—B.

pp. 312–313: *i) Label each structure shown on the illustrations:* 1. Red blood cell, 2. Platelet, 3. Cytoplasm, 4. Nucleus, 5. Heme, 6. Globin protein strand, 7. Iron ion, 8. Macrophage, 9. Monocyte, 10. Neutrophil, 11. Lymphocyte, 12. Basophil, 13. Eosinophil.

ii) Label each structure shown on the illustrations: 1. Platelets, 2. Leukocytes (white blood cells), 3. Erythrocytes (red blood cells), 4. Leukocyte (neutrophil), 5. Smooth muscle, 6. Endothelium, 7. Fenestrations, 8. Fenestrated capillary, 9. Continuous capillary, 10. Capillary, 11. Arteriole, 12. Artery, 13. Vein, 14. Venule, 15. Valves closed, 16. Intima, 17. Muscularis, 18. Adventitia.

pp. 314–315: 1. reticulocytes, 2. lymphocyte, 3. multilobed, 4. spectrin, 5. fibrinogen; albumen, 6. buffy coat, 7. mast, 8. spleen, 9. diapedesis, 10. spherocytosis, 11. alpha; beta, 12. heme, 13. oxyhemoglobin, 14. carbaminohemoglobin, 15. hemolysis, 16. positive chemotaxis, 17. ameboid, 18. degranulation, 19. differential, 20. acidophils.

p. 316: 1—B, 2—A, 3—E, 4—D, 5—C, 6—F.

p. 317: 1—E, 2—D, 3—B, 4—A, 5—C.

p. 320: 1. True, 2. False—The erythroid CFU gives rise to red blood cells, 3. True, 4. True, 5. False—Megakaryocyte production is under the control of thrombopoietin from the liver, 6. True, 7. False—Folate and vitamin B12 are the most important vitamins for red blood cell production, 8. True, 9. False—Basophil granulocytes are involved in allergic skin reactions, 10. True.

p. 321: 1—A, 2—E, 3—D, 4—C, 5—D, 6—A, 7—E.

p. 322: 1. pluripotency, 2. neutropenia, 3. leukocytosis, 4. leucopenia, 5. erythroid, 6. marrow stromal compartment, 7. thrombocytopenia, 8. granulocyte-macrophage, 9. colony-stimulating factors, 10. lymphoblast, 11. polycythemia rubra vera, 12. myelofibrosis, 13. hematopoiesis, 14. macrophages.

p. 323: 1—A, 2—E, 3—D, 4—B, 5—C.

p. 326: 1. False—The first event in hemostasis is vasoconstriction, 2. True, 3. True, 4. True, 5. False—Platelet deficiency usually gives rise to tiny petechial hemorrhages, 6. True, 7. False—Most of the clotting factors (e.g. fibrinogen, vitamin K dependent clotting factors) are made by the liver, 8. False—The extrinsic pathway is usually activated by external injury to a blood vessel, 9. True, 10. True.

pp. 328–329: 1—C, 2—D, 3—A, 4—A, 5—D, 6—E, 7—D, 8—C, 9—A, 10—D.

p. 330: 1. universal donors, 2. erythroblastosis fetalis, 3. plasmin, 4. von Willebrand factor, 5. thrombotic thrombocytopenic purpura, 6. laminin; collagen, 7. tissue plasminogen activator, 8. fibrin clot, 9. endothelin, 10. warfarin; dicumarol, 11. calcium, 12. hemophiliac, 13. thrombus, 14. coronary artery.

p. 331: 1—B, 2—A, 3—E, 4—C, 5—D.

p. 336: 1. False—Most lymph nodes are located alongside visceral organs, 2. True, 3. False—Most thoracic lymph nodes are in the posterior and superior mediastinum, 4. True, 5. False—The brain has no draining lymph nodes, 6. True, 7. True, 8. False—Infected lymph nodes are usually hot and tender, but not rubbery, 9. True, 10. False—Lymph nodes that are invaded by cancer tend to become firm or rubbery, and not tender unless there is additional infection.

p. 337: 1—C, 2—B, 3—A, 4—A, 5—C, 6—E, 7—A.

p. 338: 1. subcapsular sinus, 2. left internal jugular vein; left subclavian vein, 3. right lymphatic duct, 4. cisterna chyli, 5. sentinel node, 6. tracheobronchial, 7. hilum, 8. macrophage; plasma, 9. plasma, 10. trabeculae.

p. 339: 1– A, 2—E, 3—D, 4—C, 5—B.

pp. 340–341: *i) Label each structure shown on the illustration:* 1. Retroauricular nodes, 2. Parotid nodes, 3. Axillary nodes, 4. Apical axillary nodes, 5. Lateral group, 6. Anterior group, 7. Parasternal nodes, 8. Posterior intercostal nodes, 9. Cubital nodes, 10. Palmar and dorsal plexus, 11. Plantar and dorsal plexus, 12. Popliteal nodes (posterior side of the knee), 13. Inguinal and femoral nodes, 14. Internal iliac nodes, 15. External iliac nodes, 16. Common iliac nodes, 17. Cisterna chyli, 18. Posterior mediastinal nodes, 19. Thoracic duct, 20. Cervical nodes, 21. Buccal nodes.

ii) Add numbers to the boxes to match each label to the correct part of the illustrations: 1. Lymphatic endothelial cell, 2. Closed valve, 3. Medullary sinus, 4. Follicle of cortex, 5. Blood capillary, 6. Valve, 7. Efferent lymphatic vessel, 8. Artery, 9. Vein, 10. Subcapsular sinus, 11. Capsule, 12. Trabeculae, 13. Afferent lymphatic vessels.

p. 344: 1. False—Although the spleen has a capsule, it is quite soft, 2. True, 3. True, 4. False—The white pulp is the immune system component of the spleen, 5. True, 6. True, 7. False—Most T lymphocyte development takes place in the cortex, 8. True, 9. False—The left atrium is on the opposite side of the heart from the thymus, 10. True.

p. 345: 1—D, 2—B, 3—E, 4—A, 5—D, 6—B, 7—C.

p. 346: 1. lienorenal ligament, 2. superior border; pancreas, 3. pancreas, 4. diaphragmatic, 5. red pulp, 6. white pulp, 7. ectodermal; endodermal, 8. Hassall's corpuscles, 9. cortex, 10. apoptosis, 11. nine; eleven, 12. reticular, 13. anterior; superior, 14. trabeculae.

p. 347: 1—A, 2—E, 3—C, 4—B, 5—D.

pp. 348–349: *i) Label each structure shown on the illustration:* 1. Lingual tonsil, 2. Pharyngeal tonsil (adenoid).

ii) Use the key to color these structures: refer to i), above, for answers.

iii) Label each structure shown on the illustrations: 1. Right lobe, 2. Left lobe, 3. Capsule, 4. White pulp nodule, 5. Red pulp, 6. Trabecular arteries, 7. Venous sinusoids, 8. Splenic vein, 9. Splenic artery (terminal branches), 10. Impression of the kidney, 11. Impression of the colon (left colic flexure), 12. Impression of the stomach, 13. Notch in superior border, 14. Superior border.

p. 352: 1. True, 2. True, 3. False—Complement molecules are made in the liver, 4. False—Plasma cells usually reside in lymphoid tissue, 5. False—B lymphocytes give rise to plasma cells, 6. True, 7. True, 8. False—The thymus regresses after puberty, 9. False—The spleen plays a key role in clearing old red blood cells and in immune responses, 10. False—HIV usually infects T lymphocytes.

p. 353: 1—D, 2—E, 3—A, 4—B, 5—E, 6—C, 7—E.

p. 354: 1. IgM, 2. memory, 3. antigens, 4. blood pressure; airways, 5. histamine; mast cells, 6. T lymphocyte; natural killer cell, 7. nasopharynx, 8. Waldeyer's ring, 9. tonsillitis, 10. human immunodeficiency virus (HIV, mainly subtypes B and C), 11. inguinal, 12. venous, 13. internal jugular.

p. 355: 1—A, 2—D, 3—B, 4—E, 5—C.

Index

A

abdomen 14, 142
abdominal aorta 218, 224, 238
abdominal cavity 16
abdominal organs 196, 197
abdominopelvic cavity 16
abductor pollicis longus 62
ABO blood groups 325
accessory pancreatic duct 196, 208, 266
acromioclavicular joint 40
acromion 38, 40
acute lymphoblastic leukemia 320
Addison's disease 271
adductor longus 64
adductor magnus 64
adductor tubercle 42
adenohypophysis (anterior pituitary) 254, 256
adipose tissue 20, 290
adrenal cortex 266
adrenal glands 218, 224, 250, 266, 272
  anterior view 225
  coronal section, left adrenal gland 271
  cross-sectional view, left adrenal gland 267
  microstructure 267, 271
adrenal medulla 266
afferent arteriole 226
amnion 298
ampulla of ductus deferens 282
ampullae 107
anatomical planes 18, 19
anconeus 62
anemia (sickle cell disease) 315
ankle/tarsus 14
anterior border 42
anterior cerebral artery 136
anterior chamber 106
anterior communicating artery 136
anterior corticospinal tract 82
anterior fontanelle 302
anterior group lymph nodes 334
anterior inferior cerebellar artery 136
anterior intercondylar area 42
anterior median fissure 82
anterior nasal (piriform) aperture 38
anterior orientation 18
anterior pituitary (adenohypophysis) 254, 256
anterior radicular artery 82
anterior ramus of spinal nerve 82
anterior spinal artery 82
anterior spinal vein 82
anterior tibial artery 134
antibody 324, 350
antigen 324, 350
anus 196, 201
aorta 134
aortic arch 122, 134, 170
aortic valve 122, 124
apex of fibula 42
apical axillary nodes 334
appendix 186, 196
appetite regulation 103
arachnoid 82
arcuate artery 224, 226

arcuate vein 226
areola 290
arm/brachium 14
armpit/axilla 14
arteries 134–145, 153, 342
articular facet for talus 42
articular facet of medial malleolus 42
articular surface with head of fibula 42
ascending aorta 122
ascending colon 186, 196
atherosclerosis 145
atlas (C1) 38
atrium 122, 124, 170
auditory (acoustic) meatus, external 302
auditory association area 92
autoimmune disease 353
axial skeleton 38–39
axillary artery 134
axillary (circumflex) nerve 80
axillary nodes 334
axillary vein 146
axis (C2) 38
axon 82, 109, 256
azygos vein 146

B

B lymphocyte 350
back 16, 75
ball-and-socket joint 59
basal lamina 226, 308, 342
basilar artery 136
basilic vein 146
basophil 308
basophilic erythroblast 318
biceps brachii 62
biceps femoris 64
big toe/hallux 14
bile ducts 208, 214
  anterior view 212
  common 196
bladder 218, 238, 282
  carcinoma of 245
  female, anterior view 239, 243
  male, anterior view 239, 243
  male, posterior view 283
bladder calculi (stones) 246, 247
bladder (detrusor) muscle 282
bladder lining 238
bladder stones (calculi) 246, 247
blastocyst 298
blood
  blood cells 27, 308–317, 350
  components of 34, 309, 313
  production 318–323
  blood groups 324–325, 328–329
  ABO blood groups 324, 325
  Type A 324
  Type AB 324
  Type B 324
  Type O 324
blood vessel 350
body
  anatomical planes 18, 19
  anterior view 15
  cavities 16
  major arteries 135
  major veins 147

overview 12–13
  posterior view 17
  regions 14–17
  symmetry and situs inversus 19
  view orientation 18
body of gallbladder 208
body of pancreas 196, 208, 266
bone formation 53
bone tissue 20
Bowman's capsule, parietal layer 226
Bowman's gland (olfactory gland) 109
Bowman's space 226
brachial artery 134
brachial plexus 80, 170
brachial vein 146
brachialis 62
brachiocephalic artery (trunk) 122
brachiocephalic vein 122, 146, 170
brachioradialis 62
brain 90–103
  arteries 136–139
  functional areas 97, 98
  lateral view 91
  lobes 93, 98
  sagittal view 91
brainstem 90, 104
broad ligament 274
Broca's area (motor speech) 92
bronchi
  bronchial tree 177
  cross-section 175, 177
  lobar 160
  primary 160
  in situ 176
bronchial tree 177
buccal nodes 334
bulb of penis 238, 282
bulbourethral (Cowper's) gland 238, 282
buttock/gluteus 16

C

calcaneus 42
calcarine branch 136
calf/sura 16
callosomarginal artery 136
cancer, causes 29
capillaries 157
capillary bed 156
capsular artery 266
capsular vein 342
capsule 266, 342
carcinoma of the bladder 245
carcinoma of the larynx 162
carcinoma of the pancreas 210
cardiac branch of vagus nerve 170
cardiac muscle 21
cardioesophageal (gastroesophageal) junction 186
carotid artery 122, 134, 136, 170
carpal bones 40
cartilage tissue 21
cataract 111
cauda equina 80
cecum 186, 196
celiac trunk 224
cell body of podocyte 226
cell structure 24, 26
cells of the immune system 351

cellular/humoral response 350–355
central canal 82
central vein 208
cephalic vein 146
cerebellar artery 136
cerebellum 90
cerebral aneurysms 139
cerebral artery 136
cerebral cortex 95
cerebrospinal fluid 101
cerebrum 90
cervical enlargement of spinal cord 80
cervical nerve 80
cervical nodes 334
cervical vertebrae 38, 55
cervix 290
cheek (buccal) 14
chest/thorax (thoracic) 14
chin (mental) 14
cholesterol 317
chordae tendineae 122
chorionic villi 298
choroid 106
chronic renal failure 237
cilia 109
cilia function, defects in 23
ciliary body 106
ciliary muscle 106
Circle of Willis 136, 137
circulatory system 12, 152
cisterna chyli 334
clavicles 38
clitoris 290
coccyx 38
cochlea 107, 119
cochlear duct 107
cochlear nerve 107
cochlear (round) window 107
collagen 20
collecting duct 226
colliculi 256
colliculus seminalis 238
committed cell 318
common bile duct 196
common carotid artery 122, 134, 136, 170
common fibular nerve 80
common hepatic duct 196, 208
common iliac artery 134, 218, 238
common iliac nodes 334
common iliac vein 146, 238
communicating artery 136
components of blood 34, 309, 313
conception 298–301
congenital T cell deficiency (DiGeorge syndrome) 347
connecting tubule 226
connective tissue 20
connective tissue septum 342
coracoid process 40
cornea 106
corona glandis 238
coronal suture 302
coronary ligament 208
coronary sinus 122
corpus albicans 274
corpus callosum 90, 256
corpus cavernosum 238, 282
corpus luteum 274

corpus spongiosum 238, 282
cortex 224, 342
corti 117
corticospinal tract 82
costal cartilage 38
costodiaphragmatic recess 170
cotyledon 298
Cowper's (bulbourethral) gland 238, 282
cranial cavity 16
cribriform plate of ethmoid bone 109
cricoid cartilage 170
cricothyroid muscle 170
cubital nodes 334
cuneate fasciculus 82
Cushing's disease 268
Cushing's syndrome 268
cyclical changes in ovary 277
cystic artery 208
cystic duct 196, 208
cytotoxic (or killer) T cell 350

D

deep fibular nerve 80
deep muscles of the head 74
deep muscles of the neck 74
deep plexus 266
deep vein thrombosis 149
deltoid 60, 62
dense connective tissue 20
depressor anguli oris 60
descending aorta 122, 134
detrusor (bladder) muscle 282
diabetes mellitus 273
diaphragm 16, 160, 170, 183, 196
DiGeorge syndrome (congenital T cell deficiency) 347
digestive system 13, 186–187
digital nerve 80
digits/fingers (digital or phalangeal) 14
discharging follicle 274
distal convoluted tubule 226
distal orientation 18
distal straight tubule (thick ascending limb of loop of Henle) 226
dorsal arch 134
dorsal branch to corpus callosum 136
dorsal cavity 16
dorsal funiculus 82
dorsal horn 82
dorsal orientation 18
dorsal rootlets 82
dorsal spinocerebellar tract 82
dorsal surface 18
dorsal venous arch 146
dorsolateral sulcus 82
ductus deferens (vas deferens) 282
duodenojejunal junction 196
duodenum 186, 206
dura mater 82

E

ear 14, 107, 114
efferent arteriole 226
ejaculatory duct 238, 282
elastic cartilage 21
elbow (antecubital) 14
elbow/olecranon 16

ellipsoidal joint 57
embryo 298, 299
endocrine system overview 13, 250–255
    endocrine function 274
    endocrine glands 256, 266
    female 251, 274, 275
    male 251, 274
endometrium 298
endoneurium 82
endothelial cell 226
eosinophil 308
epididymis 279, 282
epiglottis 108, 169
epilepsy 102
epineurium 82
epithelial tissue 20
erythroblast 318
erythrocytes (red blood cells) 308, 309, 318, 319, 324
erythroid colony-forming unit (CFU) 318
erythropoietin 318
esophagus 186
eustachian (auditory) tube 107
extensor digitorum 62
extensor digitorum longus 64
extensor pollicis brevis 62
extensor retinaculum 62, 64
external abdominal oblique 60
external auditory (acoustic) meatus 302
external ear canal (meatus) 107
external genitalia 294
external iliac artery 134, 218, 238
external iliac nodes 334
external iliac vein 146, 218, 238
external urethral orifice 238
eye 14, 106

F

face 14, 96, 134
facial artery 134
facial nerve branches 96
falciform ligament 208
fallopian (uterine) tube 274, 290
false ribs (pairs 8–10) 38
fascia penis 282
female bladder 239, 243
female endocrine system 251
female pelvis 45
female reproductive system 290–297
    cross-sectional view 291
    organs, anterior view 291
female urethra 245
female urinary system 219, 242
femoral artery 134
femoral head 42, 48
femoral neck 42, 45
femoral nerve 80
femoral vein 146
femur 42, 43
fenestrated capillary 157
fertilization 298
fetal development 298, 302–305
    changes in uterus and fetus 305
    early stages of pregnancy 299
    skull 303
fibrocartilage 21

fibrous flexor sheath 62
fibula 42, 43
fibular artery 134
fibular nerve 80
fibular notch 42
fibularis (peroneus) longus 64
fila olfactoria 109
filiform papillae 108
fingers/digits (digital or phalangeal) 14
first capillary venule 342
first rib 170
flexor carpi ulnaris 62
flexor digitorum superficialis 62
flexor retinaculum 62
floating ribs (pairs 11 & 12) 38
foot/pes (pedal) 14
foot process of podocytes 226
forearm/antebrachium 14
forehead (frontal) 14
fornix 90
fovea capitis 42
fractured neck of femur 45
frontal bone 38, 302
frontal (coronal) plane 18
frontal lobe 92, 109
frontalis 60
functional areas of brain 97, 98
functional cortical area 93
fundus of gallbladder 208
fungiform papillae 108

G

galea aponeurotica 60
gallbladder 186, 196, 208, 209
    anterior view 212
    body of 208
gas exchange 181
gastrocnemius 64
gastroesophageal (cardioesophageal) junction 186
glans penis 238, 282
glenoid cavity 40
glenoid fossa 40
gliding (plane) joint 56
glioma 95
globin protein strand 308
glomerular tuft of capillaries 226
glomerulus 226
gluteus maximus 64
gluteus medius 64
gracile fasciculus 82
gracilis 64
Graves' disease 264
great saphenous vein 146
greater duodenal papilla 196
greater trochanter 42
greater tubercle of humerus 40
grey ramus communicans 82
groin/inguen (inguinal) 14
gyrus 90, 92

H

hair 30
hallux 14
hand 14, 51, 66
Hassall's corpuscle 342
head 14
    deep muscles 74

lymphoid organs in 348
    superficial arteries 153
    superficial muscles 61
    superficial veins 153
    surface arteries 139
head of femur 42, 48
head of fibula 42
head of humerus 40
head of pancreas 196, 208, 266
heart 122–133, 134
    anterior view 131, 179
    cross-section 123
    cross-sectional view 130
heart valves 124–125
heartbeat 132
heel/calcaneus (calcaneal) 16
helicotrema 107
helper T cell 350
heme 308
hemoglobin (Hb) 318
hemophilia 331
hemostasis 326–327
hepatic artery (branch) 208
hepatic duct 196, 208
hepatic (right colic) flexure 196
hepatocyte 208
hiatus hernia 199
hinge joint 59
hormones
    mechanisms of action 253
    regulation of daily cycles 263
horseshoe kidney 221
humerus 38, 40
humoral response 350–355
hyaline tissue 21
hypophyseal artery 256
hypophyseal portal system 256
hypothalamus 90, 103, 256
hypothenar muscles 62

I

ileum 186, 196, 207
iliac artery 134, 218, 238
iliac crest 60
iliac nodes 334
iliac vein 146, 218, 238
iliopsoas 64
iliotibial tract 64
ilium 38
immune system
    cells 351
        response to viral invasion 351
        tissue 338
immune system tissue 21
impression of the colon (left colic flexure) 342
impression of the kidney 342
impression of the stomach 342
incus 107
inferior articular surface 42
inferior border of liver 208
inferior colliculi 256
inferior extensor retinaculum 64
inferior orientation 18
inferior pulmonary vein 122
inferior thyroid vein 170
inferior vena cava 122, 146, 218, 224, 238
inguinal ligament 64

insula cortex 90
intercondylar eminence 42
intercondylar fossa 42
intercostal arteries 134
intercostal muscles 183
intercostal nerve 80
intercostal nodes 334
interlobular artery 224, 226
interlobular bile duct 208
interlobular vein 226
internal carotid artery 136
internal iliac artery 134, 218, 238
internal iliac nodes 334
internal iliac vein 146, 218, 238
internal jugular vein 146, 170
internal thoracic vein 170
interosseous border 42
intestinal jejunum 200, 203
intestinal villus 201
intestines 200, 205
iron ion 308
ischium 38
islets of Langerhans 269, 270

J
jejunum 186, 196, 200, 203, 207
joints 40–46, 56–59
jugular vein 146, 170

K
kidney nephron 227
kidneys 218, 222, 224, 226, 266, 318
    anterior view 225, 230
    coronal view 230
    macroscopic features 224–225
    microscopic features 226–227
killer (or cytotoxic) T cell 350
kneecap/patella (patellar) 14

L
labium majus 290
labium minus 290
labyrinthine artery 136
lacrimal apparatus 166
lactiferous duct 290
lactiferous sinus 290
lambdoid suture 302
larynx 160, 162, 163, 168
lateral border of scapula 40
lateral condyle 42
lateral corticospinal tract 82
lateral epicondyle 42
lateral femoral cutaneous nerve 80
lateral fissure 92
lateral funiculus 82
lateral group lymph nodes 334
lateral head of gastrocnemius 64
lateral head of triceps brachii 62
lateral malleolus 42
lateral orientation 18
lateral reticulospinal tract 82
lateral surface 42
lateral vestibulospinal tract 82
latissimus dorsi 60
leaflet/cusp of mitral valve 122, 124
leaflet/cusp of tricuspid valve 122,
    124
left adrenal gland 224
left atrium 122, 124

left brachiocephalic vein 122, 170
left colic flexure (impression of the
    colon) 342
left common carotid artery 122
left hepatic duct 208
left inferior pulmonary vein 122
left lobe of liver 196, 208
left lobe of thymus 342
left lower limb 43
left primary bronchus 160
left pulmonary artery 122
left renal artery 224
left subclavian artery 122
left superior pulmonary vein 122
left upper limb 41
left ventricle 122, 124
leg/crus (crural) 14
lens 106
lesser trochanter 42
lesser tubercle 40
leukemia 321
leukocytes (white blood cells) 308,
    310
levator labii superioris 60
ligamentum teres 208
limbic system, and smell 113
lingual tonsil 108
lipoproteins 317
liver 186, 208
    inferior border 208
    left lobe 196
    microstructure 211
    right lobe 196, 208
liver lobule 208, 209
liver plate 208
lobar bronchi 160
lobes of the brain 93, 98
lobules of mammary gland 290
long head of triceps brachii 62
loop of Henle 226
loose connective tissue 20
lower back (lumbar) 16
lower limb 16
    arteries 144
    bones and joints 42–43
    major nerves 89
    muscles 64–65
    veins 148
lower lobar bronchus 160
lower lobes (left/right lung) 170
lower teeth 38
lumbar vertebra 38, 55
lumbosacral enlargement of spinal
    cord 80
lumbosacral plexus 80
lung disease 77
lungs 77, 131, 170–171, 179
lymph nodes 22, 334–341
lymphatic system 12, 335, 340
lymphatic vessel 338, 341
lymphedema 337
lymphoblastic leukemia, acute 320
lymphocyte 308
lymphoid organs 342, 348
lymphoma 339

M
macrophage 308, 350
macula densa 226

main pancreatic duct 196, 208, 266
major arteries of the body 135
major calyx 224
major nerves of lower limb 89
major nerves of upper limb 88
major veins of the body 147
male bladder 239, 243
male endocrine system 251
male reproductive system 282–289
    anterior view 286
    cross-sectional view 283
    organs, anterior view 283
    sagittal view 286
male urethra 241
male urinary system 219, 242
malignant neoplasm, metastasis by
    355
malleus 107
mammary glands 290, 291, 295
mammillary body 256
mandible 38, 302
masseter 60
mastoid fontanelle 302
mastoid part of temporal bone 302
maternal blood vessels 298
mature ovum 274
maxilla 38, 302
meatus (opening) of left ureter 238
mechanisms of action of anterior
    pituitary 254
mechanisms of action of hormones
    253
mechanisms of potassium regulation
    229
medial border of scapula 40
medial condyle 42
medial epicondyle 42
medial frontal branches 136
medial frontobasal artery 136
medial malleolus 42
medial occipital artery 136
medial orientation 18
medial reticulospinal tract 82
medial striate artery 136
medial vestibulospinal tract 82
median antebrachial vein 146
median nerve 80
median plane 18
median sulcus 108
mediastinum 16
medulla 266, 342
medulla oblongata 80
medullary plexus 266
medullary plexus of peritubular
    capillaries 226
medullary vein 266
membranous urethra 282
memory B cell 350
memory T cell 350
Ménière's disease 117
metacarpal bone 40
metastasis by a malignant neoplasm
    355
metatarsal bones 42
metopic suture 302
mid-sagittal plane 18
middle cerebral artery 136
middle lobar bronchus 160
middle lobe (right lung) 170

minor calyx 224
mitochondrial disease 25
mitral cell 109
mitral valve 122, 124, 125
monocyte 308
mons pubis 290
motor speech area (Broca's) 92
mouth
    anterior view 192
    sagittal view 191
mouth (oral) 14
muscle(s)
    fiber 68
    of the hand 66
    of lung disease 77
    muscular system 12
    of the neck 76
    shapes 69
    tissue 21
    of ventilation 77
muscular system 12
musculocutaneous nerve 80
myelin sheath of Schwann cell 82
myocardial infarction 127
myometrium 298

N
nail 30
nasal bone 302
nasal cavity 160
nasal (piriform) aperture 38
neck (cervical) 14
    muscles 61, 74, 76
    superficial arteries 153
    superficial veins 153
    surface arteries 139
neck of bladder 238
neck of femur 42, 45
neck of fibula 42
neck of gallbladder 208
neck of pancreas 208, 266
nephritis 233
nephron 226, 231
nervous system 12, 81
neural tissue 20
neuron 26, 85
neurosecretory cells 256
neutrophil 308
nipple 290
node of Ranvier 82
normal vein, structure 150
nose 14, 115
notch in superior border of spleen 342
nucleus of endothelial cell 308

O
obturator artery 134
obturator nerve 80
occipital bone 38, 302
occipital lobe 92
occipitalis 60
olfactory apparatus 109
olfactory bulb 109
olfactory gland (Bowman's gland) 109
olfactory mucosa 109
olfactory nerve cell 109
olfactory system 115
olfactory tract 109
opening (meatus) of left ureter 238

opening of coronary sinus 122
opening of ejaculatory duct 238
opening of ureters 238
optic chiasm 256
optic nerve 106
oral cavity and salivary glands 195
orbicularis oculi 60
orbicularis oris 60
orbit 38
organ of Corti 117
organs of female reproduction 291
organs of male reproduction 283
orifice of urethra 290
orthochromatic erythroblast 318
ossicles 107
osteogenesis imperfecta 33
ovarian artery 218, 238
ovarian vein 218, 238
ovaries 250, 274, 290, 295
    cross-sectional view 275
    cyclical changes in 277
    endocrine function 275
    structure and function 293
ovum 274, 298

P

painful arc syndrome 71
palatine tonsil 108
palatoglossus muscle and arch 108
palatopharyngeus muscle and arch
    108
palm (palmar) 14
palmar and dorsal plexus 334
palmar arterial arches 134
palmar surface 18
palmar venous arch 146
palmaris brevis 62
palmaris longus 62
pancreas 186, 208, 209, 250, 266
    anterior view 213, 267, 270
    body 196, 208, 266
    carcinoma of 210
    exocrine cells 215
    head 196, 208, 266
    neck 266
    tail 266
    uncinate process 208, 266
pancreatic duct 196, 208, 266
papillary muscle 122
para-median plane 18
para-sagittal plane 18
paracentral artery 136
paranasal sinuses 164, 166
paraplegia 86, 87
parasternal nodes 334
parathyroid glands 250, 256, 257, 261
    microstructure 258, 261
parietal bone 38, 302
parietal layer of Bowman's capsule
    226
parietal lobe 92
parietal tuber 302
parietooccipital branch 136
parotid nodes 334
patella 42
patellar surface 42
pectineus 64
pectoralis major 60, 62, 170, 290
pelvic cavity 16

pelvic floor muscles (female) 73
pelvic wall, arteries 143
pelvis 14, 45
penis 282, 283
pericallosal artery 136
pericardial cavity 16
pericardium 122, 170
perineurium 82
perirenal fat 224
peritubular capillaries 226
perivitelline space 298
pernicious anemia 323
phalanges 40, 42
pharynx 160, 167
pia mater 82
pineal gland 90, 256, 257, 260
piriform aperture 38
pituitary adenomas 259
pituitary gland 250, 254, 256
    location 257
    structure 257, 260
pituitary stalk 256
pivot joint 58
placenta 298
    anterior view 301
    cross-sectional view 300
placenta previa 304
plantar and dorsal plexus 334
plantar arch 134
plantar surface 18
plantar venous arch 146
plasma B cell 350
platelets 308, 317
pneumonia 172
pneumothorax 175
polar body 298
polar frontal artery 136
pollex 14
polychromatophilic erythroblast 318
polycystic kidney 223
polycystic ovarian syndrome 292
polyribosomes 318
popliteal archery 134
popliteal nodes 334
portal system (digestive function) 142,
    212
portal vein (branch) 196, 208
postcentral gyrus 92
posterior auricular artery 134
posterior cerebral artery 136
posterior chamber 106
posterior communicating artery 136
posterior fontanelle 302
posterior inferior cerebellar artery 136
posterior labial commissure 290
posterior mediastinal nodes 334
posterior orientation 18
posterior pituitary (neurohypophysis)
    256
posterior ramus of spinal nerve 82
posterior spinal artery 82
posterior spinal vein 82
posterior tibial artery 134
potassium regulation, mechanisms of
    229
precentral gyrus (motor cortex) 92
precuneal artery 136
prefrontal cortex 92

pregnancy 298–301
    early stages 299
    rhesus (Rh) factor 328–329
prepuce (foreskin) 238, 282
primary auditory cortex 92
primary bronchi 160
primary motor cortex 92
primary somatosensory cortex 92
primary visual cortex 92
process of renal regulation 236
production of red blood cells 319
proerythroblast 318
promontary covering first coil of
    cochlea 107
pronator teres 62
prostate gland 238, 282, 283
prostatic urethra 282
prostatic utricle 238
proteins 350
proximal convoluted tubule 226
proximal orientation 18
proximal straight tubule (thick
    descending limb of loop of Henle)
    226
pterion 302
pubis (pubic) 14
pulmonary artery 122, 129
pulmonary circulation 127
pulmonary valve 122, 124
pulmonary vein 122
pyelonephritis 235
pyloric sphincter 196
pylorus 186, 196

R

radial artery 134
radial nerve 80
radicular artery 82
radius 40
ramus of spinal nerve 82
reading comprehension area 92
rectum 186, 196, 201
rectus abdominis 60
red blood cells (erythrocytes) 308,
    309, 318, 319, 324
red bone marrow 318
red pulp 342
regulation of daily cycles by hormones
    263
renal artery 134, 143, 224
renal calculi (stones) 240
renal column 224
renal failure, chronic 237
renal papilla 224
renal pelvis 224
renal pyramid (medulla) 224
renal regulation, process of 236
renal sinus 224
renal vein 146, 224
reproductive system 13
    cross-sectional view 283, 291
    female 290–297
    male 282–289
    organs, anterior view 283, 291
respiratory muscles 183
respiratory system 13, 160–169
    anterior view 161, 182
    diaphragm 183
    intercostal muscles 183

    upper part 161, 183
reticulocyte 318
retina 106, 110
retroauricular nodes 334
rhesus negative 324
rhesus positive 324
rhesus (Rh) factor 328–329
rheumatic heart disease 128
ribs
    false (pairs 8–10) 38
    first 170
    floating (pairs 11 & 12) 38
    true (pairs 1–7) 38
right adrenal gland 224
right anterior cerebral artery 136
right atrium 122, 124, 170
right brachiocephalic vein 122, 170
right colic (hepatic) flexure 196
right hepatic duct 208
right inferior pulmonary vein 122
right lobe of liver 196, 208
right lobe of thymus 342
right lower limb 43
right primary bronchus 160
right pulmonary artery 122
right upper limb 41
right ventricle 122, 124, 170
rotator cuff and painful arc syndrome
    71
rotator cuff muscles 71
ruptured spleen 345

S

saccular macula 107
saccule 107
sacral nodes 334
sacrum 38
saddle joint 56
sagittal fissure 90
sagittal (mid-sagittal or median) plane
    18
sagittal (para-sagittal or para-median)
    plane 18
sagittal plane 18
sagittal suture 302
salivary gland calculi (sialolithiasis)
    188
salivary glands 189, 195
saphenous vein, small 146
scalenus anterior 170
scalp 30
scapula 38, 40
Schwann cell, myelin sheath 82
sciatic nerve 80
scrotum 238, 274, 282
segmental artery 224
semicircular canals 107
semimembranosus 64
seminal vesicle 282
semitendinosus 64
sensory speech area (Wernicke's) 92
serratus anterior 60
shaft (diaphysis) of femur 42
shoulder (acromial) 16
    cross-section 46
    dislocation 47
    joint 40, 41
shoulder joint 40, 41

sialolithiasis (salivary gland calculi) 188
sickle cell disease (anemia) 315
sigmoid colon 186, 196
sinusitis 165
sinusoid 208
sinusoidal vessel 266
situs inversus 19
skeletal muscle 21
skeletal system 12
skin 12, 30, 33
skull/cranium (cranial) 14, 49
small saphenous vein 146
smell and the limbic system 113
smooth muscle 21
sole (plantar) 16
soleus 64
somatosensory association area 92
sperm 287, 289, 298
spinal anesthesia 105
spinal artery 82
spinal canal 16
spinal cord 82–83, 90
  anterior view 105
  cervical enlargement of 80
  cross-sectional view 83
  lumbosacral enlargement of 80
spinal (dorsal root) ganglion 83
spinal gray matter 83
spinal nerves 105
spinal vein 82
spine of scapula 40
spinothalamic tract 83
spinous processes 38
spleen 186, 196, 342–349
  location 344
  microstructure 343, 349
  ruptured 345
  visceral surface 349
splenic artery 342
splenic vein 342
squamous part of occipital bone 302
squamous part of temporal bone 302
stapes 107
stapes footplate covering vestibular (oval) window 107
sternocleidomastoid 60
sternohyoid 60
sternum 38
stomach 186, 196
  anterior view 200, 205, 213
  internal structure 199, 213
  and intestines 200, 205
structure of a normal vein 150
styloid process 302
subcapsular plexus 266
subclavian artery 122, 134, 170
subclavian vein 146, 170
subscapular fossa 40
sulcus 90, 92
superficial arteries of the head and neck 153
superficial dorsal vein 282
superficial fibular nerve 80
superficial inguinal nodes 334
superficial muscles of the back 75
superficial muscles of the head 61
superficial muscles of the lower limb 65
superficial muscles of the neck 61
superficial muscles of the trunk 61
superficial muscles of the upper limb 63

superficial veins of the head and neck 153
superior and inferior colliculi 256
superior articular surfaces (medial and lateral facets) 42
superior border of scapula 40
superior border of spleen 342
superior cerebellar artery 136
superior extensor retinaculum 64
superior lobar bronchus 160
superior mesenteric artery 224
superior orientation 18
superior pulmonary vein 122
superior vena cava 122, 146, 170
suppressor T cell 350
suprarenal artery 266
sural nerve 80
surface arteries of the head and neck 139
suspensory ligaments 106
sweat gland 28
syncytiotrophoblast 298

T

T cell *see* T lymphocytes
T lymphocytes 342
tail of pancreas 208, 266
talus 42
tarsal bones 42
teeth 194
temporal bone 38, 302
temporal lobe 92
temporalis 60
tendon of flexor carpi ulnaris 62
tendon of palmaris longus 62
tendon of triceps brachii 62
tendons of extensors of the digits 62
teres major 60
teres minor 60
terminal sulcus 108
testes 250, 274, 282
  endocrine function 274
  function 274, 279
  posterior view 284, 287
  structure 279, 287
  torsion of 285
testicular artery 218, 238
testicular vein 218, 238
thalamus 90, 256
thenar muscles 62
thick ascending limb of loop of Henle (distal straight tubule) 226
thick descending limb of loop of Henle (proximal straight tubule) 226
thigh (femoral) 14
thigh muscle injury 67
thin ascending limb of loop of Henle 226
thin descending limb of loop of Henle 226
thoracic cavity 16
thoracic duct 334
thoracic vein 170
thoracic vertebra 38, 55
thoracolumbar fascia 60
thumb/pollex (pollical) 14
thymus 170, 250, 342–349, 352
  anterior view 349
  left lobe 342
  location 344
  microstructure 343
  right lobe 342
thyroid gland 170, 250, 256, 257

anterior view 261
  microstructure 258, 261
thyrotoxicosis 264
tibia 42, 43
tibial artery 134
tibial nerve 80
tibial tuberosity 42
tibialis anterior 64
tissues 20–21
toes/digits (digital or phalangeal) 14
tongue 108, 112, 186, 194
torsion of testis 285
torso 14, 61
trabecular arteries 342
trachea 160, 170
  cross-section 172, 177
  in situ 176
transverse (axial) plane 18
transverse colon 186, 196
transverse processes 38
trapezius 60
traumatic neck injury 50
triceps brachii 60, 62
tricuspid valve 122, 124, 125
trigone 238
true ribs (pairs 1–7) 38
trunk/torso 14, 61
tympanic duct 107
tympanic membrane (eardrum) 107
tympanic ring of temporal bone 302

U

ulna 40
ulnar artery 134
ulnar nerve 80
umbilical artery 298
umbilical cord 298
umbilical vein 298
umbilicus (umbilical) 14
uncinate process of pancreas 208, 266
universal donors 324
universal recipients 324
upper limb 16
  arteries 141
  bones and joints 40–41
  major nerves 88
  muscles 62–63
  veins 154
upper lobar bronchus 160
upper lobe (left/right lungs) 170
upper part of the respiratory system 161, 183
upper teeth 38
ureter 218, 224, 238, 282
urethra 238, 282, 290
  female 245
  male 241
urethral orifice, external 238
urinary pole 226
urinary systems 13, 218–219
  anterior view 219, 231
  female 219, 242
  male 219, 242
  sagittal view 242
  urinary tract 220, 231, 242
uterine (fallopian) tube 274, 290
uterus 274, 290
  anterior view 294
  changes in 305
  posterior view 294
  structure 296
utricle 107

V

vagina 274, 290
vagus nerve, cardiac branch 170
vallate papillae 108
vallecula 108
varicose veins 150
vas deferens (ductus deferens) 282
vasa recta 226
vascular pole 226
veins 146–147, 256
  head and neck 153
  lower limb 148
  major 147
  structure 150
  upper limb 154
  varicose 150
venous sinusoids 342
venous stasis ulcers 155
ventilation, muscles of 77
ventral funiculus 83
ventral horn 83
ventral rootlets 83
ventral spinocerebellar tract 83
ventricle 122, 124, 170
ventricular diastole 124
ventricular systole 124
vertebral artery 136
vertebral column 54
vestibular duct 107
vestibular nerve branches 107
view orientation 18
viral invasion, immune response to 351
viruses 350, 351
visual association area 92
vitreous body 106
vulva 290, 291

W

wall of vagina 290
white blood cells (leukocytes) 308, 310
white pulp nodule 342
white ramus communications 83
wrist (carpal) 14, 40, 88

Z

zona fasciculata 266
zona glomerulosa 266
zona pellucida 298
zona reticularis 266
zygomatic bone 38, 302
zygomatic process of temporal bone 302
zygomaticus major 60
zygote 295, 298, 299